AF699748

# Calculus and Analytic Geometry

# Calculus and Analytic Geometry

**Harold Schachter**

*Assistant Professor of Mathematics*
*Queensborough Community College*
*City University of New York*

**McGraw-Hill Book Company**

*New York St. Louis San Francisco Düsseldorf Johannesburg*
*Kuala Lumpur London Mexico Montreal New Delhi*
*Panama Rio de Janeiro Singapore Sydney Toronto*

**Calculus and Analytic Geometry**

Copyright © 1972 by McGraw-Hill, Inc. All rights reserved. Printed in the United States of America. No part of this publication may be reproduced, stored in a retrieval system, or transmitted, in any form or by any means, electronic, mechanical, photocopying, recording, or otherwise, without the prior written permission of the publisher.

*Library of Congress Catalog Card Number* 75-174626

07-055056-5

234567890KPKP798765

This book was set in Modern by The Maple Press Company, and printed and bound by Kingsport Press. The designer was Richard Paul Kluga; the drawings were done by Danmark & Michaels, Inc. The editors were Bob Flowers, Janet Wagner, and David Damstra. James Lee supervised production.

*To my parents*
*and to my wife, Iris,*
*daughter, Nofie*
*and son, Michael*

# Contents

# Preface

This textbook is designed for nonmathematics students, with majors in such fields as electrical technology, mechanical technology, social science, business, and, in fact, for any students who desire some basic knowledge of calculus and analytic geometry. Because this book is not designed for engineering students or mathematics majors, there is not a strong emphasis on mathematical rigor. Very often, results are presented on an intuitive basis. This, however, does not imply that mathematical terms are not clearly and carefully defined. It is the opinion of this author that complete and accurate definitions are essential if the student is to understand the fundamental concepts and successfully perform the appropriate applications.

An attempt has been made to present the material in a clear and concise way with major emphasis on having a readable textbook. Once a new idea is presented and defined precisely, the emphasis is shifted to applications and illustrative examples to help indicate methods and procedures for problem solving.

For the student who desires a more thorough and rigorous discussion of certain basic concepts, special sections, indicated by an arrowhead, have been included at the ends of some chapters. These sections contain additional proofs and discussion and may be completely omitted without affecting the continuity of the chapters.

Chapter 12 on Conic Sections is independent of the other chapters in the text, but the instructor may wish to cover this chapter before Chapter 9 so as to familiarize the students with the figures with which they will be working.

This textbook presupposes a working knowledge of plane geometry, intermediate algebra, and trigonometry (eleventh-year high school mathematics). There are many review books available for the student who feels that he needs to review some of these subjects.

Many students make the mistake of expecting to understand a mathematics textbook on the first reading and are disappointed to find that such textbooks do not read like novels (unless perhaps like mysteries). This author strongly recommends that the student read the textbook two or three times before taking the final examination in the course.

I am very grateful to my colleagues Professor Morris Goldberg and Mrs. Juliana Corn of Queensborough Community College and to Professor Robert B. Sackett of Erie County Community College, Buffalo, New York, Dr. Cy Lawrence of New York City Community College, City University of New York, Professor Toby R. Shook of Asheville-Bucombe Technical Institute, Asheville, North Carolina, and Professor Edwin C. Conry of Orange Coast Junior College, Costa Mesa, California, for their helpful suggestions. But most of all I wish to thank my wife, Iris, who spent countless hours typing and illustrating the entire manuscript and without whose help this book would not have been possible.

*Harold Schachter*

# 1
# Analytic Geometry

## 1-1 RECTANGULAR COORDINATES

Analytic geometry was introduced in the seventeenth century by the French mathematician and philosopher, René Descartes. The main advantage of analytic geometry is that it combines geometry and algebra, using algebraic methods to solve geometric problems and giving geometric interpretations to algebraic equations.

Until the seventeenth century geometry dealt chiefly with points and figures composed of points; algebra was concerned mainly with numbers. In combining the two, Descartes associated with each point in the plane an ordered pair of real numbers which were called the *coordinates* of the point. Analytic geometry is sometimes referred to as coordinate geometry and sometimes as cartesian geometry (where *Cartesius* is a Latinized form of Descartes' name).

The simplest type of coordinate system is a rectangular-coordinate system. A rectangular-coordinate system has two perpendicular axes, one horizontal and one vertical. The horizontal axis is generally called the $X$ axis and the vertical axis is generally called the $Y$ axis. The point of intersection of the two axes is called the *origin*. From the origin we also choose a positive and negative direction for each axis. For the $X$ axis the positive direction is chosen to the right and the negative direction to the left. For the $Y$ axis the positive direction is chosen upward and the negative direction, downward.

The two perpendicular axes divide the plane into four parts, called *quadrants*. The position of a point in the plane is determined by its directed distances from the two perpendicular axes. If we are given a point $A$ in a plane, then its $x$ coordinate would be the perpendicular distance from the point to the $Y$ axis (taking into account the quadrant), and its $y$ coordinate would be the perpendicular distance from the point to the $X$ axis. In writing the coordinates the $x$ coordinate is always given first. Thus, for a point in the first quadrant, if the distance from the point to the $Y$ axis is 3 and the distance from the point to the $X$ axis is 5, the coordinates are written (3,5) (Fig. 1-1).

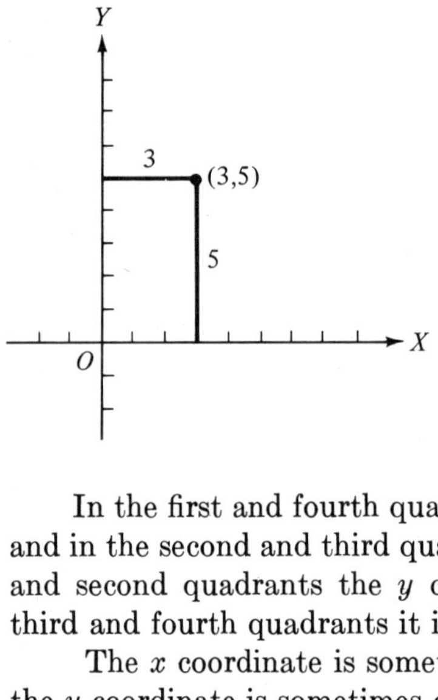


**Fig. 1-1**

In the first and fourth quadrants the $x$ coordinate is always positive, and in the second and third quadrants it is always negative. In the first and second quadrants the $y$ coordinate is always positive, and in the third and fourth quadrants it is always negative (Fig. 1-2).

The $x$ coordinate is sometimes called the *abscissa* of the point, and the $y$ coordinate is sometimes called the *ordinate* of the point. The process of locating points in the plane by using the coordinates of the point is known as *plotting*. When accuracy is desired, it is recommended that rectangular-coordinate graph paper be used for plotting and sketching.

**Example 1** To plot the point whose coordinates are (3,2), we first count off 3 units in the positive direction on the $X$ axis and then 2 units in the positive direction on the $Y$ axis (Fig. 1-3).

**Example 2** The coordinates of three vertices of a rectangle are given as (0,0), (3,0), and (0,5). We wish to find the coordinates of the fourth vertex.

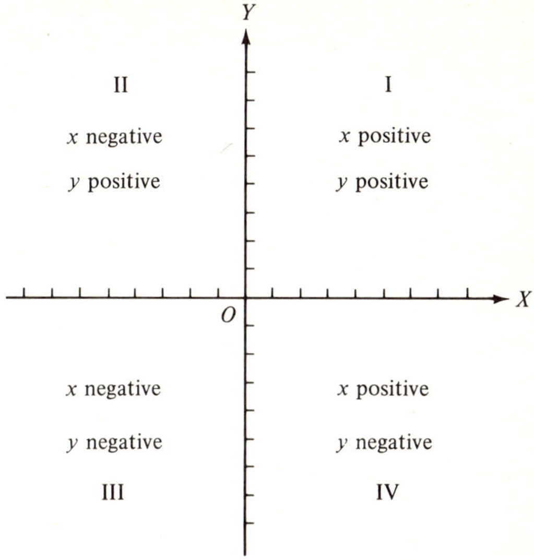


**Fig. 1-2**

*Solution* We can label the coordinates of the fourth vertex $(a,b)$, and label the four vertices as $A$, $B$, $C$, and $D$. We note that points $A$ and $B$ have the same $y$ coordinate and that points $C$ and $D$ have the same $y$ coordinate. Similarly, point $A$ has the same $x$ coordinate as point $D$, and point $B$ has the same $x$ coordinate as point $C$. Having a rectangle, we know that the distance from the point $C$ to the $Y$ axis is the same as the distance from the point $B$ to the $Y$ axis, namely 3. Similarly, the distance from the point $C$ to the $X$ axis is the same as the distance from the point $D$ to the $X$ axis, namely 5. Thus, the point $(a,b)$ must be the point (3,5) (Fig. 1-4).

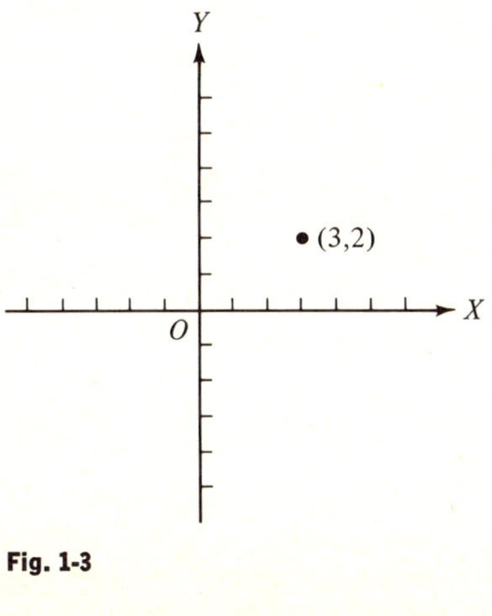


**Fig. 1-3**

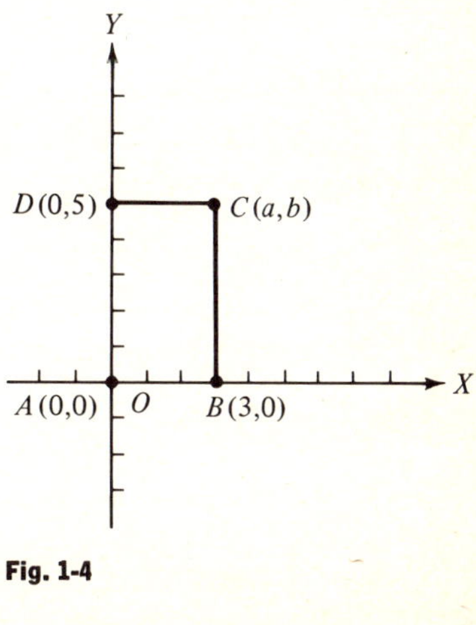


**Fig. 1-4**

## 1-2 THE DISTANCE BETWEEN TWO POINTS

If we wish to find the length of a line segment connecting two points whose $y$ coordinates are the same, we merely have to take the positive difference of their $x$ coordinates. This will give us the length of the segment connecting the two points. Similarly, if the $x$ coordinates are the same, the positive difference of the $y$ coordinates will give us the length of the segment. If, however, the two points do not lie on a line parallel to one of the axes, both the ordinates and the abscissas of the two points will be different and a formula for distance will have to be developed.

Let $A$ and $B$ be two points with coordinates $(x_1,y_1)$ and $(x_2,y_2)$, respectively.† From the point $B$ we will draw a line perpendicular to the $X$ axis, calling the point of intersection on the $X$ axis $D$. From the point $A$ we will draw a line perpendicular to the segment $\overline{BD}$, calling the point of intersection $C$ (Fig. 1-5).

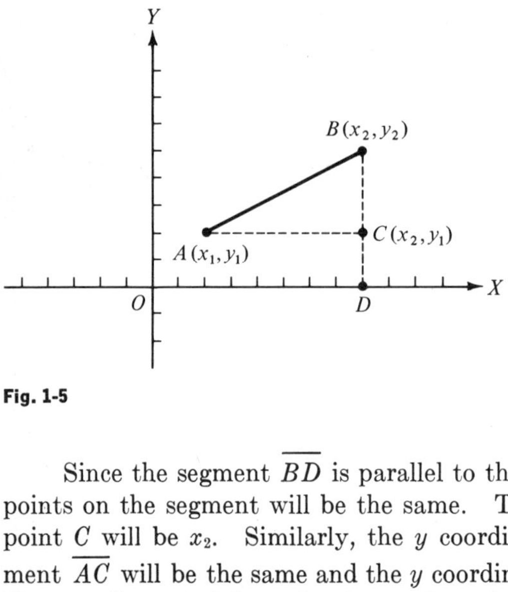


**Fig. 1-5**

Since the segment $\overline{BD}$ is parallel to the $Y$ axis, the $x$ coordinate of points on the segment will be the same. Thus, the $x$ coordinate of the point $C$ will be $x_2$. Similarly, the $y$ coordinate of all points on the segment $\overline{AC}$ will be the same and the $y$ coordinate of the point $C$ will be $y_1$. The coordinates of the point $C$ are, then, $(x_2,y_1)$.

Since the $y$ coordinate is the same for the points $A$ and $C$, the length of the segment $\overline{AC}$ can be found by taking the difference of the $x$ coordinates, $x_2 - x_1$. Similarly, the length of the segment $\overline{BC}$ would be the difference of the $y$ coordinates, $y_2 - y_1$.

† Remark: For simplicity, the points $A$ and $B$ have been chosen in the first quadrant, with $x_2$ greater than $x_1$ and $y_2$ greater than $y_1$. It can be shown that our results will hold for a choice of points in any quadrant.

We can now consider the right triangle $ABC$ in which we know the lengths of the sides $\overline{AC}$ and $\overline{BD}$. By the pythagorean theorem the length of side $\overline{AB}^2$ can be found by the formula $\overline{AB}^2 = \overline{AC}^2 + \overline{BC}^2$. Thus,

$$\overline{AB}^2 = (x_2 - x_1)^2 + (y_2 - y_1)^2 \quad \text{and} \quad \overline{AB} = \sqrt{(x_2 - x_1)^2 + (y_2 - y_1)^2}$$

**Example 1** Find the distance between the point $A(3,3)$ and the point $B(7,6)$.

*Solution* Using the formula,

$$\begin{aligned} \overline{AB} &= \sqrt{(x_2 - x_1)^2 + (y_2 - y_1)^2} \qquad & x_2 - x_1 = 7 - 3 \\ & & y_2 - y_1 = 6 - 3 \\ &= \sqrt{(7-3)^2 + (6-3)^2} \\ &= \sqrt{(4)^2 + (3)^2} \\ &= \sqrt{16 + 9} \\ &= \sqrt{25} \\ &= 5 \end{aligned}$$

Thus, segment $\overline{AB} = 5$.

**Example 2** Find the coordinates of the point whose $y$ coordinate is 4 and which is equidistant from the points (1,1) and (7,5).

*Solution* Since the $y$ coordinate is given as 4, we have only to find the $x$ coordinate. We will label the coordinates of the point $(a,4)$ (Fig. 1-6).

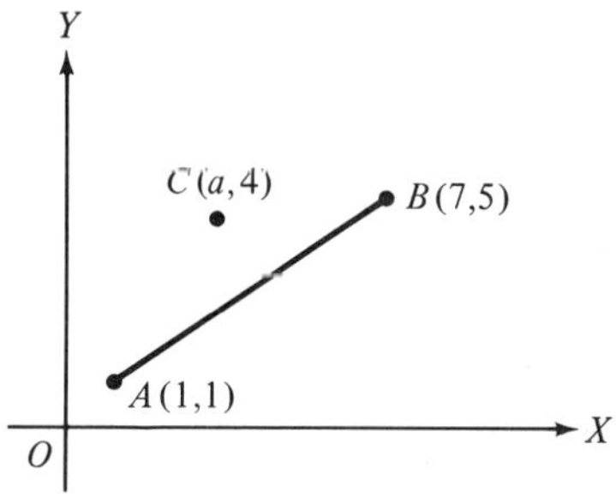


**Fig. 1-6**

The distance from $A$ to $C$ must equal the distance from $B$ to $C$.

$$\overline{AC} = \overline{BC}$$

But,

$$\begin{aligned} \overline{AC} &= \sqrt{(a-1)^2 + (4-1)^2} \\ \overline{BC} &= \sqrt{(a-7)^2 + (4-5)^2} \end{aligned}$$

Therefore,
$$\begin{aligned}
\sqrt{(a-1)^2+(4-1)^2} &= \sqrt{(a-7)^2+(4-5)^2} \\
\sqrt{(a-1)^2+9} &= \sqrt{(a-7)^2+1} \\
(a-1)^2+9 &= (a-7)^2+1 \\
a^2-2a+1+9 &= a^2-14a+49+1 \\
a^2-2a+10 &= a^2-14a+50 \\
12a &= 40 \\
a &= \tfrac{40}{12} = \tfrac{10}{3}
\end{aligned}$$

Thus, the coordinates of the point are $(\frac{10}{3},4)$.

### EXERCISES

1. Find the lengths of the line segments connecting the following sets of points:
   (a) $(-1,3)$, $(2,7)$ (b) $(-5,2)$, $(3,-4)$
   (c) $(3,8)$, $(-6,8)$ (d) $(4,0)$, $(-1,5)$
   (e) $(-4,1)$, $(-4,7)$ (f) $(\frac{7}{2},-\frac{1}{2})$, $(\frac{1}{2},\frac{7}{2})$
2. Find the lengths of the sides of the triangle with vertices $A(-3,1)$, $B(0,5)$, $C(5,7)$.
3. Find the lengths of the sides of the quadrilateral with vertices $A(0,0)$, $B(4,0)$, $C(10,8)$, and $D(3,4)$.

## 1-3 MIDPOINT OF A LINE SEGMENT

We will now consider the question of finding the coordinates of the midpoint of a line segment, given the coordinates of the endpoints.

Let $A$ and $B$ be points with coordinates $(x_1,y_1)$ and $(x_2,y_2)$, respectively. Let $M$ be the midpoint of segment $\overline{AB}$. We will temporarily label the point $M$ as $(x,y)$ and attempt to find $x$ and $y$ in terms of the given coordinates.

From the point $B$ draw a line perpendicular to the $X$ axis. Label the point of intersection $C$. Similarly, from the point $M$ draw a line perpendicular to the $X$ axis, labeling the point of intersection $D$. From the point $A$ draw a line perpendicular to the segment $\overline{BC}$, labeling the points of intersection on $\overline{MD}$ and $\overline{BC}$, $E$ and $F$, respectively. The coordinates of $E$ are $(x,y_1)$ and the coordinates of $F$ are $(x_2,y_1)$ (Fig. 1-7). Thus,

$$\overline{AE} = x - x_1 \qquad \overline{AF} = x_2 - x_1$$
$$\overline{ME} = y - y_1 \qquad \overline{BF} = y_2 - y_1$$

We note that triangle $AME$ is similar to triangle $ABF$ and therefore the corresponding sides are proportional. In particular:

$$\frac{\overline{AM}}{\overline{AB}} = \frac{\overline{AE}}{\overline{AF}}$$

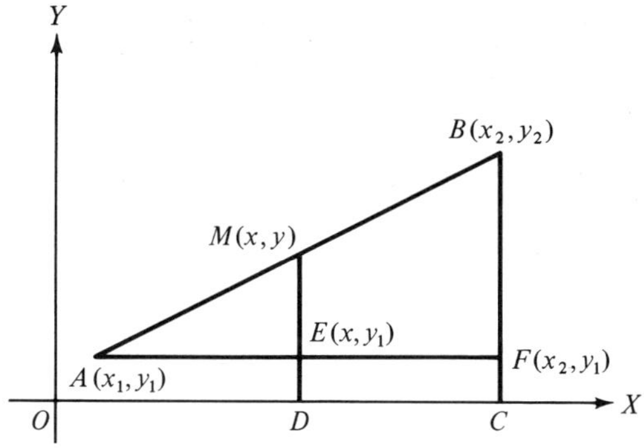


**Fig. 1-7**

Since $M$ is the midpoint of $\overline{AB}$, $\overline{AM} = \frac{1}{2}(\overline{AB})$ and $\dfrac{\overline{AM}}{\overline{AB}} = \frac{1}{2}$. But,

$$\frac{\overline{AE}}{\overline{AF}} = \frac{x - x_1}{x_2 - x_1}$$

and, therefore,

$$\frac{x - x_1}{x_2 - x_1} = \frac{1}{2}$$
$$x - x_1 = \tfrac{1}{2}(x_2 - x_1)$$
$$x = x_1 + \tfrac{1}{2}(x_2 - x_1)$$
$$= \tfrac{1}{2}x_1 + \tfrac{1}{2}x_2$$
$$= \frac{x_1 + x_2}{2}$$

Using the fact that $\dfrac{\overline{AM}}{\overline{AB}} = \dfrac{\overline{ME}}{\overline{BF}}$, we could similarly show that

$$y = \frac{y_1 + y_2}{2}$$

Thus, the midpoint $M$ of segment $\overline{AB}$ has coordinates $(x,y)$ with

$$x = \frac{x_1 + x_2}{2} \qquad \text{and} \qquad y = \frac{y_1 + y_2}{2}$$

**Example 1** Find the midpoint of the segment connecting the points (3,7) and (9,1).

*Solution* For the midpoint

$$x = \frac{x_1 + x_2}{2} \qquad y = \frac{y_1 + y_2}{2}$$

Therefore
$$x = \frac{3+9}{2} = \frac{12}{2} = 6$$
$$y = \frac{7+1}{2} = \frac{8}{2} = 4$$

The midpoint is (6,4).

**Example 2** Let the coordinates of triangle $ABC$ be (2,1), (8,9), and (6,3), respectively. Find the length of the median from vertex $C$.

*Solution* A median of a triangle is a line segment joining a vertex of a triangle to the midpoint of the opposite side. We must first find the coordinates of the midpoint $D$ (Fig. 1-8). To find the coordinates of $D$:

$$x = \frac{2+8}{2} = \frac{10}{2} = 5 \qquad y = \frac{1+9}{2} = \frac{10}{2} = 5$$

The coordinates of $D$ are (5,5). We now have the coordinates of both $C$ and $D$, the endpoints of the desired median. The length of the median is

$$\begin{aligned}\overline{CD} &= \sqrt{(6-5)^2 + (3-5)^2} \\ &= \sqrt{1+4} \\ &= \sqrt{5}\end{aligned}$$

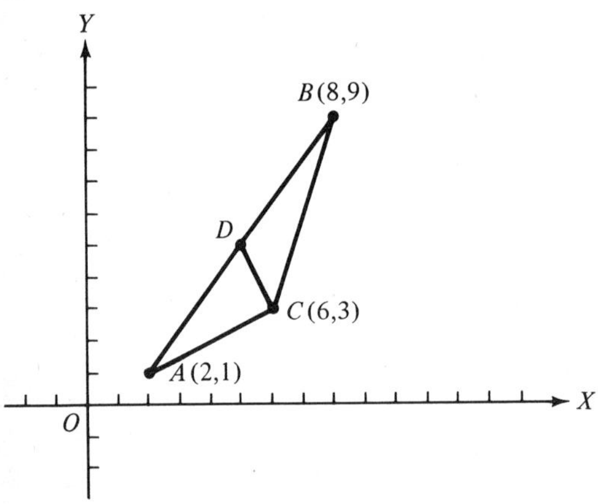


**Fig. 1-8**

**EXERCISES**

1. Find the coordinates of the midpoint for the following sets of points:
   (*a*) (−2,3), (8,7) (*b*) (−4,0), (6,4)
   (*c*) (−1,−1), (7,−7) (*d*) (3,4), (−4,−3)
   (*e*) (6,−6), (−2,8)

**2.** Find the coordinates of the midpoints of the sides of the triangle whose vertices are $A(1,4)$, $B(-3,2)$, and $C(7,2)$.

**3.** In triangle $ABC$ with vertices $A(2,7)$, $B(5,5)$, and $C(-7,1)$, find the length of the median from vertex $A$ on side $\overline{BC}$.

**4.** Find the midpoints of the sides of the quadrilateral with vertices $A(0,0)$, $B(1,3)$, $C(3,7)$, and $D(5,0)$.

**5.** Determine whether the line segments joining the consecutive midpoints of the sides of the rectangle $ABCD$ with vertices $A(0,0)$, $B(0,3)$, $C(5,3)$, and $D(5,0)$ form a parallelogram.

## 1-4 DIRECTION OF A LINE

The direction of a straight line is often given in terms of the *angle of inclination* of the line, which is defined as the smallest angle measured counterclockwise from the positive $X$ axis to the line (Fig. 1-9).

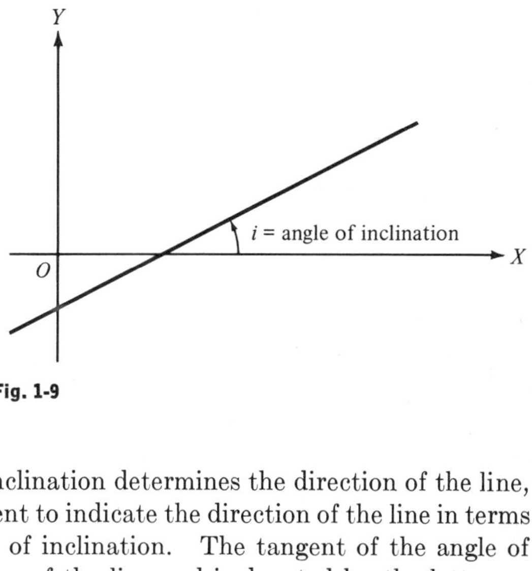


**Fig. 1-9**

Though the angle of inclination determines the direction of the line, it is generally more convenient to indicate the direction of the line in terms of the tangent of the angle of inclination. The tangent of the angle of inclination is called the *slope* of the line and is denoted by the letter $m$.

$$\tan i = m = \text{slope}$$

We recall from trigonometry that $\tan i = \dfrac{\sin i}{\cos i}$. When angle $i = 0$, then $m = 0$. When angle $i = 90°$, the slope is not defined. We also note that when the angle of inclination is acute, the tangent is positive and thus the slope is positive, and when the angle of inclination is obtuse, the slope is negative (Fig. 1-10).

**Example 1**

(*a*) Angle of inclination $i = 45°$ (Fig. 1-10*a*).
(*b*) Angle of inclination $i = 120°$ (Fig. 1-10*b*).

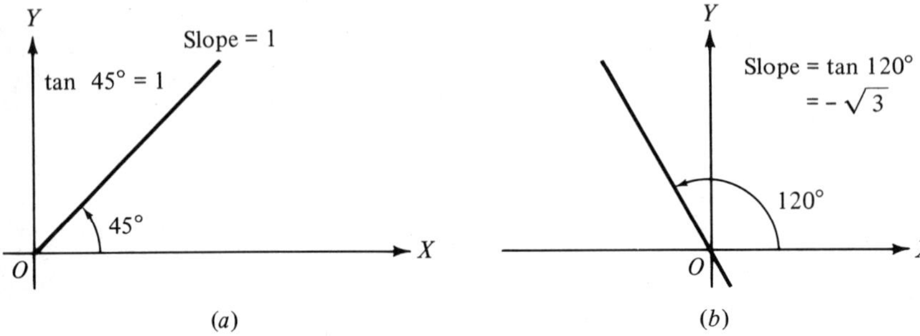


**Fig. 1-10**

In plane geometry we found that a knowledge of two points on a line was enough to determine the line. In coordinate geometry a knowledge of the coordinates of two points on a line is sufficient to determine the slope of the line.

Let the coordinates of two points $A$ and $B$ be $(x_1,y_1)$ and $(x_2,y_2)$, respectively. As before, we can drop perpendiculars to form a right triangle with right angle at vertex $C$ with coordinates $(x_2,y_1)$ (Fig. 1-11).

$$\text{Slope } m = \frac{\overline{BC}}{\overline{AC}} = \frac{y_2 - y_1}{x_2 - x_1}$$

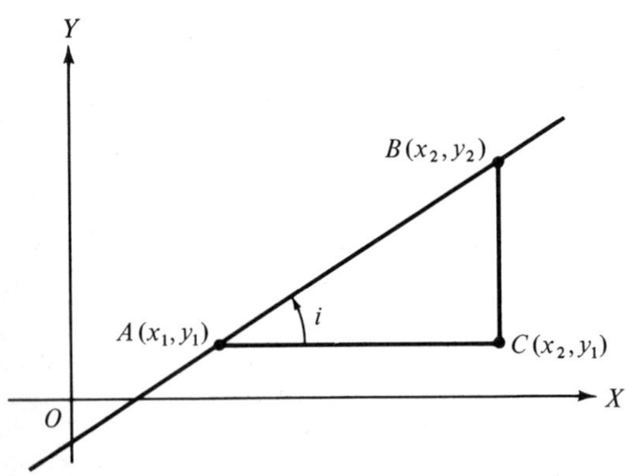


**Fig. 1-11**

**Example 2** If point $A$ has coordinates (7,9) and point $B$ has coordinates (10,15), then

$$m = \frac{15 - 9}{10 - 7} = \frac{6}{3} = 2$$

In applying the formula for the slope of a line given the coordinates of two points, it is immaterial which point we choose as $(x_1,y_1)$ and which point we choose as $(x_2,y_2)$. In the above example, if we interchange the two points, we get

$$m = \frac{9 - 15}{7 - 10} = \frac{-6}{-3} = 2 \qquad \text{As before}$$

Having considered the direction of a line, we now turn to the question of perpendicularity, parallelism, and, in general, the determination of the angle between two lines whose slopes are given. Let $l_1$ and $l_2$ be two lines with angles of inclination $i_1$ and $i_2$, respectively.† Let $\theta$ be the angle from $l_1$ to $l_2$ (Fig. 1-12).

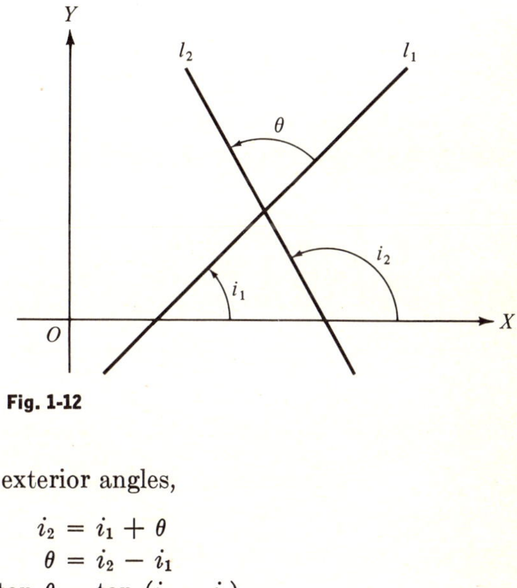


**Fig. 1-12**

Using the property of exterior angles,

$$i_2 = i_1 + \theta$$
$$\theta = i_2 - i_1$$
$$\tan \theta = \tan (i_2 - i_1)$$

Since $$\tan (A - B) = \frac{\tan A - \tan B}{1 + \tan A \tan B}$$

$$\tan \theta = \frac{\tan i_2 - \tan i_1}{1 + \tan i_2 \tan i_1}$$

† For simplicity we shall consider the case where $i_2 > i_1$.

Using the slope notation:

$$\tan \theta = \frac{m_2 - m_1}{1 + m_2 m_1}$$

In a similar way we can show that $\cot \theta = \dfrac{1 + m_2 m_1}{m_2 - m_1}$. If $l_1$ and $l_2$ are parallel, then

$$\theta = 0° \qquad \tan 0° = 0 \qquad \text{and} \qquad m_2 - m_1 = 0$$

$$m_1 = m_2\dagger$$

If $l_1$ and $l_2$ are perpendicular, then $\theta = 90°$. Tan 90° is not defined; however, cot 90° = 0. Thus

$$1 + m_2 m_1 = 0$$
$$m_2 m_1 = -1$$
$$m_1 = \frac{-1}{m_2}$$

Thus, if two lines are parallel, then their slopes are equal, and if two lines are perpendicular, then the slope of one line is the negative reciprocal of the slope of the other. Conversely, if

$$m_1 = m_2$$

then $$\tan i_1 = \tan i_2$$

Since $0° \le i_1 < 180°$ and $0° \le i_2 < 180°$, $i_1 = i_2$ and $l_1$ is parallel to $l_2$. Similarly, if

$$m_2 = -\frac{1}{m_1}$$

$$\tan i_2 = -\frac{1}{\tan i_1} = -\cot i_1 = \tan (90° + i_1)$$

Thus, $$i_2 = 90° + i_1$$

and $l_1$ is perpendicular to $l_2$.

**Example 3** Prove that the line segment joining the points (−3,6) and (9,1) is perpendicular to the line segment joining the points (0,0) and (5,12).

*Solution* We shall call the first line $l_1$ and the second line $l_2$. The corresponding slopes of $l_1$ and $l_2$ are

$$ml_1 = \frac{6 - 1}{-3 - 9} = -\frac{5}{12} \qquad ml_2 = \frac{12 - 0}{5 - 0} = \frac{12}{5}$$

The slope of $l_2$ is the negative reciprocal of the slope of $l_1$, and therefore $l_1$ is perpendicular to $l_2$.

† **Remark:** We could also have obtained this result by noting that if $l_1$ and $l_2$ are parallel, then their angles of inclination are equal and therefore $\tan i_1 = \tan i_2$ and $m_1 = m_2$.

**Example 4** Show that (2,3), (4,7), and (8,5) are the coordinates of a right triangle.

*Solution: Method* 1 We shall call the three points $A$, $B$, and $C$, respectively.

$$\overline{AB} = \sqrt{(4-2)^2 + (7-3)^2} = \sqrt{4+16} = \sqrt{20}$$
$$\overline{BC} = \sqrt{(8-4)^2 + (5-7)^2} = \sqrt{16+4} = \sqrt{20}$$
$$\overline{AC} = \sqrt{(8-2)^2 + (5-3)^2} = \sqrt{36+4} = \sqrt{40}$$

Thus, $\overline{AB}^2 + \overline{BC}^2 = (\sqrt{20})^2 + (\sqrt{20})^2 = 40$

and $\overline{AB}^2 + \overline{BC}^2 = \overline{AC}^2$

Therefore, triangle $ABC$ is a right triangle. (It is, in fact, an isosceles right triangle.)

*Solution: Method* 2

$$m\overline{AB} = \frac{7-3}{4-2} = \frac{4}{2} = 2$$

$$m\overline{BC} = \frac{5-7}{8-4} = \frac{-2}{4} = \frac{-1}{2}$$

$$m\overline{AC} = \frac{5-3}{8-2} = \frac{2}{6} = \frac{1}{3}$$

Since the slope of $\overline{AB}$ is the negative reciprocal of the slope of $\overline{BC}$, side $\overline{AB}$ is perpendicular to side $\overline{BC}$ and triangle $ABC$ is a right triangle.

### EXERCISES

**1.** Find the slope of the line segments connecting each of the following sets of points:

(*a*) (3,9) and (9,3) (*b*) (4, −7) and (7, −1)

(*c*) (3,6) and (12,4) (*d*) (3, −4) and (7,0)

(*e*) (−2,4) and (5, −3) (*f*) (5, −1) and (3,4)

**2.** Find the slope of the line whose angle of inclination is 45°.

**3.** Show that the line segment joining the points $A(-5,-2)$ and $B(4,2)$ is parallel to the line segment joining the points $C(1,0)$ and $D(10,4)$.

**4.** Find the slope of a line perpendicular to the line passing through the points $A(4,-3)$ and $B(7,2)$.

**5.** Determine whether the points $A(-2,3)$, $B(2,5)$, $C(7,0)$, and $D(3,-2)$ are the vertices of a parallelogram.

**6.** Determine whether the points $A(-2,0)$, $B(2,4)$, $C(5,0)$, and $D(1,-4)$ are the vertices of a rectangle.

**7.** Determine whether the points $A(-2,1)$, $B(1,4)$, $C(5,0)$, and $D(2,-3)$ are the vertices of a rectangle.

**8.** Show that the line segment joining the points $A(-2,3)$ and $B(7,1)$ is perpendicular to the line segment joining the points $C(5,5)$ and $D(3,-4)$.

**9.** Determine whether the points $A(1,3)$, $B(-1,-1)$, and $C(3,7)$ all lie on the same line.

## 1-5 EQUATION OF A STRAIGHT LINE

When we speak of the equation of a straight line, we are referring to the relationship between the $x$ and $y$ coordinates for all points on the line.

If the line is parallel to the $Y$ axis and is $x_1$ units away from the $Y$ axis, the points on the line can best be described in terms of their $x$ coordinates, and the equation is given as $x = x_1$. Similarly, if the line is parallel to the $X$ axis and is $y_1$ units away, the equation is given as $y = y_1$ (Fig. 1-13).

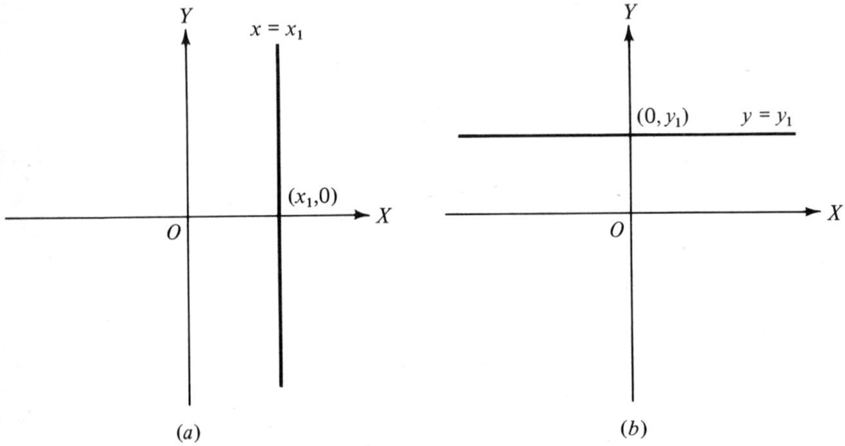


**Fig. 1-13**

Let us now consider the case in which the line is not parallel to either axis. We shall derive an equation for a straight line where we are given one point $(x_1,y_1)$ and the slope $m$. Let $(x,y)$ be the coordinates of any other point on the line (Fig. 1-14). We shall now find the relationship between $x$ and $y$. Using the formula for the slope,

$$m = \frac{y - y_1}{x - x_1}$$

$$y - y_1 = m(x - x_1)$$

This is known as the *point-slope form* of the equation of the straight line.

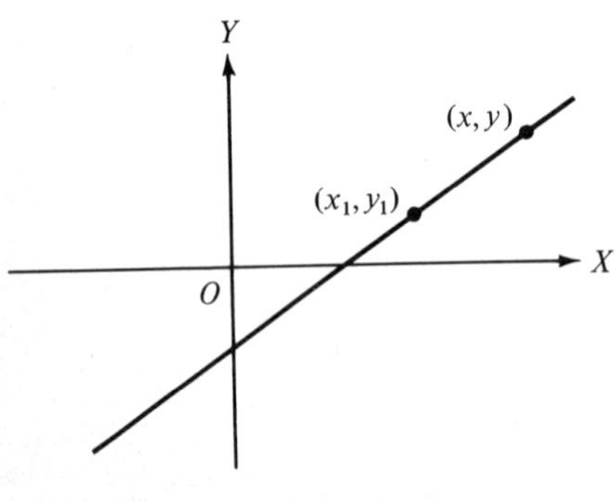


**Fig. 1-14**

**Example 1** Find an equation for the straight line whose slope is 2 and which passes through the point $(2,-1)$.

*Solution* Using the point-slope form,

$$y - y_1 = m(x - x_1)$$
$$y + 1 = 2(x - 2)$$

If we have as our known point a point on the $Y$ axis, say at $(0,b)$, the equation then becomes

$$y - b = m(x - 0)$$
$$y - b = mx$$
$$y = mx + b$$

This is known as the *slope-intercept* form. This form is most useful when we are given the slope $m$ and the $y$ intercept $y = b$ (Fig. 1-15).

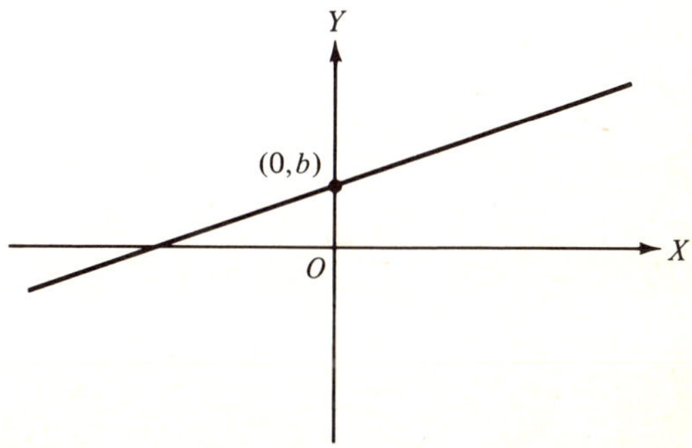


**Fig. 1-15**

**Example 2** Find the equation of the straight line whose slope is 3 and whose $y$ intercept is 1.

*Solution* From the given information, we know that $m = 3$ and $b = 1$. Thus, the equation of the line is

$$y = 3x + 1$$

**Example 3** Find the equation of the straight line passing through the points (2,3) and (6,9).

*Solution* We shall use the point-slope form, but first we must find the slope.

$$m = \frac{9 - 3}{6 - 2} = \frac{6}{4} = \frac{3}{2}$$
$$y - 3 = \tfrac{3}{2}(x - 2)$$

**Example 4** Find the equation of the line passing through the points $(a,0)$ and $(0,b)$.

*Solution*

$$m = -\frac{b}{a}$$

$$y - b = -\frac{b}{a}x$$

$$y = -\frac{b}{a}x + b$$

Though the equation may be left in this form, we shall continue:

$$ay = -bx + ab$$
$$bx + ay = ab$$

Now, dividing each term by $ab$,

$$\frac{x}{a} + \frac{y}{b} = 1$$

This is known as the *intercept* form of the equation of a straight line. It is a convenient form to use when we are given the $x$ intercept $x = a$ and the $y$ intercept $y = b$.

The various forms of the equation of the straight line are often changed to the form $Ax + By + C = 0$, which is known as the general form of the equation of the straight line.

**EXERCISES**

**1.** Find the equation of each of the following lines:
   (*a*) Slope $m = -2$ and passing through $(-2,4)$
   (*b*) Passing through the points $(2,-3)$ and $(7,5)$
   (*c*) Passing through the points $(0,4)$ and $(-1,-2)$

**2.** Find the equation of each of the following lines:
   (*a*) Passing through the points $(6,7)$ and $(-2,3)$
   (*b*) Passing through the points $(3,5)$ and $(3,-2)$
   (*c*) Passing through the points $(-2,-4)$ and $(7,-4)$

**3.** Find the slope of the line $7x - 4y = 8$.

**4.** Find the equation of the line that passes through the point $(2,3)$ and is parallel to the line $3x - 2y = 10$.

**5.** Find the equation of the line that passes through the point $(-1,-1)$ and is perpendicular to the line $x + 2y = 4$.

**6.** Find the equation of the line passing through the point $(2,7)$ and having an angle of inclination $i = 45°$.

**7.** Find the equation of the line passing through the point $(5,2)$ and parallel to the line $2x - 5y = 2$.

**8.** Find the equation of the line passing through the point $(1,6)$ and perpendicular to the line $3x - y = 7$.

**9.** Find the equation of the perpendicular bisector of the line segment joining the points $(2,3)$ and $(8,5)$.

**10.** Find the equation of the perpendicular bisector of the line segment joining the points $(-1,2)$ and $(7,6)$.

## 1-6 PROVING THEOREMS ANALYTICALLY

With the aid of the formulas we have developed, and the use of our coordinate system, we can now prove some of the theorems of plane geometry in a much simpler way. In proving theorems analytically, it is very important to place the figure in the simplest position so as to make use of our coordinate system. The labeling of the coordinates of the points will also be very important.

**Example 1** Prove analytically that the midpoint of the hypotenuse of any right triangle is equidistant from the three vertices.

*Solution* First we construct our coordinate axes. Then we try to place the figure in the simplest position. We can take one vertex at the origin, and, since we have a right triangle, the two legs of the triangle can lie on the two axes. We then label the three vertices as $A$, $B$, $C$. The point $A$ at the origin has coordinates $(0,0)$. The point $B$ on the $X$ axis has a $y$ coordinate equal to zero, and an $x$ coordinate which we can label $a$. The point $C$ on the $Y$ axis has an $x$ coordinate equal to zero and a $y$ coordinate which we can label $b$.

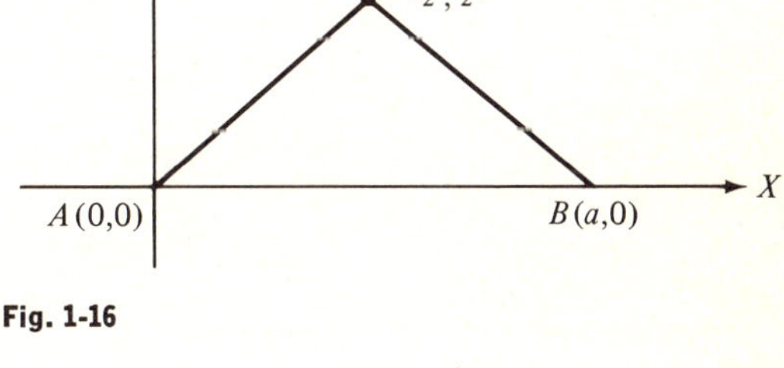


**Fig. 1-16**

We are interested in the midpoint of $\overline{BC}$, which we shall label as the point $D$. The $x$ coordinate of $D$ is $x = \dfrac{a+0}{2}$ and the $y$ coordinate is $y = \dfrac{b+0}{2}$. Thus $D$ is the point $\left(\dfrac{a}{2}, \dfrac{b}{2}\right)$ (Fig. 1-16). We must now show that the distance from

$A$ to $D$ is the same as the distances from $B$ to $D$ and from $C$ to $D$ (that is, $\overline{AD} = \overline{BD} = \overline{CD}$). Using the formula,

$$\overline{AD} = \sqrt{\left(\frac{a}{2}\right)^2 + \left(\frac{b}{2}\right)^2} = \sqrt{\frac{a^2}{4} + \frac{b^2}{4}}$$

$$\overline{BD} = \sqrt{\left(a - \frac{a}{2}\right)^2 + \left(\frac{b}{2}\right)^2} = \sqrt{\frac{a^2}{4} + \frac{b^2}{4}}$$

$$\overline{CD} = \sqrt{\left(\frac{a}{2}\right)^2 + \left(\frac{b}{2} - b\right)^2} = \sqrt{\frac{a^2}{4} + \frac{b^2}{4}}$$

Thus, $\overline{AD} = \overline{BD} = \overline{CD}$, and the theorem is proved.

**Example 2** Prove analytically that the line segment joining the midpoints of two sides of a triangle is parallel to the third side and equal to half of its length.

*Solution* We first note that since we are not given a right triangle, we cannot use the two axes as two of the sides of the triangle. However, we can still use one of the axes, and we shall place the triangle on the $X$ axis with one vertex at the origin. We will label the triangle $ABC$ with vertex $A$ at the origin and vertex $B$ on the $X$ axis. We will label the point $B$ as $(2a,0)$ and the point $C$ as $(2b,2c)$. We have used $2a$, $2b$, $2c$ rather than $a$, $b$, $c$ so that the coordinates of the midpoints will not involve fractions. We will label the midpoint of $\overline{AC}$ as $D$, and the midpoint of $\overline{BC}$ as $E$. Using the midpoint formula, $D = (b,c)$ and $E = (a + b, c)$ (Fig. 1-17).

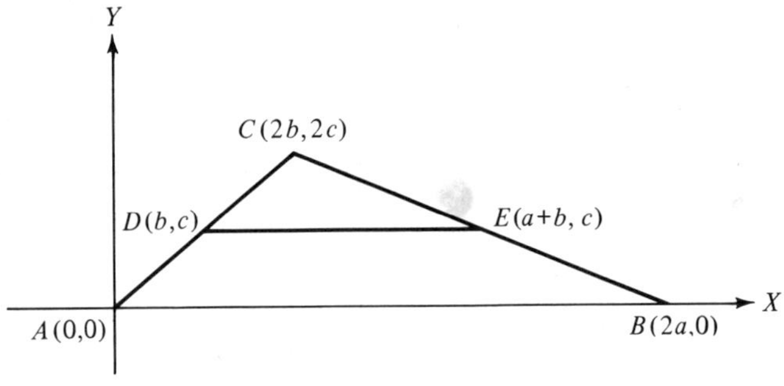


**Fig. 1-17**

We must show that $\overline{DE}$ is parallel to $\overline{AB}$ and $\overline{DE}$ is $\frac{1}{2}\overline{AB}$. Since

$$m\overline{AB} = 0 \qquad \text{and} \qquad m\overline{DE} = \frac{c - c}{(a + b) - b} = 0$$

$\overline{DE}$ is parallel to $\overline{AB}$.

The length of $\overline{AB}$ is $2a$, and, using the distance formula,

$$\overline{DE} = \sqrt{(a + b - b)^2 + (c - c)^2} = \sqrt{a^2} = a$$

Thus $$\overline{DE} = \tfrac{1}{2}\overline{AB}$$

**Example 3** Prove analytically that the line segments joining the consecutive midpoints of the sides of any quadrilateral form a parallelogram.

*Solution* We will label the quadrilateral as $ABCD$ with $A = (0,0)$, $B = (2a,0)$, $C = (2b,2c)$, and $D = (2d,2e)$. We will label the corresponding midpoints $E$, $F$, $G$, and $H$, and, by the midpoint formula, $E = (a,0)$, $F = (a + b,\ c)$, $G = (b + d,\ c + e)$, and $H = (d,e)$ (Fig. 1-18). We must now show that $\overline{EF}$ is parallel to $\overline{HG}$ and $\overline{FG}$ is parallel to $\overline{EH}$.

$$m\overline{EF} = \frac{c - 0}{a + b - a} = \frac{c}{b} \qquad m\overline{HG} = \frac{c + e - e}{b + d - d} = \frac{c}{b}$$

$$m\overline{FG} = \frac{c + e - c}{b + d - a - b} = \frac{e}{d - a} \qquad m\overline{EH} = \frac{e - 0}{d - a} = \frac{e}{d - a}$$

Therefore, $\overline{EF} \| \overline{HG}$ and $\overline{FG} \| \overline{EH}$

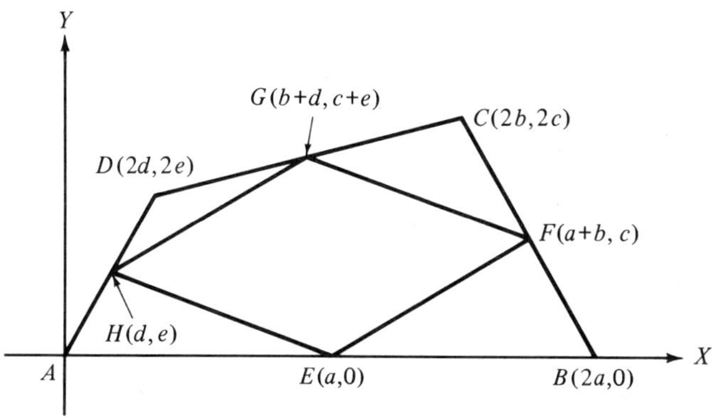


**Fig. 1-18**

**EXERCISES**

Prove the following theorems analytically.

1. The diagonals of any parallelogram bisect each other.
2. If the diagonals of a parallelogram are equal, the figure is a rectangle.
3. The diagonals of a rhombus are perpendicular to each other.
4. The line segment joining the midpoints of the nonparallel sides of a trapezoid is parallel to the bases.
5. The line segments joining the midpoints of the successive sides of a rhombus form a rectangle.
6. The line segments joining the midpoints of the successive sides of a rectangle form a rhombus.
7. If the diagonals of a parallelogram are perpendicular to each other, the parallelogram is a rhombus.
8. The sum of the squares of the lengths of the sides of any parallelogram is equal to the sum of the squares of the lengths of its diagonals.

# 2

# Equations of Curves and Curve Sketching

## 2-1 DISCUSSION OF AN EQUATION

One of the fundamental problems in analytic geometry is to determine the geometric representation of a given equation. This geometric representation is generally referred to as the *graph* of the equation. Thus, the graph consists of all those points whose coordinates satisfy the given equation. Generally, it is impossible to plot all points whose coordinates satisfy the equation, since there may be an infinite number of such points. After plotting several points, the general procedure is to draw a smooth curve through these plotted points. In following this procedure, we are assuming that this smooth curve actually gives us a good approximation of the appearance of the curve. This assumption is not always correct.

We can often increase the accuracy of our sketch and decrease the labor involved by first making an algebraic examination of the given equation. This algebraic examination may bring to light certain properties of the curve which will enable us to sketch the curve by plotting only a few well-chosen points.

The algebraic examination is generally called a discussion of the equation. Our discussion will consist of four items: intercepts, symmetry, extent, and horizontal and vertical asymptotes.

#### INTERCEPTS

The points where the curve crosses the $X$ axis are known as the $x$ intercepts, and the points where the curve crosses the $Y$ axis are known as the $y$ intercepts. Since the $y$ coordinate for all points on the $X$ axis is 0, to find the $x$ intercepts we have merely to set $y = 0$ in the equation and then solve for $x$. Similarly, to find the $y$ intercepts, we set $x = 0$ and solve for $y$.

**Example 1** Find the intercepts of the graph of $y = x^2 - 4$.

*Solution* To find the $x$ intercept, set $y = 0$:

$$0 = x^2 - 4$$
$$x^2 = 4$$
$$x = \pm 2$$

Similarly, to find the $y$ intercept, set $x = 0$:

$$y = -4$$

Thus, the $x$ intercepts are $\pm 2$, and the $y$ intercept is $-4$.

#### SYMMETRY

We shall be concerned mainly with symmetry with respect to the $X$ and $Y$ axes and with respect to the origin. We will first consider some basic definitions, and then construct a test for symmetry.

**Definition** *A point $Q$ is said to be the* symmetric point of a point $P$ with respect to a line $L$ *if $L$ is the perpendicular bisector of the segment* $\overline{PQ}$ (*Fig.* 2-1).

**Definition** *A point $Q$ is said to be the* symmetric point of a point $P$ with respect to a point $O$ *if the point $O$ is the midpoint of the segment* $\overline{PQ}$ (*Fig.* 2-2).

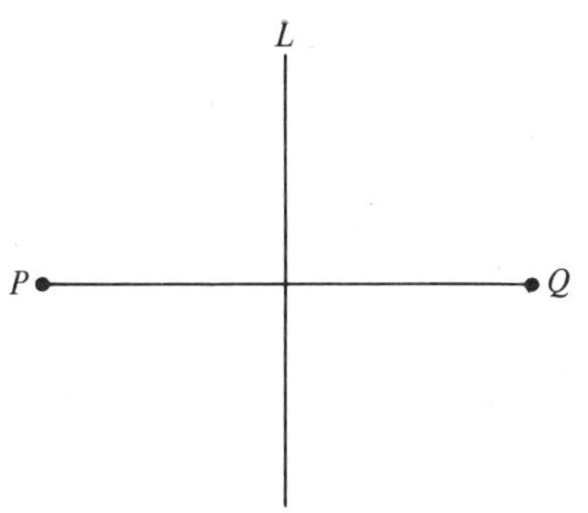

**Fig. 2-1**

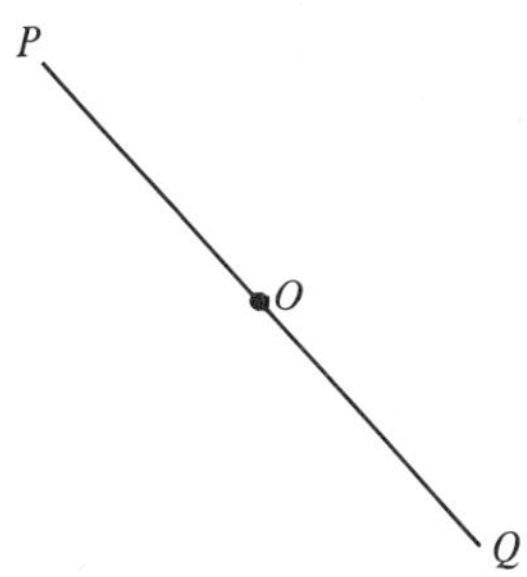

**Fig. 2-2**

**Definition** *A curve is said to be symmetric with respect to a line L (or a point O) if for every point on the curve its symmetric point with respect to L (or O) is also on the curve (Fig. 2-3).*

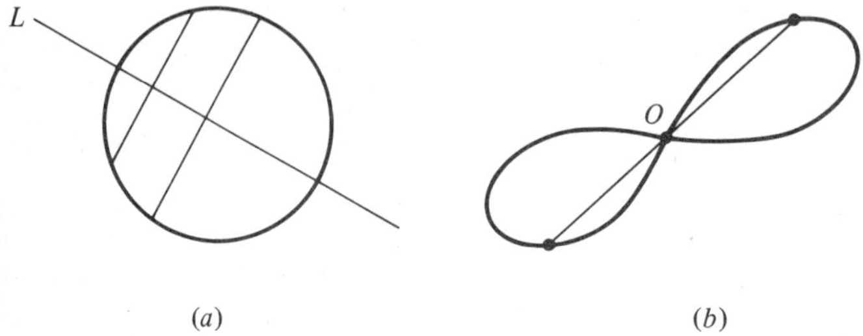


**Fig. 2-3**

We shall now consider symmetry with respect to the coordinate axes and the origin. Let $P$ be any point on the curve, and let the coordinates of $P$ be $(x,y)$. The symmetric point of $P$ with respect to the $X$ axis must have coordinates $(x,-y)$. Thus, if the curve is symmetric with respect to the $X$ axis for every point $(x,y)$ on the curve, the point $(x,-y)$ must also be on the curve and must satisfy the equation of the curve. Conversely, if for every set of coordinates $(x,y)$ satisfying the equation of the curve, the coordinates $(x,-y)$ also satisfy the equation, then the curve is symmetric with respect to the $X$ axis. Similarly, if the curve is symmetric with respect to the $Y$ axis, for each point $(x,y)$ the point $(-x,y)$ must also be on the curve and must also satisfy the equation of the curve. Also, if the curve is symmetric with respect to the origin, for each point $(x,y)$ the point $(-x,-y)$ must also be on the curve and must also satisfy the equation of the curve.

We summarize the above discussion with the following tests for symmetry (Fig. 2-4):

1. To test the equation for symmetry with respect to the $X$ axis, replace $y$ by $-y$. If, and only if, the resulting equation is equivalent to the original equation, the curve is symmetric with respect to the $X$ axis.
2. To test for symmetry with respect to the $Y$ axis, replace $x$ by $-x$. If the equation remains equivalent to the original equation, the curve is symmetric with respect to the $Y$ axis.
3. To test for symmetry with respect to the origin, replace $x$ by $-x$ and $y$ by $-y$. If the equation remains equivalent to the original equation, the curve is symmetric with respect to the origin.

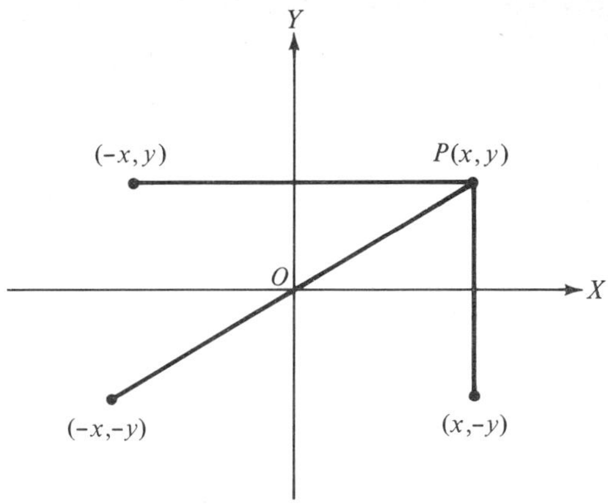


Fig. 2-4

**Example 2** Test the following equations for symmetry:

(a) $x^2 + y^2 = 4$
(b) $y = x^2 - 4$
(c) $xy = 4$

*Solution*

| | | |
|---|---|---|
| (a) | i. $(-x)^2 + y^2 = 4$ | Curve is symmetric with respect to the $Y$ axis |
| | $x^2 + y^2 = 4$ | |
| | ii. $x^2 + (-y)^2 = 4$ | Symmetric with respect to $X$ axis |
| | $x^2 + y^2 = 4$ | |
| | iii. $(-x)^2 + (-y)^2 = 4$ | Symmetric with respect to origin |
| | $x^2 + y^2 = 4$ | |
| (b) | i. $y = (-x)^2 - 4$ | Symmetric with respect to $Y$ axis |
| | $y = x^2 - 4$ | |
| | ii. $-y = x^2 - 4$ | Not symmetric with respect to $X$ axis |
| | iii. $-y - (-x)^2 - 4$ | Not symmetric with respect to origin |
| | $-y = x^2 - 4$ | |
| (c) | i. $(-x)y = 4$ | Not symmetric with respect to $Y$ axis |
| | $-xy = 4$ | |
| | ii. $x(-y) = 4$ | Not symmetric with respect to $X$ axis |
| | $-xy = 4$ | |
| | iii. $(-x)(-y) = 4$ | Symmetric with respect to origin |
| | $xy = 4$ | |

**EXTENT**

When given the equation for $y$ in terms of $x$, it will often occur that for certain values of $x$ the corresponding values of $y$ are imaginary or perhaps nonexistent. A careful analysis of the equation of the curve will often reveal to us the sets of values permissible for $x$ and for $y$, and the corresponding regions of the plane that will either be included or excluded.

**Example 3** Consider the equation $x^2 + y^2 = 25$. If we solve for $y$ in terms of $x$, we get

$$y = \pm\sqrt{25 - x^2}$$

For $-5 \leq x \leq 5$ (less than or equal to $+5$ but greater than or equal to $-5$), the corresponding $y$ values will be real. However, for $x > 5$ or $x < -5$, the corresponding $y$ values will be imaginary. Similarly, in solving for $x$ in terms of $y$,

$$x = \pm\sqrt{25 - y^2}$$

and only values of $y$ for which $-5 \leq y \leq 5$ will give corresponding real values for $x$. Thus, we know that the entire curve lies within the region from $x = -5$ to $x = 5$ and $y = -5$ to $y = 5$ (Fig. 2-5).

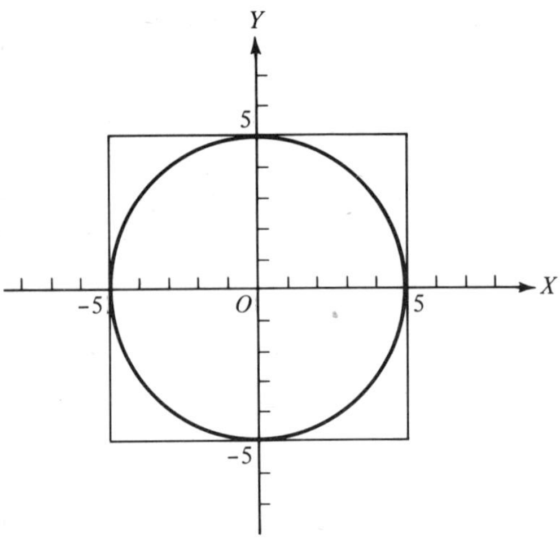


**Fig. 2-5**

**Example 4** Consider the equation $xy = 1$. When we solve for $y$ in terms of $x$, we get

$$y = \frac{1}{x}$$

In this case for no value of $x$ will the corresponding $y$ value be imaginary. However, for $x = 0$ there is no corresponding $y$ value. (Why?) Similarly, $x = \frac{1}{y}$ and for $y = 0$ there is no corresponding $x$ value. Thus, all $x$ and all $y$ values are permissible except $x = 0$ and $y = 0$. A careful examination of the equation will also reveal that when $x$ is positive $y$ is also positive and when $x$ is negative $y$ is negative. We can thus conclude that the curve will appear only in the first and third quadrants where the signs of $x$ and $y$ are both the same (Fig. 2-6).

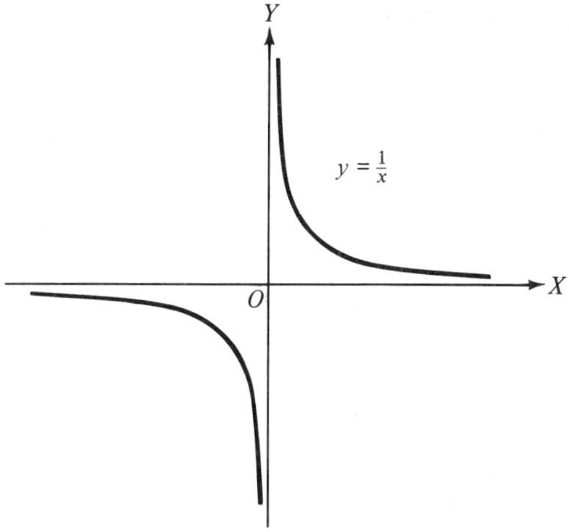


Fig. 2-6

#### HORIZONTAL AND VERTICAL ASYMPTOTES

An asymptote may be defined as a line such that as the $x$ or $y$ values of the curve get infinitely large, the distance between the curve and the line approaches 0. We shall presently consider only vertical and horizontal asymptotes.

**Example 5** We have already considered the equation $y = \frac{1}{x}$ (Fig. 2-6). We wish to examine the behavior of the curve when the values of $x$ get infinitely large and also when the values of $y$ get infinitely large. When the values of $x$ get infinitely large, the values of $\frac{1}{x}$ get very small and close to 0. Thus, as $x$ gets infinitely large, the distance between the curve $y = \frac{1}{x}$ and the line $y = 0$ approaches 0, and by definition the line $y = 0$ is an asymptote. Similarly, we can rewrite the equation in the form $x = \frac{1}{y}$. As the values of $y$ get infinitely large, the values of $\frac{1}{y}$ get very small and close to 0, and the line $x = 0$ is an asymptote.

As a practical procedure in determining the asymptotes of a curve, it is often convenient to detect the asymptotes by finding the values of $x$ such that as $x$ approaches $x_1$, $y$ becomes infinitely large, and the values of $y$ such that as $y$ approaches $y_1$, $x$ becomes infinitely large. As in this example, when our equation is in the form of a fraction, the values of the variable that make the denominator equal to 0 will give us a clue as to what the asymptote will be.

For the curve $y = \frac{1}{x}$, the denominator will equal 0 when $x = 0$. When $x > 0$ but very close to 0, the value of $y$ becomes very large and is positive. The closer $x$ comes to 0, the larger the corresponding value of $y$. This is generally described by saying that $y$ approaches infinity ($y \to \infty$) as $x$ approaches 0. When $x < 0$ but is very close to 0, then the value of $y$ becomes very large but negative. The closer $x$ comes to 0, the more negative the value of $y$ (that is, $y \to -\infty$). Thus, $x = 0$ is a vertical asymptote. Similarly, $y = 0$ is a horizontal asymptote.

**Example 6** Find the vertical and horizontal asymptotes for the graph of the equation $xy + 2y - 4 = 0$.

*Solution* Solving for $y$ in terms of $x$, we get

$$y(x + 2) = 4$$

$$y = \frac{4}{x + 2}$$

For $x > -2$, as $x \to -2$, $y \to \infty$.
For $x < -2$, as $x \to -2$, $y \to -\infty$.

Thus, $x = -2$ is a vertical asymptote.
Solving for $x$ in terms of $y$, we get:

$$xy = 4 - 2y$$

$$x = \frac{4 - 2y}{y}$$

For $y > 0$, as $y \to 0$, $x \to \infty$.
For $y < 0$, as $y \to 0$, $x \to -\infty$.

Thus, $y = 0$ is the horizontal asymptote (Fig. 2-7).

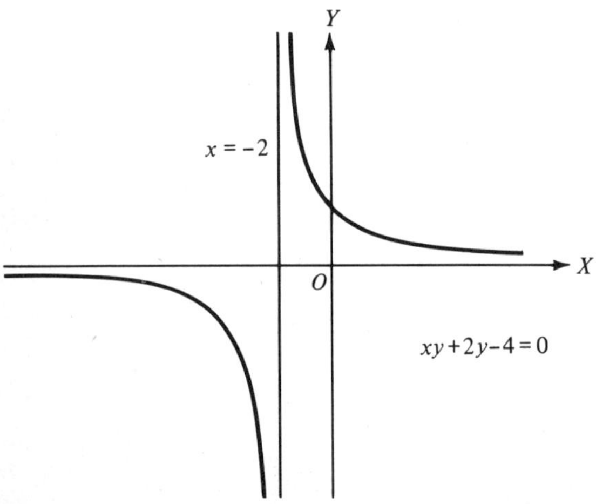


Fig. 2-7

## 2-2 CURVE SKETCHING

**Example 1** Give a complete discussion of the equation $x^2y - 4y + 8 = 0$ and sketch the curve.

*Solution* We shall determine intercepts, symmetry, extent, and asymptotes, and we will use this information to sketch the curve.

(*a*) *Intercepts* When $x = 0$,

$$\begin{aligned} -4y + 8 &= 0 \\ y &= 2 \end{aligned}$$

When $y = 0$ we get the erroneous statement $8 = 0$. Thus, there is no $x$ intercept.

(*b*) *Symmetry* If we replace $x$ by $-x$, we get

$$\begin{aligned} (-x)^2y - 4y + 8 &= 0 \\ x^2y - 4y + 8 &= 0 \end{aligned}$$

The equation remains unchanged, and we have symmetry with respect to the $Y$ axis.

If we replace $y$ by $-y$, we get

$$\begin{aligned} x^2(-y) - 4(-y) + 8 &= 0 \\ -x^2y + 4y + 8 &= 0 \end{aligned}$$

The equation is changed and there is no symmetry with respect to the $X$ axis. There is also no symmetry with respect to the origin.

(*c*) *Extent* Solving for $y$ in terms of $x$,

$$y = \frac{-8}{x^2 - 4}$$

For $x = \pm 2$ there is no corresponding $y$ value. Thus, all $x$ values can be used except $x = \pm 2$. Solving for $x$ in terms of $y$, we get

$$\begin{aligned} x^2 &= \frac{4y - 8}{y} \\ x &= \pm\sqrt{\frac{4y - 8}{y}} \\ &= \pm 2\sqrt{\frac{y - 2}{y}} \end{aligned}$$

For $y = 0$ there is no corresponding $x$ value. For $y < 0$ the corresponding $x$ values are real. For $y > 2$ the corresponding $x$ values are real. However, for $0 < y < 2$ the corresponding $x$ values are imaginary. Thus, all $y$ values can be used except $0 \leq y < 2$.

(*d*) *Horizontal and vertical asymptotes* Solving for $y$ in terms of $x$,

$$y = \frac{-8}{x^2 - 4} = \frac{-8}{(x + 2)(x - 2)}$$

When $x > 2$ and $x \to 2$, $y \to -\infty$.
When $x < 2$ and $x \to 2$, $y \to \infty$.
When $x > -2$ and $x \to -2$, $y \to \infty$.
When $x < -2$ and $x \to -2$, $y \to -\infty$.

Thus, $x = 2$ and $x = -2$ are the vertical asymptotes.
Solving for $x$ in terms of $y$,

$$x = \pm 2\sqrt{\frac{y-2}{y}}$$

We recall that there are no $y$ values between 0 and 2.

When $y < 0$ and $y \to 0$, $x \to \infty$ and $x \to -\infty$.

Thus, $y = 0$ is a horizontal asymptote. Using this information, we can then sketch (Fig. 2-8).

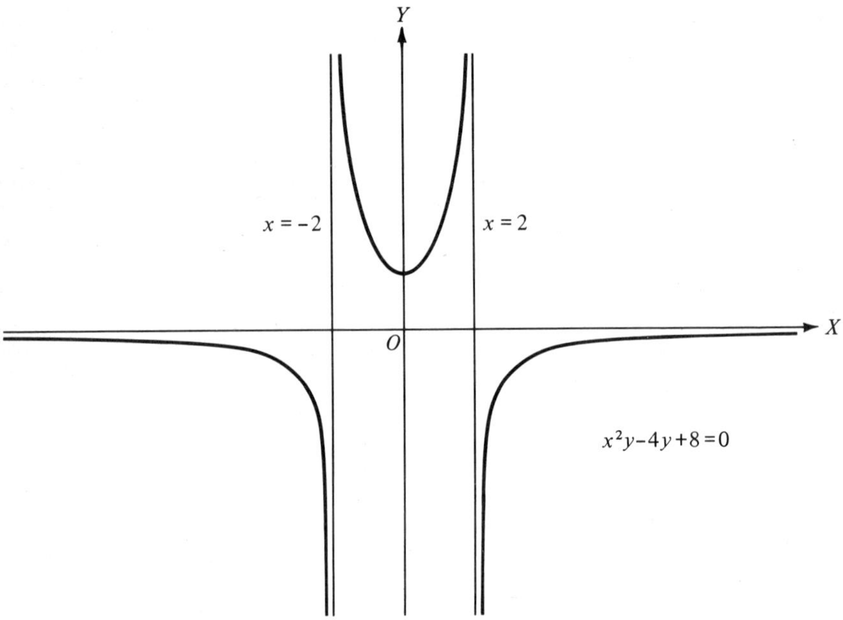


Fig. 2-8

**EXERCISES**

For the following equations give a discussion of the equation (including intercepts, symmetry, extent, and asymptotes) and sketch the curve.

**1.** $x^2 + y^2 = 36$

**2.** $4x^2 - 9y^2 = 36$

**3.** $y^2 = 4x$

**4.** $xy = 16$

**5.** $xy - 3y - 9 = 0$

**6.** $9x^2 + 16y^2 = 144$

**7.** $x^2y - 4y - 2 = 0$

**8.** $y = 4 - x^2$

**9.** $y = x^3$

**10.** $xy - y - 1 = 0$

**11.** $y = x^2 - 6x + 9$

**12.** $x^2y = 1$

# 3 Functions

## 3-1 REPRESENTING A FUNCTION

In this chapter we shall consider the question of the relationship among various quantities and in particular the concept of a correspondence between two specific quantities. There may, for example, be a correspondence between a person's income and his level of education, a person's wages and the number of hours worked, a person's weight and the number of calories consumed daily, the cost of an airplane ticket and the distance to be traveled. We shall now give the necessary definitions to characterize that important relationship between quantities known as a *function.*

**Definition** *A* variable *is a symbol that may take on any value in some specified set of numbers.*

**Definition** *A* constant *is a symbol that retains the same value throughout.*

*Remark* Letters from the beginning part of the alphabet are often used to denote constants (for example, $a$, $b$, $c$), while letters from the latter part of the alphabet are often used to denote variables (for example, $x$, $y$, $z$).

**Definition** *A* function *is a set of ordered pairs* $(x,y)$, *where the set consisting of all the first elements is called the* domain *of the function, the set consisting of all the second elements is called the* range *of the function, and each element of the domain is the first element of exactly one ordered pair.*

**Example 1** The set of ordered pairs of numbers $\{(1,2), (3,6), (5,10), (7,14), (9,18)\}$ is a function. The domain of the function is the set of numbers $\{1, 3, 5, 7, 9\}$ and the range of the function is the set $\{2, 6, 10, 14, 18\}$.

In representing a function as a set of ordered pairs of numbers in the form $(x,y)$, we are using the variable $x$ to represent the numbers in the domain of the function and the letter $y$ to represent the numbers in the range of the function.

A function is generally denoted by a single letter such as $F$, $G$, $f$, etc. The ordered pair $(a,b)$ for the function $F$ can be denoted as $F(a) = b$ (read "$F$ of $a$ equals $b$"), and we say that $b$ is the value of the function $F$ at $a$.

In practice, it is usually not convenient, and sometimes not possible, to list all the ordered pairs of a function. Instead, the function can be completely determined by giving the domain of the function together with a rule to find the corresponding function values in the range of the function.

**Example 2** List the ordered pairs $(x,y)$ of the functions $f$ whose rule of correspondence is $y = x^2$ and whose domain is the set of integers $\{1, 2, 3, 4, 5\}$.

*Solution*

$$\{(1,1), (2,4), (3,9), (4,16), (5,25)\}$$

The rule of correspondence for the above function is sometimes denoted as $f(x) = x^2$ and also $(x,x^2)$. No matter what notation is used to denote a function, the variable used to represent the domain of the function is called the *independent variable*. The variable used to represent the range of the function is called the *dependent variable*. Thus, when $y = f(x) = x^2$, $x$ is the independent variable and $y$ is the dependent variable.

Since a function is a set of ordered pairs $(x,y)$, we can associate with each ordered pair $(x,y)$ a point in the coordinate plane where the $x$ coordinate of the point is the $x$ value in the ordered pair $(x,y)$, and the $y$ coordinate is the $y$ value in $(x,y)$. Thus, we can graph functions just as we did equations. This will be especially useful for those functions whose domain and range consist of an infinite set of numbers.

**Example 3** Sketch the graph of the function whose domain is the set of real numbers $1 \leq x \leq 3$, with rule of correspondence $f(x) = x + 2$.

*Solution* There are an infinite number of $x$ values between $x = 1$ and $x = 3$. Since the corresponding $y$ values are always 2 more than the $x$ values, the range also contains an infinite set of numbers and goes from $y = 3$ to $y = 5$ (Fig. 3-1).

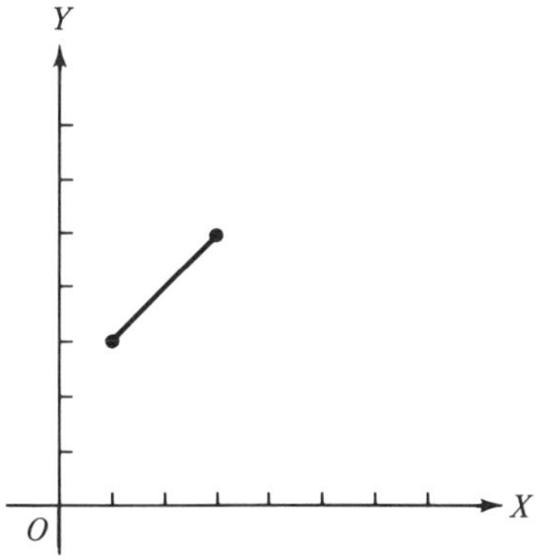


**Fig. 3-1**

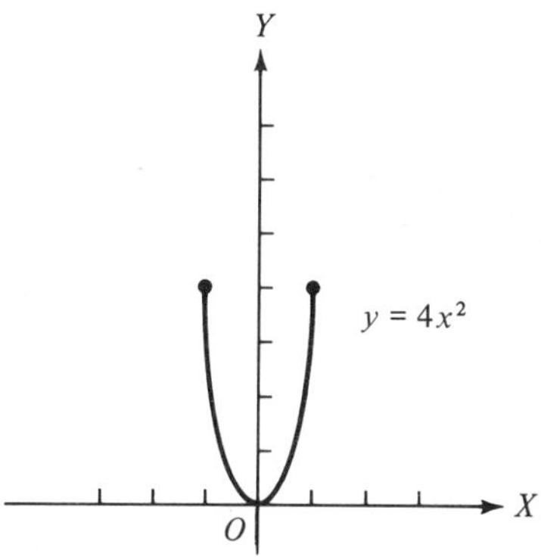


**Fig. 3-2**

**Example 4** Sketch the graph of the function whose domain is the set of real numbers $-1 \leq x \leq 1$, with rule of correspondence $y = 4x^2$.

*Solution* The range of the function is $0 \leq y \leq 4$ (Fig. 3-2).

When an equation is used as a rule of correspondence to describe a function, and one wishes to include all possible values in the domain of the function, the specification of the domain is often omitted and it is understood to be all values for which the equation can be used.

**Example 5** If $y = \dfrac{x}{x+2}$ is used as a rule of correspondence for a function, the domain is understood to be all values of $x$, except $x = -2$, for which the equation is not defined. If we wish to find the range of the function we can solve for $x$ in terms of $y$.

$$\begin{aligned} xy + 2y &= x \\ xy - x &= -2y \\ x(y - 1) &= -2y \\ x &= \frac{-2y}{y - 1} \end{aligned}$$

Only $y = 1$ will not appear in the range.

**EXERCISES**

For the functions in Exercises 1 to 4 determine the domain and range of the function.

**1.** $F = \{(1,3), (2,4), (3,5), (5,7), (6,9)\}$

**2.** $G = \{(1,1), (2,2), (3,5), (6,7)\}$

**3.** $f = \{(1,2), (3,4), (5,6), (7,8), (9,10), (11,12)\}$

**4.** $h = \{(2,1), (3,2), (4,1), (5,0)\}$

**5.** Determine whether the following sets of ordered pairs of numbers form a function. If they do, determine the domain and the range.

(*a*) $\{(1,2), (2,3), (3,4), (4,5), (5,2)\}$
(*b*) $\{(6,5), (4,3), (2,1), (0,0), (4,2)\}$
(*c*) $\{(5,2), (4,2), (3,2), (2,2), (1,2)\}$
(*d*) $\{(1,2), (1,3), (1,4), (1,5)\}$

**6.** Sketch the functions whose rules of correspondence are:

(*a*) $y = 2x,\ 0 \le x \le 3$
(*b*) $y = \frac{2}{3}\sqrt{9 - x^2},\ -3 \le x \le 3$
(*c*) $y = x^2 + 1,\ -1 \le x \le 1$

**7.** Sketch the function whose rule of correspondence is:

$$y = \begin{cases} 1 & 0 \le x < 1 \\ 2 & 1 \le x \le 2 \end{cases}$$

**8.** For the following functions determine the possible set of values that can be used for $x$ in the domain, and find the possible set of values for $y$ in the range.

(*a*) $y = \dfrac{x - 4}{x + 1}$ (*b*) $y = \sqrt{x}$

(*c*) $y = x^3$ (*d*) $y = \dfrac{2}{x^2 + 2}$

**9.** Given the function $f(x) = x^2$, find $f(2)$, $f(3)$, $f(h)$, $f(x + h)$.

**10.** Given the function $f(x) = \dfrac{1}{x}$, find $f(1)$, $f(3)$, $f\left(\dfrac{1}{x}\right)$, $f(x + h)$.

**11.** Given the function $f(x) = 1 - 2x^2$, find $f(0)$, $f(\frac{1}{2})$, $f(a)$, $f(a + 1)$.

**12.** Given the function $f(x) = \dfrac{1}{x + 1}$, find $f(0)$, $f(1)$, $f(x + h)$, $[f(x + h) - f(x)]$.

## 3-2 DETERMINING THE RULE OF CORRESPONDENCE FOR FUNCTIONS GIVEN BY VERBAL STATEMENTS

Many verbal problems involving functions can be solved by determining the rule of correspondence for the function and working with the algebraic equation. In this section we shall be concerned with finding a formula in mathematical symbols to represent the information given in a verbal statement.

**Example 1** If one leg of a right triangle has a length of 10 in., express the length of the other leg as a function of the hypotenuse.

*Solution* We shall label the hypotenuse $h$ and the unknown leg $x$ (Fig. 3-3). By the pythagorean theorem:

$$h^2 = x^2 + 100$$
$$x^2 = h^2 - 100$$
$$\dagger x = \sqrt{h^2 - 100}$$

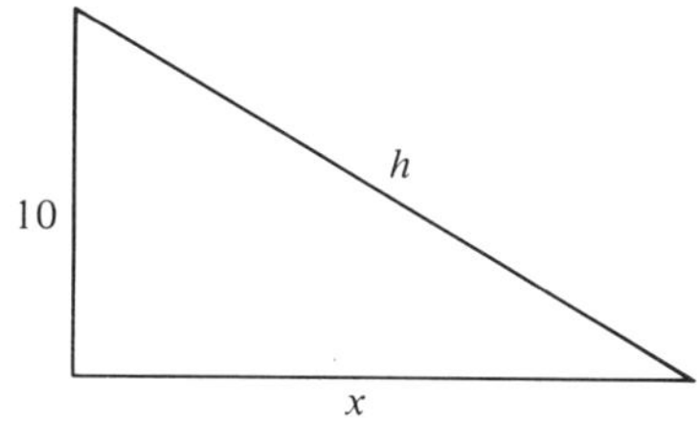


**Fig. 3-3**

**Example 2** A merchant estimates that he can sell each of his 50 tomatoes at the current price of 10 cents per tomato. For every day that he waits, the price of tomatoes will increase by 1 cent, but 2 tomatoes will spoil each day. If all tomatoes are sold the same day, express the total income if the merchant waits $x$ days before selling.

*Solution*

$$\begin{aligned}\text{Total income} &= (\text{number of articles sold})\ (\text{price per article}) \\ &= (50 - 2x)(10 + x) \\ &= 500 + 30x - 2x^2\end{aligned}$$

**EXERCISES**

**1.** Express the area of a circle as a function of the radius $r$.

**2.** For a right circular cylinder of radius 5, express the volume as a function of the height $h$.

**3.** If 40 ft of wire are available for the construction of a rectangular fence, express the area $A$ as a function of the length $x$ of one side.

**4.** An open box is to be made from a piece of cardboard 20 in. long and 10 in. wide by cutting out a square from each corner and turning up the sides. Express the volume of the box as a function of the side of the cut-out square.

**5.** If a rectangular area is to be enclosed by 80 ft of fence, express the area $A$ as a function of the length $x$ of one side.

**6.** A magazine publisher finds that he can get 10,000 subscribers at a yearly rate of \$4.00 each. He will get 200 more subscribers for each 15-cent decrease in the rate.

(*a*) Express the number of subscribers as a function of $x$, if the price has been reduced 15 cents $x$ number of times.

(*b*) Express the total income as a function of $x$.

**7.** A fence is needed to enclose a rectangular area consisting of 2,000 $yd^2$ in a beach area. Only three sides of fence are necessary, since no fence is needed at the water's edge. Express the amount of fence needed as a function of the length $x$ of one side.

† Note: Since $x$ represents a length, its value is positive and therefore $x$ equals the positive square root.

### 3-3 INCREMENTS

The term *increment of x* refers to the change in the value of $x$. If we change the value of $x$ from $x = 1$ to $x = 1.3$, then the increment of $x$ is 0.3. If we change the value of $x$ from $x = 2$ to $x = 1.98$, then the increment of $x$ is $-0.02$. The symbol used to represent the increment of $x$ is $\Delta x$, which is read "delta $x$." It is important to remember that $\Delta x$ is one symbol and cannot be thought of as $\Delta$ times $x$.

If we have a rule of correspondence for a function which gives us $y$ in terms of $x$, $y = f(x)$, then a change in the value of $x$ will produce a corresponding change in the value of $y$ which we will refer to as $\Delta y$. Graphically, if $A$ is the point $(x_1,y_1)$ and we change the value of $x$ from $x_1$ to $x_1 + \Delta x$, the corresponding $y$ value will change from $y_1$ to $y_1 + \Delta y$ (Fig. 3-4). The change in $x$, $\Delta x$, is represented in the graph by $\overline{AC}$; the corresponding change in $y$, $\Delta y$, is represented by $\overline{BC}$.

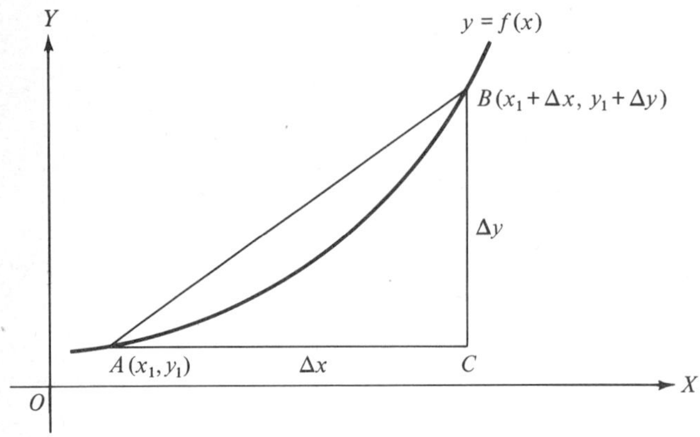


**Fig. 3-4**

**Example 1** If $y = f(x) = x^2$, then when $x = 1$, $y = 1$. Suppose we change the value of $x$ from 1 to 1.2. What is $\Delta x$ and what is $\Delta y$?

*Solution*

$$\Delta x = 0.2$$

$\Delta y$ will be the change in the value of $y$. Therefore,

$$\Delta y = (1.2)^2 - (1)^2 = 1.44 - 1 = 0.44$$

In the above example the change in the value of $y$ was found by taking the difference between the new value of $y$ and the original value of $y$. We will now use this same reasoning in forming a general expression for $\Delta y$ in terms of $x$ and $\Delta x$. If $y = f(x)$, and we change the value of $x$ from $x_1$ to $x_1 + \Delta x$, then the new value of $y$ can be thought of as $f(x_1 + \Delta x)$ (read "$f$ of $x_1 + \Delta x$"). Thus, the change in the value of $y$ would be the new value of $y$ minus the original value of $y$, or

$$\Delta y = f(x_1 + \Delta x) - f(x_1)$$

**Example 2** If $y = x^2$, write an expression for $\Delta y$ in terms of $x$ and $\Delta x$.

*Solution* We know that

$$\Delta y = f(x + \Delta x) - f(x)$$

It is important that we understand the functional notation. For the above equation, we note that:

$$\begin{aligned} f(3) &= (3)^2 = 9 \\ f(5) &= (5)^2 = 25 \\ f(a) &= (a)^2 = a^2 \\ f(a+1) &= (a+1)^2 = a^2 + 2a + 1 \\ f(x + \Delta x) &= (x + \Delta x)^2 = x^2 + 2x\,\Delta x + \Delta x^2 \end{aligned}$$

Thus,

$$\begin{aligned} \Delta y &= f(x + \Delta x) - f(x) \\ &= (x + \Delta x)^2 - x^2 \\ &= x^2 + 2x\,\Delta x + \Delta x^2 - x^2 \\ &= 2x\,\Delta x + \Delta x \end{aligned}$$

As a check, if we consider the case where $x = 1$ and $\Delta x = 0.2$:

$$\begin{aligned} \Delta y &= 2(1)(0.2) + (0.2)^2 \\ &= 0.4 + 0.04 = 0.44 \end{aligned}$$

This checks with the result in our previous example.

**Example 3** If $y = 2x^2 - x + 1$, write an expression for $\Delta y$ in terms of $x$ and $\Delta x$.

*Solution*

$$\begin{aligned} \Delta y &= f(x + \Delta x) - f(x) \\ &= [2(x + \Delta x)^2 - (x + \Delta x) + 1] - (2x^2 - x + 1) \\ &= [2(x^2 + 2x\,\Delta x + \Delta x^2) - (x + \Delta x) + 1] - (2x^2 - x + 1) \\ &= 2x^2 + 4x\,\Delta x + 2\Delta x^2 - x - \Delta x + 1 - 2x^2 + x - 1 \\ &= 4x\,\Delta x + 2\Delta x^2 - \Delta x \end{aligned}$$

**Example 4** If $y = x - 2x^2$, find $\Delta y$ when $x = 3$ and $\Delta x = -0.02$.

*Solution*

$$\begin{aligned} \Delta y &= f(x + \Delta x) - f(x) \\ &= [(x + \Delta x) - 2(x + \Delta x)^2] - (x - 2x^2) \\ &= x + \Delta x - 2x^2 - 4x\,\Delta x - 2\Delta x^2 - x + 2x^2 \\ &= \Delta x - 4x\,\Delta x - 2\Delta x^2 \\ &= (-0.02) - 4(3)(-0.02) - 2(-0.02)^2 \\ &= -0.02 + 0.24 - 0.0008 \\ &= 0.2192 \end{aligned}$$

In addition to finding $\Delta y$ we will also be interested in finding the quotient $\frac{\Delta y}{\Delta x}$. To find this quotient, we first find $\Delta y$ as in the previous examples and then we divide by $\Delta x$.

**Example 5** If $y = 2x^2 - x + 1$, find $\frac{\Delta y}{\Delta x}$.

*Solution*

$$\frac{\Delta y}{\Delta x} = \frac{f(x + \Delta x) - f(x)}{\Delta x}$$

We have already found

$$\Delta y = 4x\,\Delta x + 2\Delta x^2 - \Delta x$$

$$\frac{\Delta y}{\Delta x} = \frac{4x\,\Delta x + 2\Delta x^2 - \Delta x}{\Delta x} = 4x + 2\Delta x - 1$$

Geometrically, if $A$ is the point $(x_1,y_1)$ on the curve $y = f(x)$, and $B$ is the point $(x_1 + \Delta x,\ y_1 + \Delta y)$, then $\frac{\Delta y}{\Delta x}$ represents the slope of $\overline{AB}$ (Fig. 3-5).

$$m = \text{slope } \overline{AB} = \frac{\overline{BC}}{\overline{AC}} = \frac{\Delta y}{\Delta x}$$

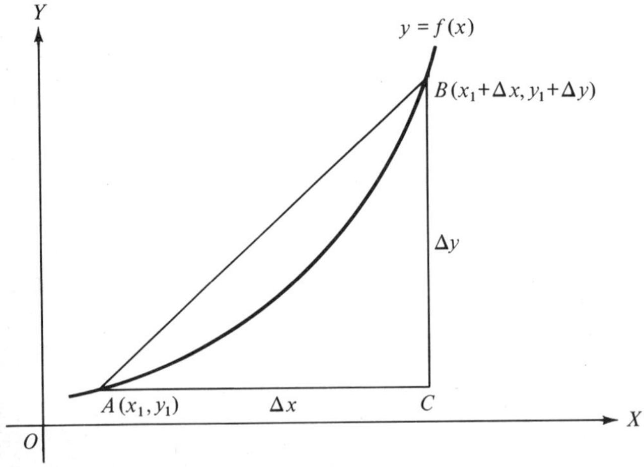


**Fig. 3-5**

**EXERCISES**

**1.** If $y = x^2 + x$, find $\Delta y$ when $x = 2$ and $\Delta x = 0.1$.
**2.** If $y = 3x - 7$, find $\Delta y$ when $x = 5$ and $\Delta x = -0.3$.

In Exercises 3 to 14 find (*a*) $\Delta y$ and $\frac{\Delta y}{\Delta x}$ in terms of $x$ and $\Delta x$ and (*b*) $\Delta y$ and $\frac{\Delta y}{\Delta x}$ when $x = 2$ and $\Delta x = 0.1$.

**3.** $y = x - x^2$

**4.** $y = x^2 + 3x + 2$

**5.** $y = \frac{1}{x}$

**6.** $y = x^3$

**7.** $y = \frac{x}{x+1}$

**8.** $y = 1 - \frac{1}{x^2}$

**9.** $y = 2x^2 + 4x - 5$

**10.** $y = ax^2 + 7$

**11.** $y = 7x - x^2$

**12.** $y = \frac{x+1}{x+2}$

**13.** $y = 3 - 2x^3$

**14.** $y = bx + c$

## 3-4 LIMITS

One of the most fundamental concepts of the calculus is that of a limit of a function. Let us consider the function $y = 3x$. We shall take values of $x$ close to 1 and observe the corresponding values of $y$.

| $x$ | $y$ |
|---|---|
| 0.9 | 2.7 |
| 0.99 | 2.97 |
| 0.999 | 2.997 |
| 1.1 | 3.3 |
| 1.01 | 3.03 |
| 1.001 | 3.003 |

We can see that as the values of $x$ get closer to 1, whether we get closer to 1 from values less than 1 or values greater than 1, the values of $y$ get closer and closer to 3. We describe this situation by saying that the limit of $f(x)$ as $x$ approaches 1 is equal to 3. This is denoted by writing

$$f(x) \to 3 \qquad \text{when } x \to 1$$

or, more commonly,

$$\lim_{x \to 1} f(x) = 3$$

In general, we may say: "If when $x$ becomes arbitrarily close to some number $a$, $f(x)$ becomes arbitrarily close to some number $L$, then $\lim_{x \to a} f(x) = L$ [or $f(x) \to L$ as $x \to a$]." Our definition tells us that when $x$ gets close to the number $a$, then $f(x)$ gets close to the number $L$. (A more rigorous definition is given in Sec. 3-8.)

It is important to note that in considering the $\lim_{x \to a} f(x)$, we are concerned with what happens to $f(x)$ when $x$ gets close to the value $a$, using values of $x$ larger than $a$ and values of $x$ smaller than $a$. We are *not* concerned with what happens to $f(x)$ when $x = a$.

**Example 1**

(a) $\lim_{x \to 3} x^2 + 1 = 10$

(b) $\lim_{x \to 4} 2x - 3 = 5$

(c) $\lim_{x \to 4} \frac{1}{x} = \frac{1}{4}$

To aid us in the evaluation of limits, we will make use of the following theorems:

### Theorems on limits

**1.** *If* $\lim_{x \to a} f(x) = A$, *and* $k$ *is any constant, then:*

(a) $\lim_{x \to a} [f(x) + k] = \lim_{x \to a} f(x) + k = A + k$

(b) $\lim_{x \to a} k \cdot f(x) = k \cdot \lim_{x \to a} f(x) = k \cdot A$

**2.** *If* $\lim_{x \to a} f(x) = A$ *and* $\lim_{x \to a} g(x) = B$, *then:*

(a) $\lim_{x \to a} [f(x) + g(x)] = \lim_{x \to a} f(x) + \lim_{x \to a} g(x) = A + B$

(b) $\lim_{x \to a} [f(x) - g(x)] = \lim_{x \to a} f(x) - \lim_{x \to a} g(x) = A - B$

(c) $\lim_{x \to a} [f(x) \cdot g(x)] = \lim_{x \to a} f(x) \cdot \lim_{x \to a} g(x) = A \cdot B$

(d) $\lim_{x \to a} \frac{f(x)}{g(x)} = \frac{\lim_{x \to a} f(x)}{\lim_{x \to a} g(x)} = \frac{A}{B}$ *where* $g(x) \neq 0$, $B \neq 0$

**3.** (a) $\lim_{x \to a} [f(x)]^n = [\lim_{x \to a} f(x)]^n = A^n$

(b) $\lim_{x \to a} [f(x)]^{1/n} = \lim_{x \to a} \sqrt[n]{f(x)} = \sqrt[n]{\lim_{x \to a} f(x)} = \sqrt[n]{A}$ *n a positive integer*

**Example 2**

(a) $\lim_{x \to 3} x^3 + 4x^2 - 2x + 3 = \lim_{x \to 3} x^3 + \lim_{x \to 3} 4x^2 - \lim_{x \to 3} 2x + \lim_{x \to 3} 3 = 27 + 36 - 6 + 3 = 60$

(*b*) $\lim_{x\to 2} x^2 + x - \frac{1}{x} = \lim_{x\to 2} x^2 + \lim_{x\to 2} x - \lim_{x\to 2} \frac{1}{x} = 4 + 2 - \frac{1}{2} = \frac{11}{2}$

**Example 3**

(*a*) Find $\lim_{x\to 0^+} \frac{1}{x}$ (*b*) Find $\lim_{x\to 0^-} \frac{1}{x}$

*Solution* (*a*) The symbol $x \to a^+$† means $x$ approaches $a$ from the right side, i.e., using values of $x$ greater than $a$. Thus, the symbol $x \to 0^+$ means $x$ approaches 0 from the positive side. We note that as $x \to 0^+$, the value of our fraction $\frac{1}{x}$ gets larger and larger, and does not approach a limit (Fig. 3-6*a*). However, it is sometimes convenient to describe the behavior of the function by writing

$$\lim_{x\to 0^+} \frac{1}{x} = \infty \qquad \text{(read infinity)}$$

It is important to realize that the symbol $\infty$ is not a number and is being used here to denote the fact that the function is getting larger and larger (infinitely large).

(*b*) The symbol $x \to 0^-$ means $x$ approaches 0 from the negative side, i.e., using only negative values. As $x \to 0^-$, $\frac{1}{x}$ becomes large and negative and does not approach a limit (Fig. 3-6*b*). Here, again, we can describe the behavior of the function by writing

$$\lim_{x\to 0^-} \frac{1}{x} = -\infty$$

† Note: If $\lim_{x\to a} f(x)$ exists, then the two one-sided limits, $\lim_{x\to a^+} f(x)$ and $\lim_{x\to a^-} f(x)$, both exist and are equal.

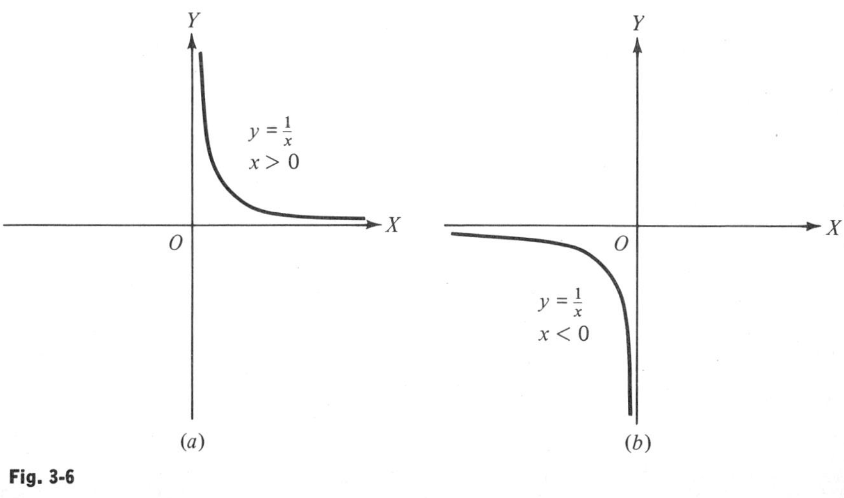


**Fig. 3-6**

We shall also be interested in examining the behavior of functions as the values of $x$ get larger and larger. We shall use the $\infty$ symbol to describe this situation. The symbol $+\infty$ is sometimes used to denote the concept of large and positive. The symbol $-\infty$ is used to denote the concept of large and negative.

**Example 4**

(a) $\lim_{x \to 1^+} \dfrac{x}{x-1} = \infty$

(b) $\lim_{x \to \infty} \left(2 + \dfrac{1}{x}\right) = 2$

(c) $\lim_{x \to 0^-} \left(4 + \dfrac{1}{x}\right) = -\infty$

**EXERCISES**

Evaluate the following limits.

1. $\lim_{x \to 1} \dfrac{x}{x+1}$
2. $\lim_{x \to 3} (2x^2 - x + 7)$
3. $\lim_{x \to \infty} \left(3 + \dfrac{1}{x^2}\right)$
4. $\lim_{x \to 2^+} \dfrac{x+3}{x-2}$
5. $\lim_{x \to a} \dfrac{x}{x+1}$
6. $\lim_{x \to -2} (x^2 - 4x + 1)$
7. $\lim_{x \to \infty} \left(2 - \dfrac{1}{x^2-4}\right)$
8. $\lim_{h \to 0} (2x^2 - 4h + 3)$
9. $\lim_{x \to -1^+} \left(2 + \dfrac{4}{x+1}\right)$
10. $\lim_{h \to 0} \dfrac{x^2 + 2xh - 4}{(x+h)(x+h+1)}$
11. $\lim_{x \to -2} \left(2 - \dfrac{3}{x-2}\right)$
12. $\lim_{x \to \infty} (x^2 - 7)$

## 3-5 INDETERMINATE FORMS: $\dfrac{0}{0}, \dfrac{\infty}{\infty}$

Let us consider the following limit:

$$\lim_{x \to 4} \frac{x^2 - 16}{x - 4}$$

If we apply our quotient rule for limits, then $\lim_{x \to 4} \dfrac{x^2-16}{x-4} = \dfrac{0}{0}$, which is an indeterminate form. What we are actually saying is that in the limit we end up with a very small number divided by a very small number, the quotient of which could be anything. In many cases, however, it will be possible to factor the equation defining the function $f(x)$ to get it into a form where the limit can be evaluated.

**Example 1** Evaluate $\lim_{x \to 4} \dfrac{x^2 - 16}{x - 4}$.

*Solution*

$$\begin{aligned}\lim_{x \to 4} \frac{x^2 - 16}{x - 4} &= \lim_{x \to 4} \frac{(x + 4)(x - 4)}{(x - 4)} \\ &= \lim_{x \to 4} (x + 4) \\ &= 8\end{aligned}$$

In the above solution we have canceled $(x - 4)$ from both numerator and denominator, or equivalently we have divided by $(x - 4)$. Since we are considering the limits as $x$ approaches 4, this raises the question: "Are we dividing by 0?" If $x = 4$, then $x - 4$ does equal 0 and it would seem that we are dividing by 0.

We have already pointed out that the symbol $\lim_{x \to a} f(x)$ examines the behavior of $f(x)$ when $x$ gets *close to* the value $a$, but we are not concerned with what happens when $x = a$. Therefore, in considering

$$\lim_{x \to 4} \frac{x^2 - 16}{x - 4}$$

we are concerned with what happens when $x$ gets close to 4 but we do not consider what happens when $x = 4$. Thus, since $x$ does not equal 4, $(x - 4)$ does not equal 0 and the division is permissible.

**Example 2** Evaluate $\lim_{x \to 2} \dfrac{(2x^2 - 4x)}{x - 2}$.

*Solution*

$$\begin{aligned}\lim_{x \to 2} \frac{2x^2 - 4x}{x - 2} &= \lim_{x \to 2} \frac{2x(x - 2)}{x - 2} \\ &= \lim_{x \to 2} 2x \\ &= 4\end{aligned}$$

**Example 3** Evaluate $\lim_{x \to \infty} \dfrac{2x^2 - 4x + 7}{3x^2 + 2x + 3}$.

*Solution* If we use our theorem on the limit of a quotient, we obtain the result $\dfrac{\infty}{\infty}$ (a very large number divided by a very large number), which again is an indeterminate form. Here we shall not attempt to factor, but shall apply another technique to solve limit problems which end up in the form $\frac{\infty}{\infty}$. We shall divide each term by $x$ raised to the highest power that it appears in the denominator. In our example this would be $x^2$. Thus

$$\lim_{x\to\infty}\frac{2x^2-4x+7}{3x^2+2x+3}=\lim_{x\to\infty}\frac{\frac{2x^2}{x^2}-\frac{4x}{x^2}+\frac{7}{x^2}}{\frac{3x^2}{x^2}+\frac{2x}{x^2}+\frac{3}{x^2}}$$

$$=\lim_{x\to\infty}\frac{2-\frac{4}{x}+\frac{7}{x^2}}{3+\frac{2}{x}+\frac{3}{x^2}}$$

$$=\frac{2}{3}$$

**Example 4** Evaluate $\lim_{x\to\infty}\frac{3x^3-x+7}{x^3+4x^2+x-3}$.

*Solution*

$$\lim_{x\to\infty}\frac{3x^3-x+7}{x^3+4x^2+x-3}=\lim_{x\to\infty}\frac{3-\frac{1}{x^2}+\frac{7}{x^3}}{1+\frac{4}{x}+\frac{1}{x^2}-\frac{3}{x^3}}$$

$$=\frac{3}{1}$$

$$=3$$

**Example 5** Evaluate $\lim_{x\to\infty}\frac{2x^2+4x-7}{x^3+3x^2-5}$.

*Solution*

$$\lim_{x\to\infty}\frac{2x^2+4x-7}{x^3+3x^2-5}=\lim_{x\to\infty}\frac{\frac{2}{x}+\frac{4}{x^2}-\frac{7}{x^3}}{1+\frac{3}{x}-\frac{5}{x^3}}$$

$$=\frac{0}{1}$$

$$=0$$

**EXERCISES**

Evaluate the following limits.

**1.** $\lim_{x\to3}\frac{x^2-9}{x-3}$

**2.** $\lim_{x\to8}\frac{x^2-8x}{x-8}$

**3.** $\lim\limits_{x\to\infty} \dfrac{2x^2 + 7x - 5}{5x^2 + 3x + 7}$

**4.** $\lim\limits_{x\to -1} \dfrac{x + 1}{x^2 - 1}$

**5.** $\lim\limits_{x\to\infty} \dfrac{x + 7}{3x^2 - 4x + 7}$

**6.** $\lim\limits_{h\to 0} \dfrac{2hx^2 - 4hx + h}{2h}$

**7.** $\lim\limits_{x\to\infty} \dfrac{x^3 - 4x^2 + x - 7}{x^2 + 9x + 8}$

**8.** $\lim\limits_{x\to -2} \dfrac{x + 2}{x^2 + 5x + 6}$

**9.** $\lim\limits_{x\to 3} \dfrac{4x^2 - 12x}{2x - 6}$

**10.** $\lim\limits_{x\to\infty} \dfrac{3x^2 + 7x - 7}{7x^2 - 9x + 1000}$

**11.** $\lim\limits_{x\to -4} \dfrac{x + 4}{x^2 - 16}$

**12.** $\lim\limits_{h\to 0} \dfrac{(a + h)^2 - a^2}{h}$

## 3-6 CONTINUITY

**Definition** *A function $f(x)$ is said to be* continuous *at a point $(a,b)$ if it satisfies the following three conditions:*

1. $\lim\limits_{x\to a} f(x)$ *exists.*
2. $f(a)$ *exists (i.e., the function is defined at $x = a$).*
3. $\lim\limits_{x\to a} f(x) = f(a)$.

**Example 1** Determine whether the function $y = \dfrac{1}{x - 2}$ is continuous at (*a*) $x = 0$, (*b*) $x = 1$, and (*c*) $x = 2$.

*Solution*

(*a*) $\lim\limits_{x\to 0} \dfrac{1}{x - 2} = -\dfrac{1}{2}$

$f(0) = -\frac{1}{2}$

Therefore, $\lim\limits_{x\to 0} \dfrac{1}{x - 2} = f(0) = -\frac{1}{2}$ and the function is continuous at $x = 0$.

(*b*) $\lim\limits_{x\to 1} \dfrac{1}{x - 2} = -1$

$=$

$f(1) = -1$

Therefore, the function is continuous at $x = 1$.

(*c*) $\lim\limits_{x\to 2} \dfrac{1}{x - 2} = \pm\infty$

Since the right-hand and left-hand limits are not equal, the limit does not exist and this function is not continuous at $x = 2$ (Fig. 3-7).

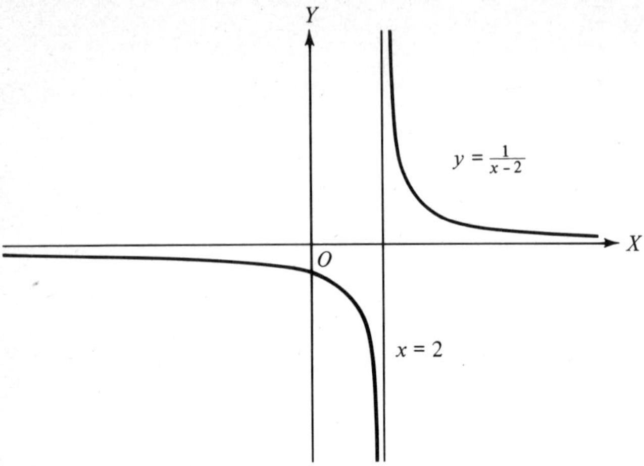


Fig. 3-7

**Example 2** Determine whether the function $y = \dfrac{x^2 - 9}{x - 3}$ is continuous at (*a*) $x = 2$ and (*b*) $x = 3$.

*Solution*

$$(a)\ \lim_{x \to 2} \frac{x^2 - 9}{x - 3} = \frac{-5}{-1} = 5$$

$$f(2) = 5$$

The function is continuous at $x = 2$.

$$(b)\ \lim_{x \to 3} \frac{x^2 - 9}{x - 3} = \frac{0}{0}$$

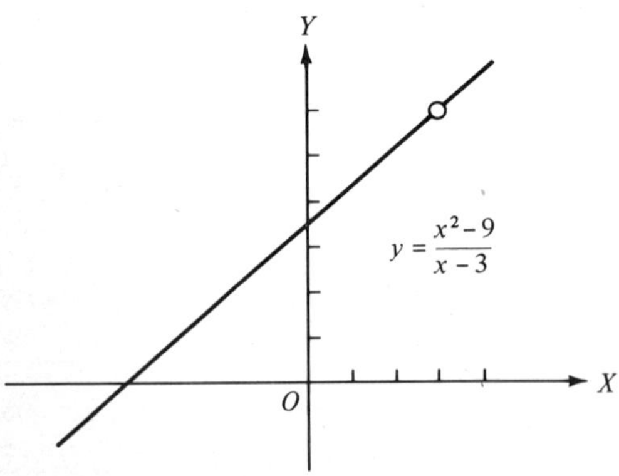


Fig. 3-8

Here we have an indeterminate form, and we shall factor $x^2 - 9$ into $(x - 3)(x + 3)$.

$$\lim_{x \to 3} \frac{x^2 - 9}{x - 3} = \lim_{x \to 3} \frac{(x - 3)(x + 3)}{x - 3} = \lim_{x \to 3} x + 3 = 6$$

We recall that in finding the limit as $x$ approaches 3 we are not concerned with what happens at $x = 3$ and therefore $x - 3$ does not equal 0.

The first condition for finding a limit is satisfied. However, $f(3)$ is not defined since here we actually do have $\frac{0}{0}$. Thus, the function is not continuous at $x = 3$ (Fig. 3-8).

**Example 3** Determine whether the function

$$f(x) = \begin{cases} 45 & \text{when } 0 < x < \frac{1}{6} \\ 55 & \frac{1}{6} \le x < \frac{3}{6} \\ 65 & \frac{3}{6} \le x < \frac{5}{6} \\ 75 & \frac{5}{6} \le x < \frac{7}{6} \\ 85 & \frac{7}{6} \le x < \frac{9}{6} \end{cases}$$

is continuous at (*a*) $x = \frac{1}{3}$ and (*b*) $x = \frac{1}{2}$.

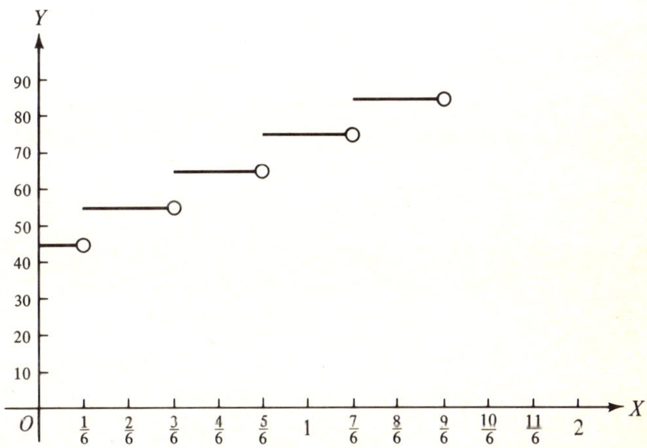


**Fig. 3-9**

*Solution* The domain of the function is $0 < x < \frac{3}{2}$. A function such as this is sometimes called a *step function* and sometimes a *jump function* (Fig. 3-9). We now test for continuity.

(*a*) $\lim_{x \to \frac{1}{3}} f(x) = 55$

If we look at the graph we can see that as $x$ approaches $\frac{1}{3}$ from values both above and below $\frac{1}{3}$, the function values approach and, in fact, equal 55. Also, $f(\frac{1}{3}) = 55$ and the function is continuous at $x = \frac{1}{3}$.

(*b*) We note that the function is defined at $x = \frac{1}{2}$ and that $f(\frac{1}{2}) = 65$. However, $\lim_{x \to \frac{1}{2}} f(x)$ does not exist. If we take values of $x$ less than $\frac{1}{2}$, $f(x) = 55$. If we take values of $f(x)$ greater than $\frac{1}{2}$, $f(x) = 65$. Since the right-hand limit does not equal the left-hand limit, the limit does not exist. We thus see that it is possible for condition 2 to be satisfied even though condition 1 fails. Since all three conditions must be satisfied, the function is not continuous at $x = 3$.

Though at first glance the above function seems quite unfamiliar, it can be used to represent the cost of a taxi ride where we must pay 45 cents for the first sixth of a mile and 10 cents for each additional third of a mile.

**Example 4** Determine whether the function

$$f(x) = \begin{cases} \dfrac{x^2 - 4}{x - 2} & \text{when } x \neq 2 \\ 6 & \text{when } x = 2 \end{cases}$$

is continuous at $x = 2$.

*Solution*

$$\lim_{x \to 2} \frac{x^2 - 4}{x - 2} = \lim_{x \to 2} \frac{(x + 2)(x - 2)}{x - 2} = \lim_{x \to 2} x + 2 = 4$$
$$f(2) = 6$$

Here, the limit as $x$ approaches 2 exists, and the function is defined at $x = 2$, but the two are not equal. Since condition 3 is not satisfied, the function is not continuous at $x = 2$ (Fig. 3-10).

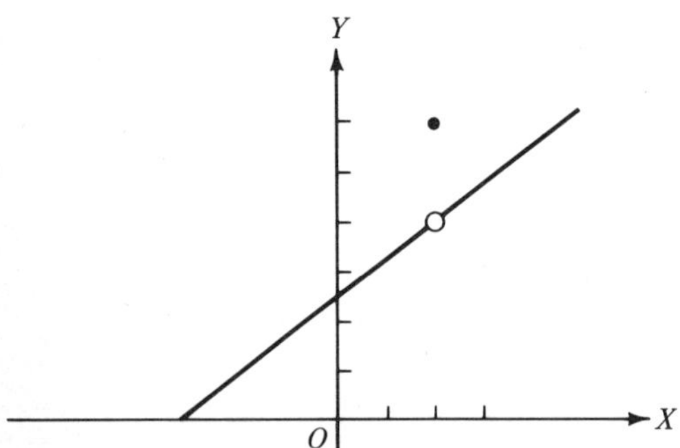


**Fig. 3-10**

The concept of continuity at a point can be extended to continuity over an interval by saying that a function is continuous over an interval if it is continuous at each point in that interval. In terms of the graph, if the function is discontinuous at a point, the curve will be broken at that point.

**EXERCISES**

**1.** Determine whether the function $y = \dfrac{x^2 - x - 6}{x - 3}$ is continuous at (a) $x = 0$, (*b*) $x = 3$, (*c*) $x = -3$.

**2.** Determine whether the function $y = \dfrac{x^2}{x}$ is continuous at (*a*) $x = 0$, (*b*) $x = -1$, (*c*) $x = 1$.

**3.** For the following functions determine the points of discontinuity.

(*a*) $y = \dfrac{x^2 + 2}{(x - 1)(x + 1)}$ (*b*) $y = \dfrac{x^2 + 4}{x - 2}$

(*c*) $y = \dfrac{2}{x^3 + x}$ (*d*) $y = \dfrac{x^2 - 3x - 4}{x^2 + 3x - 4}$

**4.** Determine whether the function

$$y = \begin{cases} \dfrac{x^2 - 16}{x - 4} & \text{when } x \neq 4 \\ 8 & \text{when } x = 4 \end{cases}$$

is continuous at $x = 4$. What is the domain of this function?

**5.** Graph and discuss the continuity of the function defined by:

$$y = \begin{cases} x & \text{when } 0 \leq x < 1 \\ x + 1 & \text{when } 1 \leq x \leq 2 \end{cases}$$

**6.** Determine whether the function defined by

$$f(x) = \begin{cases} x & \text{when } x > 0 \\ 0 & \text{when } x \leq 0 \end{cases}$$

is continuous for all values of $x$.

## ▶ 3-7 ABSOLUTE VALUES

### Definition

$$|a| = \begin{cases} a & \text{if } a \geq 0 \\ -a & \text{if } a < 0 \end{cases}$$

**Example 1**

(*a*) $|7| = 7$
(*b*) $|-7| = +7$
(*c*) $|-3| = +3$
(*d*) $|0| = 0$

**PROPERTIES OF ABSOLUTE VALUES**

1. $|a \cdot b| = |a| \cdot |b|$
2. $|a + b| \leq |a| + |b|$

3. If $a \geq 0$:
   (a) $|x| \leq a$ implies $-a \leq x \leq a$
   (b) $|x - b| \leq a$ implies $-a \leq x - b \leq a$
   or $b - a \leq x \leq b + a$
   (c) $|x| \geq a$ implies $x \leq -a$ or $x \geq a$

**Example 2** Find the possible values of $x$ if $|x| < 7$.

*Solution*

$$-7 < x < 7$$

**Example 3** Find the possible values of $x$ if $|x - 1| < 5$.

*Solution*

$$|x - 1| < 5$$

Therefore,

$$-5 < x - 1 < 5$$
$$-4 < x < 6$$

**EXERCISES**

Determine the possible values of $x$ in the following problems.

**1.** $|x - 5| < 8$
**2.** $|x + 1| < 3$
**3.** $|x - 4| < 9$
**4.** $|x + 7| < 1$
**5.** $|x - 5| < 10$
**6.** $|x + 5| < 10$

## ▶ 3-8 PROOFS OF LIMIT THEOREMS

In our previous definition of a limit, we said that if when $x$ becomes arbitrarily close to some number $a$, $f(x)$ becomes arbitrarily close to some number $L$, then $\lim_{x \to a} f(x) = L$. We will now give a more rigorous definition which avoids the ambiguous expression "$x$ gets arbitrarily close to some number $a$."

**Definition** *If for any number $\epsilon > 0$ there exists a number $\delta > 0$ such that*

$$|f(x) - L| < \epsilon \qquad \textit{whenever } 0 < |x - a| < \delta$$

*then we say that the limit of $f(x)$ as $x$ approaches $a$ is $L[\lim_{x \to a} f(x) = L]$.*

The above definition tells us that if $\lim_{x \to a} f(x) = L$, then we can get the values of $f(x)$ as close to $L$ as we wish. In fact, for any degree of accuracy desired, for any positive number $\epsilon$, the difference between $f(x)$ and $L$ will be less than $\epsilon$, so long as $x$ differs from $a$ by less than some number $\delta$. In proving that the limit of $f(x)$ exists as $x$ approaches $a$, we must show that there exists a positive number $\delta$ such that when the difference between $x$ and $a$ is less than $\delta$, then the difference between $f(x)$ and $L$ will be less than any specified positive number $\epsilon$. We will now consider $\lim_{x \to 1} 3x$ and we will prove that the limit equals 3.

*Proof* We want to show that there exists a number $\delta$ such that for any $\epsilon > 0$

$$|f(x) - L| = |3x - 3| < \epsilon \qquad \text{whenever } 0 < |x - 1| < \delta$$

Now $\quad |3x - 3| = |3(x - 1)| = |3|\,|x - 1| = 3|x - 1|$

If
$$|x - 1| < \frac{\epsilon}{3}$$
then
$$3|x - 1| < \epsilon$$
$$|3x - 3| < \epsilon$$

Thus, if we choose $\delta = \dfrac{\epsilon}{3}$,

$$|3x - 3| < \epsilon \qquad \text{whenever } 0 < |x - 1| < \delta = \frac{\epsilon}{3} \; ■$$

If, for example, $\epsilon = 0.000003$, we choose $\delta = 0.000001$, and so long as $x$ differs from 1 by less than 0.000001, then $3x$ will differ from 3 by less than 0.000003.

We will now use our definition of limits to indicate how the limit theorems can be proved. We shall prove Theorem 2*a*.

**Theorem** *If* $\lim_{x \to a} f(x) = A$ *and* $\lim_{x \to a} g(x) = B$, *then*

$$\lim_{x \to a} [f(x) + g(x)] = \lim_{x \to a} f(x) + \lim_{x \to a} g(x) = A + B$$

*Proof* We must show that there exists a number $\delta > 0$ such that for any $\epsilon > 0$

$$|[f(x) + g(x)] - (A + B)| < \epsilon \qquad \text{whenever } 0 < |x - a| < \delta$$

We are given the fact that $\lim_{x \to a} f(x) = A$ and $\lim_{x \to a} g(x) = B$. Therefore, for any $\epsilon' > 0$, there exists a $\delta' > 0$ such that

$$|f(x) - A| < \epsilon \qquad \text{whenever } 0 < |x - a| < \delta'$$

Similarly, there exists a $\delta'' > 0$ such that for any $\epsilon'' > 0$

$$|g(x) - B| < \epsilon'' \qquad \text{whenever } 0 < |x - a| < \delta''$$

We note that we have used $\delta'$ and $\delta''$ since the same number need not necessarily be used for the two different limits. Since we know that the $\lim_{x \to a} f(x)$ and $\lim_{x \to a} g(x)$ both exist, any number can be chosen for $\epsilon'$ and $\epsilon''$. We shall choose $\epsilon' = \epsilon'' = \frac{\epsilon}{2}$. We shall also choose $\delta$ to be the smaller of the two numbers $\delta'$ and $\delta''$. Thus, if

$$0 < |x - a| < \delta$$

then $|f(x) - A| < \epsilon' = \frac{\epsilon}{2}$ and $|g(x) - B| < \epsilon'' = \frac{\epsilon}{2}$.

Having set the groundwork, we shall now show that

$$|[f(x) + g(x)] - (A + B)| < \epsilon \qquad \text{whenever } 0 < |x - a| < \delta$$

$$\begin{aligned} |[f(x) + g(x)] - (A + B)| &= |[f(x) - A] + [g(x) - B]| \\ &\leq |f(x) - A| + |g(x) - B| \\ &< \frac{\epsilon}{2} + \frac{\epsilon}{2} = \epsilon \\ &\qquad \text{whenever } 0 < |x - a| < \delta \end{aligned}$$

This completes the proof.■

**EXERCISES**

1. Using the definition of limits prove that $\lim_{x \to 2} (2x + 1) = 5$.
2. Prove that $\lim_{x \to a} Kf(x) = K \lim_{x \to a} f(x)$.
3. Prove that if $\lim_{x \to a} f(x) = A$ and $\lim_{x \to a} g(x) = B$, then $\lim_{x \to a} [f(x) - g(x)] = A - B$.

# 4

# Derivatives

## 4-1 A TANGENT LINE TO A CURVE

The concept of a tangent line to a curve has become familiar when dealing with a circle. We shall now extend this concept to any curve.

**Definition** *Let $P(x_1,y_1)$ be a fixed point on the curve $y = f(x)$, and let $Q$ be a variable neighboring point on the curve* (Fig. 4-1). *We can thus consider the secant line $PQ$. Now, let the point $Q$ approach the point $P$ on the curve. As the point $Q$ moves to position $Q'$, $Q''$, etc., we get new secant lines $PQ'$, $PQ''$,* etc. *If there exists a limiting position for the secant lines, this limiting position is called the tangent line to the curve at the point $P$. To summarize, the tangent line to a curve $y = f(x)$ at a point $P$ is the limiting position of the secant line $PQ$ as the point $Q$ approaches the point $P$ along the curve, if such a limiting position exists.*

We shall now consider the slopes of the secant lines and the slope of the tangent line. We can label the point $Q$ as $(x_1 + \Delta x,\ y_1 + \Delta y)$ (Fig.

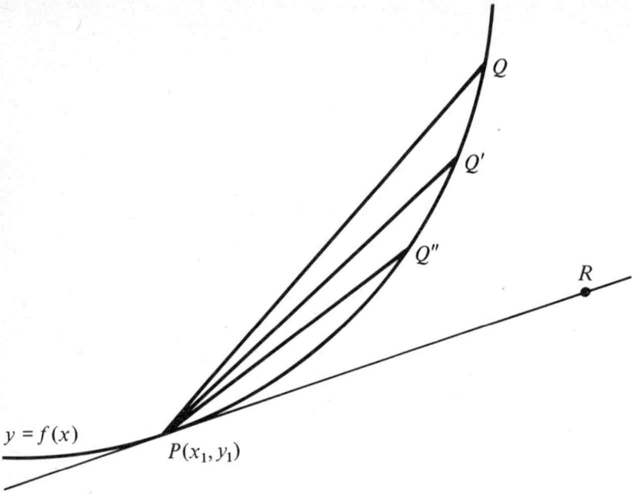


Fig. 4-1

4-2). The slope of secant $PQ$ is thus $\frac{\Delta y}{\Delta x}$. As the point $Q$ approaches the point $P$ along the curve, we note that $\Delta x$ approaches 0. The tangent line $PR$ is the limiting position of the secant lines, and the slope of the tangent line is the limit of $\frac{\Delta y}{\Delta x}$ as $\Delta x$ approaches 0. Thus, the slope of the tangent line $PR$ at the point $P = \lim_{\Delta x \to 0} \frac{\Delta y}{\Delta x}$.

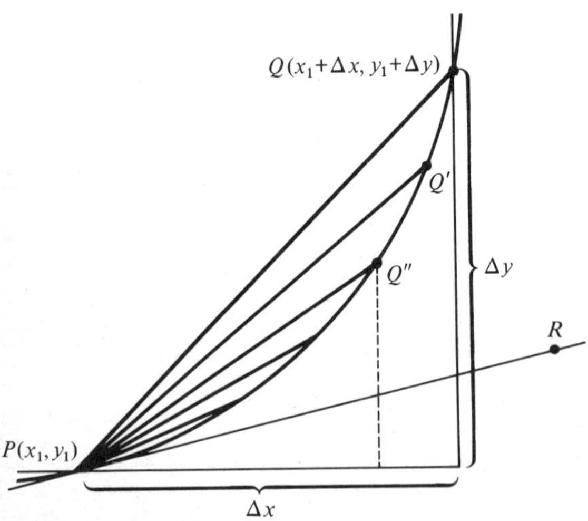


Fig. 4-2

**Example 1** Find the slope of the tangent line to the curve $y = x^2$ (Fig. 4-3) at (*a*) $x = -2$ and (*b*) $x = 1$.

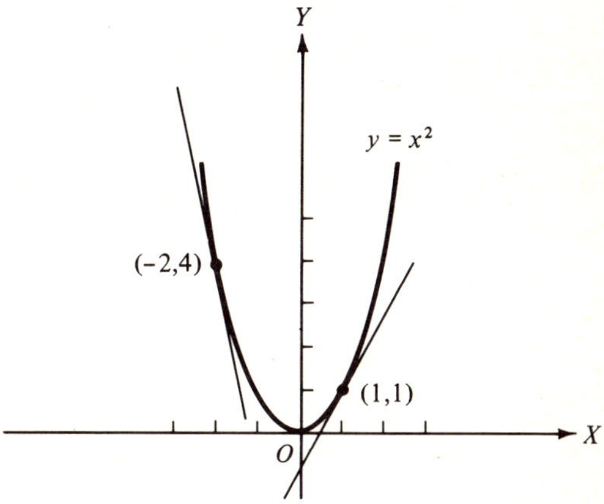


**Fig. 4-3**

*Solution*

(*a*) We must find $\lim_{\Delta x \to 0} \frac{\Delta y}{\Delta x}$.

$$\begin{aligned} \Delta y &= (x + \Delta x)^2 - x^2 \\ &= x^2 + 2x\,\Delta x + \Delta x^2 - x^2 \\ &= 2x\,\Delta x + \Delta x^2 \end{aligned}$$

$$\frac{\Delta y}{\Delta x} = 2x + \Delta x$$

$$\lim_{\Delta x \to 0} \frac{\Delta y}{\Delta x} = \lim_{\Delta x \to 0} (2x + \Delta x) = 2x$$

At $x = -2$:

$$2x = -4 \qquad \text{Slope} = -4$$

(*b*) At $x = 1$:

$$2x = 2 \qquad \text{Slope} = 2$$

**Example 2** Find the equation of the tangent line to the curve $y = 4x - x^2$ at the point (3,3) (Fig. 4-4).

*Solution* We recall that the equation of a straight line can be given using the point-slope form $y - y_1 = m(x - x_1)$. We have the point (3,3) so that $x_1 = 3$ and $y_1 = 3$. We must now find the slope $m$.

$$\begin{aligned} \Delta y &= [4(x + \Delta x) - (x + \Delta x)^2] - (4x - x^2) \\ &= 4x + 4\Delta x - x^2 - 2x\,\Delta x - \Delta x^2 - 4x + x^2 \\ &= 4\Delta x - 2x\,\Delta x - \Delta x^2 \end{aligned}$$

$$\frac{\Delta y}{\Delta x} = 4 - 2x - \Delta x$$

$$\lim_{\Delta x \to 0} \frac{\Delta y}{\Delta x} = \lim_{\Delta x \to 0} (4 - 2x - \Delta x) = 4 - 2x$$

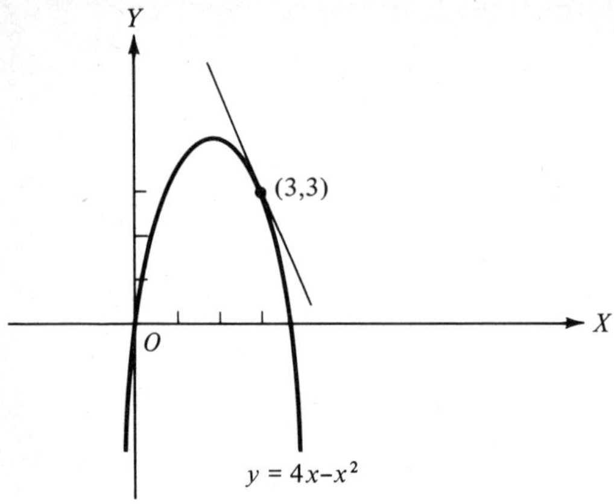


Fig. 4-4

At $x = 3$:

$$4 - 2x = 4 - 6 = -2 \qquad \text{and} \qquad m = -2$$

Therefore, the equation of the tangent line is $y - 3 = -2(x - 3)$.

**EXERCISES**

For the following curves find the slope of the tangent line at (*a*) $x = x_1$ and (*b*) $x = 1$.

**1.** $y = x^2 - 3x$

**2.** $y = x^3$

**3.** $y = \dfrac{1}{x}$

**4.** $y = 2x^2 + 1$

**5.** $y = 2x - x^2$

**6.** $y = \dfrac{x}{x + 1}$

For the following curves find the equation of the tangent line at $x = 2$.

**7.** $y = 2x^2 + 4x$

**8.** $y = 8x - x^2$

**9.** $y = \dfrac{1}{x^2}$

**10.** $y = x^3 + 1$

**11.** Find the point on the curve $y = 2x^2 - 8x$ where the slope of the tangent equals 1.

**12.** Find the point on the curve $y = 4x - x^2$ where the tangent line is horizontal.

## 4-2 THE DERIVATIVE OF A FUNCTION

We shall now consider one of the most important concepts in calculus, the derivative of a function.

**Definition** *Let $y = f(x)$ be defined over the interval from $x = a$ to $x = b$, and let $x_1$ be any $x$ coordinate such that $a < x_1 < b$. Form the difference quotient* $\dfrac{\Delta y}{\Delta x} = \dfrac{f(x_1 + \Delta x) - f(x_1)}{\Delta x}$. *Consider the limit of* $\dfrac{\Delta y}{\Delta x}$ *as $\Delta x$ approaches 0. If this limit exists, it is called the* derivative *of $y$ with respect to $x$ at $x = x_1$.*

There are various notations used to represent the derivative of $y$ with respect to $x$. Among the most common are the following:

$$D_x y \qquad \frac{dy}{dx} \qquad f'(x) \qquad y' \qquad \text{and} \qquad D_x f(x)$$

Thus,

$$\boxed{D_x y = \lim_{\Delta x \to 0} \frac{\Delta y}{\Delta x} = \lim_{\Delta x \to 0} \frac{f(x_1 + \Delta x) - f(x_1)}{\Delta x}}$$

From our discussion of the tangent line to a curve $y = f(x)$ at a point $(x_1,y_1)$, the slope of the tangent line at the point is $\lim\limits_{\Delta x \to 0} \dfrac{\Delta y}{\Delta x}$ evaluated at $(x_1,y_1)$. Thus, geometrically, the derivative can be thought of as the slope of the tangent line to the curve $y = f(x)$ at any given point $(x_1,y_1)$.

**Example 1** Find the derivative of $y$ with respect to $x$ for the function with rule of correspondence $y = x^2 + 1$ at (*a*) $x = x_1$ and (*b*) $x = 2$.

*Solution*

(*a*) We must first form the difference quotient $\dfrac{\Delta y}{\Delta x}$.

$$\begin{aligned}
\Delta y &= f(x_1 + \Delta x) - f(x_1) \\
&= [(x_1 + \Delta x)^2 + 1] - (x_1^2 + 1) \\
&= x_1^2 + 2x_1\,\Delta x + \Delta x^2 + 1 - x_1^2 - 1 \\
&= 2x_1\,\Delta x + \Delta x^2 \\
\frac{\Delta y}{\Delta x} &= 2x_1 + \Delta x \\
D_x y &= \lim_{\Delta x \to 0} \frac{\Delta y}{\Delta x} = \lim_{\Delta x \to 0} (2x_1 + \Delta x) = 2x_1
\end{aligned}$$

(*b*) We have found that when $x = x_1$, $D_x y = 2x_1$. Therefore, at $x = 2$ the derivative of $y$ with respect to $x$ is equal to $2(2) = 4$.

We also note that the slope of the tangent line to the curve $y = x^2 + 1$ at the point $(2,5)$ is equal to 4.

**Example 2** Find $D_x y$ for $y = \dfrac{1}{x}$ at (*a*) $x = x_1$ and (*b*) $x = 2$.

*Solution*

(*a*)

$$\begin{aligned}
\Delta y &= \frac{1}{x_1 + \Delta x} - \frac{1}{x_1} \\
&= \frac{x_1 - (x_1 + \Delta x)}{x_1(x_1 + \Delta x)} = \frac{-\Delta x}{x_1(x_1 + \Delta x)} \\
\frac{\Delta y}{\Delta x} &= \frac{-1}{x_1(x_1 + \Delta x)} \\
D_x y &= \lim_{\Delta x \to 0} \frac{\Delta y}{\Delta x} = \lim_{\Delta x \to 0} \frac{-1}{x_1(x_1 + \Delta x)} = \frac{-1}{x_1^2}
\end{aligned}$$

(*b*) At $x = 2$:

$$D_x y = \frac{-1}{(2)^2} = -\frac{1}{4}$$

In finding $D_x y$ we often use the letter $x$ without the subscript, and then substitute $x = x_1$ if we want the derivative at $x = x_1$.

In more advanced works it is proved that if a function has a derivative at $x = x_1$, then it is continuous at $x = x_1$. However, the converse is not true, and it is possible for a function to be continuous at $x = x_1$ and still not have a derivative at that point.

If a function has a derivative at $x = x_1$ it is said to be *differentiable* at $x = x_1$. If the function has a derivative at all points in an interval, the function is said to be differentiable in that interval.

**EXERCISES**

Find $D_x y$ at $x = x_1$.

**1.** $y = x^2 + 2x - 2$

**2.** $y = 2x^2 - 3$

**3.** $y = \dfrac{1}{x^2}$

**4.** $y = \dfrac{x}{x+1}$

**5.** $y = x^3$

**6.** $y = x - \dfrac{1}{x}$

Find $D_x y$ at $x = 2$.

**7.** $y = 2x^3 + x$

**8.** $y = \dfrac{x+1}{x+2}$

**9.** $y = \dfrac{x^2}{x+1}$

**10.** $y = 2x - x^2$

**11.** (*a*) Find the slope of the tangent line to the curve $y = 3x^2 - 1$ at the point (1,2).
(*b*) Find the equation of the tangent line to the above curve at the point (1,2).

## 4-3 PHYSICAL INTERPRETATION OF THE DERIVATIVE

We shall now consider the speed of a particle in rectilinear (straight-line) motion. If a particle moves in a straight line at a uniform speed, its speed can be found from the familiar formula (rate)(time) = distance.

**Example 1** A car moves in a straight line from a point $A$ to a point $B$ 120 miles away. If the car moves with a uniform speed and it takes 3 hr to get from $A$ to $B$, what is the speed of the car?

*Solution* Since (rate) (time) = distance,

$$\text{Speed} = \frac{\text{distance}}{\text{time}} = \frac{120}{3} = 40 \text{ mph}$$

The above procedure works well if we have rectilinear motion with a uniform speed. For motion where the speed is not uniform, we will have to define the concepts of average velocity and instantaneous velocity. The term velocity is used to indicate both speed and direction.

To introduce the above concepts we shall consider a particle moving in rectilinear motion, whose position $s$ from a fixed starting point is given as a function of $t$, the time elapsed since the particle started its motion. For this problem we shall consider the law of motion $s = 16t^2$ ($s$ in feet and $t$ in seconds) (Fig. 4-5).

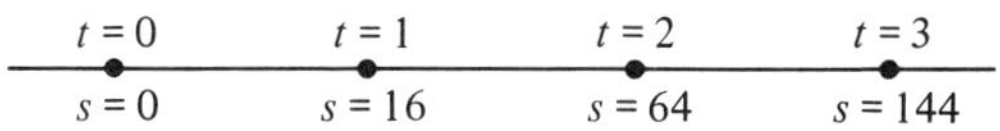


**Fig. 4-5**

We note from the graph that the motion in this problem is not uniform motion. Nevertheless, we can still examine the velocity of the particle within a given interval of time, say from $t = 1$ to $t = 3$. We define the average velocity for a given interval as the change in distance divided by the change in time. In symbols:

$$\text{Average velocity} = \frac{\Delta s}{\Delta t}$$

Thus, from $t = 1$ to $t = 3$, $\Delta t = 2$ and

$$\text{Average velocity} = \frac{144 - 16}{2} = \frac{128}{2} = 64 \text{ ft/sec}$$

This does not tell us what the velocity is when $t = 1$ or when $t = 3$. It tells us that the average velocity in the interval from $t = 1$ to $t = 3$ is 64 ft/sec. If we are more concerned with what happens at $t = 1$, we can consider the shorter interval from $t = 1$ to $t = 2$. For this interval $\Delta t = 1$ and average velocity $= \dfrac{64 - 16}{2 - 1} = \dfrac{48}{1} = 48$ ft/sec. We still cannot say what the velocity is at $t = 1$, but we are now dealing with an interval of 1 sec. If this is not yet satisfactory for us, we can cut the interval in half again, and go from $t = 1$ to $t = \frac{3}{2}$. Here $\Delta t = \frac{1}{2}$ and when $t = \frac{3}{2}$, $s = 36$.

$$\text{Average velocity} = \frac{36 - 16}{\frac{1}{2}} = \frac{20}{\frac{1}{2}} = 40 \text{ ft/sec}$$

We shall repeat the procedure once more; this time from $t = 1$ to $t = \frac{5}{4}$, with $\Delta t = \frac{1}{4}$.

$$\text{Average velocity} = \frac{25 - 16}{\frac{1}{4}} = \frac{9}{\frac{1}{4}} = 36 \text{ ft/sec}$$

Though we still do not know the velocity at $t = 1$, we now have an interval of $\frac{1}{4}$ sec. The smaller we take the interval, the closer we get to the actual velocity at $t = 1$. This leads us to the following definition:

$$v = \text{instantaneous velocity} = \lim_{\Delta t \to 0} \frac{\Delta s}{\Delta t}$$

In words, the instantaneous velocity at $t = t_1$ is defined as the limit of $\frac{\Delta s}{\Delta t}$ as $\Delta t$ approaches 0, where the interval $\Delta t$ starts at $t = t_1$.

In our problem, $\Delta s = f(t + \Delta t) - f(t)$,

$$\begin{aligned}\Delta s &= 16(t + \Delta t)^2 - 16t^2 \\ &= 16t^2 + 32t\,\Delta t + 16\Delta t^2 - 16t^2 \\ &= 32t\,\Delta t + 16\Delta t^2 \\ \frac{\Delta s}{\Delta t} &= 32t + 16\Delta t\end{aligned}$$

We note that when $t = 1$ and $\Delta t = 2$, $\frac{\Delta s}{\Delta t} = 32 + 32 = 64$.

When $t = 1$ and $\Delta t = 1$, $\frac{\Delta s}{\Delta t} = 32 + 16 = 48$.

When $t = 1$ and $\Delta t = \frac{1}{2}$, $\frac{\Delta s}{\Delta t} = 32 + 8 = 40$.

When $t = 1$ and $\Delta t = \frac{1}{4}$, $\frac{\Delta s}{\Delta t} = 32 + 4 = 36$.

These values check with what we found before. Now,

$$v = \lim_{\Delta t \to 0} \frac{\Delta s}{\Delta t} = 32t$$

At $t = 1$:

$$v = 32 \text{ ft/sec}$$

We can also find the instantaneous velocity at $t = 2$ ($v = 64$) or at any time $t$.

By now the reader should have noticed that the instantaneous velocity is actually the derivative of $s$ with respect to $t$. We shall use the term velocity to denote instantaneous velocity.

We can now go one step further and consider acceleration. As with velocity we can define average acceleration as follows:

$$\text{Average acceleration} = \frac{\Delta v}{\Delta t}$$

The average acceleration gives us the acceleration in an interval $\Delta t$. If we want the instantaneous acceleration, we must take the limit as $\Delta t \to 0$. Thus, instantaneous acceleration $= \lim_{\Delta t \to 0} \frac{\Delta v}{\Delta t}$. Here again we are using the derivative, and we can say that the derivative of $v$ with respect to $t$ gives us the instantaneous acceleration (which we shall denote by the letter $a$).

**Example 2** For the law of motion $s = 4t^2 + 8t$ find the velocity and the acceleration at $t = t_1$ and $t = 2$.

*Solution*

$$v = \lim_{\Delta t \to 0} \frac{\Delta s}{\Delta t}$$

$$\begin{aligned} \Delta s &= [4(t + \Delta t)^2 + 8(t + \Delta t)] - (4t^2 + 8t) \\ &= 4t^2 + 8t\,\Delta t + 4\Delta t^2 + 8t + 8\Delta t - 4t^2 - 8t \\ &= 8t\,\Delta t + 4\Delta t^2 + 8\Delta t \end{aligned}$$

$$\frac{\Delta s}{\Delta t} = 8t + 4\Delta t + 8$$

$$v = \lim_{\Delta t \to 0} \frac{\Delta s}{\Delta t} = 8t + 8$$

At $t = t_1$: $\quad v = 8t_1 + 8$

At $t = 2$: $\quad v = 16 + 8 = 24$

$$a = \lim_{\Delta t \to 0} \frac{\Delta v}{\Delta t}$$

$$\begin{aligned} \Delta v &= [8(t + \Delta t) + 8] - (8t + 8) \\ &= 8t + 8\Delta t + 8 - 8t - 8 \\ &= 8\Delta t \end{aligned}$$

$$\frac{\Delta v}{\Delta t} = 8$$

$$a = \lim_{\Delta t \to 0} \frac{\Delta v}{\Delta t} = 8$$

Thus, $a$ is a constant and $a = 8$ at $t = t_1$ and at $t = 2$.

**EXERCISES**

(In Exercises 1 to 6, the units for $s$ and $t$ are feet and seconds, respectively.)

**1.** For the law of motion $s = 2t^2 + 4t$ find the average velocity between $t = 1$ and $t = 3$.

**2.** For the law of motion $s = 8t^2 - 5$ find ($a$) the average velocity $\frac{\Delta s}{\Delta t}$ between $t = 1$ and $t = 3$ and ($b$) the instantaneous velocity at $t = 3$.

**3.** For a ball thrown vertically upward from the ground with an initial velocity of 64 ft/sec, its height $s$ above the ground is given by the law of motion $s = 64t - 16t^2$. Find the velocity at any time $t$, and find the velocity at $t = 1$.

**4.** For the law of motion $s = 6t - 3t^2$ find the velocity and the acceleration at $t = 2$. For what value of $t$ is the velocity equal to 0?

**5.** A ball is dropped vertically downward from a height of 1,600 ft above the ground. If its law of motion is $s = 1{,}600 - 16t^2$, find (*a*) the average velocity between $t = 0$ and $t = 2$ and (*b*) the instantaneous velocity at $t = 5$. How long will it take for the ball to reach the ground?

**6.** Find the velocity and acceleration at any time $t$ of a particle in straight-line motion whose law of motion is $s = t^3 + 1$. What was the velocity at $t = 2$? What was the acceleration at $t = 2$?

## 4-4 RATES OF CHANGE

Our discussion of instantaneous velocity and instantaneous acceleration can be generalized to rates of change between any two variables. For example, in a circle we know that the area $A$ equals $\pi$ times the square of the radius. When $r = r_1$ then $A = \pi r_1{}^2$. If we change the value of $r$ from $r_1$ to $r_1 + \Delta r$, this will produce a corresponding change in the area $A$.

$$\begin{aligned} \Delta A &= \pi(r_1 + \Delta r)^2 - \pi r_1{}^2 \\ &= \pi r_1{}^2 + 2\pi r_1\,\Delta r + \pi\,\Delta r^2 - \pi r_1{}^2 \\ &= 2\pi r_1\,\Delta r + \pi \Delta r^2 \end{aligned}$$

We can define $\dfrac{\Delta A}{\Delta r}$ as the average rate of change of area over the interval $\Delta r$.

$$\text{Average rate of change } \frac{\Delta A}{\Delta r} = (2\pi r_1 + \pi\Delta r) \text{ ft}^2 \text{ of area per foot of radius}$$

If we wish to know how fast the area is changing at any particular value of $r$, we can take $\Delta r$ smaller and smaller. This leads us to define:

$$\text{Instantaneous rate of change} = \lim_{\Delta r \to 0} \frac{\Delta A}{\Delta r}$$

Thus,

$$\text{Instantaneous rate of change} = \lim_{\Delta r \to 0} (2\pi r_1 + \pi\Delta r) = 2\pi r_1$$

We can now generalize this for any variables to say that if $y = f(x)$, then the instantaneous rate of change of $y$ per unit of change in $x$ evaluated at $x = x_1$ is given by the derivative of $y$ with respect to $x$, that is, $\displaystyle\lim_{\Delta x \to 0} \frac{\Delta y}{\Delta x}$.

**Example** A balloon in the shape of a sphere is being inflated. As the radius of the balloon increases, there is a corresponding increase in the volume, and the volume is increasing a certain number of cubic feet per foot of change of radius. Find the rate of increase of the volume when the radius is 3.

*Solution* For a sphere, $V = \frac{4}{3}\pi r^3$. Then

$$\begin{aligned}\Delta V &= \tfrac{4}{3}\pi(r + \Delta r)^3 - \tfrac{4}{3}\pi r^3 \\ &= \tfrac{4}{3}\pi(r^3 + 3r^2\,\Delta r + 3r\,\Delta r^2 + \Delta r^3) - \tfrac{4}{3}\pi r^3 \\ &= 4\pi r^2\,\Delta r + 4\pi r\,\Delta r^2 + \tfrac{4}{3}\pi\Delta r^3 \\ \frac{\Delta V}{\Delta r} &= 4\pi r^2 + 4\pi r\,\Delta r + \tfrac{4}{3}\pi\Delta r^2\end{aligned}$$

Instantaneous rate of change of $V$ with respect to $r = D_rV$.

$$D_rV = \lim_{\Delta r \to 0} \frac{\Delta V}{\Delta r} = 4\pi r^2$$

When $r = 3$, $D_rV = 36\pi$ ft$^3$ per foot of change of radius.

**EXERCISES**

**1.** If $P = 2q^2$, find the rate of change of $P$ with respect to $q$.

**2.** If the sides of a square are being increased, find the rate of increase of the area $A$ of the square, with respect to the side $x$, when $x = 10$ in.

**3.** The relationship between the pressure $P$ and the volume $V$ of a certain gas at a constant temperature is given by the formula $P = \dfrac{500}{V}\cdot$ If the volume is increasing, find the rate of change of $P$ with respect to $V$ when $V = 20$ in.$^3$

## 4-5 APPLICATIONS TO ELECTRONICS

### INSTANTANEOUS CURRENT

In electronics, the current $I$ (measured in amperes) flowing past a point in a circuit is defined as the rate of change with respect to time $t$ (measured in seconds) at which the charge $q$ (measured in coulombs) moves past the point. Thus, when the change in the charge is $\Delta q$ in an interval $\Delta t$, the average current in that time interval will be

$$I_{\text{av}} = \frac{\Delta q}{\Delta t}$$

and the instantaneous current at any time $t$ is

$$I = \lim_{\Delta t \to 0} \frac{\Delta q}{\Delta t} = D_tq$$

**Example 1** (*a*) For a given circuit the charge transferred past a certain point in the circuit is given by the equation $q = 4t^2 + 3t + 0.002$ coulombs. What was the average current between $t = 0.01$ sec and $t = 0.03$ sec?

(b) What is the instantaneous current in the circuit at any time $t$?
(c) What is the current at $t = 0.04$ sec?

*Solution*

(a) We have

$$\begin{aligned} q &= 4t^2 + 3t + 0.002 \\ \Delta q &= 4(t + \Delta t)^2 + 3(t + \Delta t) + 0.002 - (4t^2 + 3t + 0.002) \\ &= 4t^2 + 8t\,\Delta t + 4\Delta t^2 + 3t + 3\Delta t + 0.002 - 4t^2 - 3t - 0.002 \\ &= 8t\,\Delta t + 4\Delta t^2 + 3\Delta t \end{aligned}$$

$$I_{\text{av}} = \frac{\Delta q}{\Delta t} = 8t + 4\Delta t + 3$$

We wish to find the average current between $t = 0.01$ and $t = 0.03$. Thus, we start with a value of $t = 0.01$ and $\Delta t = 0.02$.

$$\frac{\Delta q}{\Delta t} = 8(0.01) + 4(0.02) + 3 = 3.16 \text{ amperes}$$

(b) The instantaneous current is $I = \lim_{\Delta t \to 0} \frac{\Delta q}{\Delta t}$:

$$I = \lim_{\Delta t \to 0} \frac{\Delta q}{\Delta t} = \lim_{\Delta t \to 0} (8t + 4\Delta t + 3) = 8t + 3 \text{ amperes}$$

(c) At $t = 0.04$:

$$I = 8(0.04) + 3 = 3.32 \text{ amperes}$$

### CURRENT IN A CAPACITOR

In a capacitor with constant capacitance $C$ (measured in farads) the charge $q$ can be expressed in terms of the capacitor voltage by the formula

$$q = CV \text{ coulombs}$$

$$\frac{\Delta q}{\Delta t} = C\frac{\Delta V}{\Delta t}$$

Since $I_{\text{av}} = \frac{\Delta q}{\Delta t}$ and $I = D_t q$,

$$I_{\text{av}} = C\frac{\Delta V}{\Delta t}$$

$$I = C\,D_t V$$

**Example 2**

(a) If a voltage of $V = 100t^2$ volts is applied to a 0.5-microfarad capacitor (1 microfarad = 0.000001 farad), what is the average current in the capacitor from $t = 2$ to $t = 5$ sec?
(b) What is the current at any time $t$?
(c) What is the current at $t = 2$ sec?

*Solution*

(*a*) We have

$$\begin{aligned} V &= 100t^2 \\ \Delta V &= 100(t + \Delta t)^2 - 100t^2 \\ &= 100t^2 + 200t\,\Delta t + 100\Delta t^2 - 100t^2 \\ &= 200t\,\Delta t + 100\Delta t^2 \\ \frac{\Delta V}{\Delta t} &= 200t + 100\Delta t \end{aligned}$$

We wish to find the average current from $t = 2$ to $t = 5$. Thus, we start with a value of $t = 2$ and $\Delta t = 3$.

$$\begin{aligned} \frac{\Delta V}{\Delta t} &= 200(2) + 100(3) \\ &= 700 \\ I_{av} &= C(700) \\ &= 0.0000005(700) \\ &= 0.00035 \text{ ampere} \end{aligned}$$

(*b*) The instantaneous current $I = C\, D_tV$.

$$D_tV = \lim_{\Delta t \to 0} \frac{\Delta V}{\Delta t} = \lim_{\Delta t \to 0} (200t + 100\Delta t) = 200t$$

$$\begin{aligned} I &= C\, D_tV \\ &= 0.0000005(200t) \\ &= 0.0001t \text{ amperes} \end{aligned}$$

(*c*) At $t = 2$

$$\begin{aligned} I &= 0.0001(2) \\ &= 0.0002 \text{ ampere} \end{aligned}$$

**INSTANTANEOUS POWER**

In a circuit the power $P$ (measured in watts) is defined as the rate of change of work (measured in joules) with respect to time $t$ (measured in seconds). Thus, the instantaneous power $P$ is represented by

$$P = \lim_{\Delta t \to 0} \frac{\Delta W}{\Delta t} = D_tW$$

**Example 3** (*a*) The work done by a certain current is given by the equation $W = 8t^2 - 2t$ joules. Find the power in the circuit at any time $t$.
(*b*) Find the current when $t = 2$ sec.

*Solution*

(*a*) We have

$$\begin{aligned} W &= 8t^2 - 2t \\ \Delta W &= 8(t + \Delta t)^2 - 2(t + \Delta t) - (8t^2 - 2t) \\ &= 8t^2 + 16t\,\Delta t + 8\Delta t^2 - 2t - 2\Delta t - 8t^2 + 2t \\ &= 16t\,\Delta t + 8\Delta t^2 - 2\Delta t \\ \frac{\Delta W}{\Delta t} &= 16t + 8\Delta t - 2 \end{aligned}$$

$$P = \lim_{\Delta t \to 0} \frac{\Delta W}{\Delta t} = \lim_{\Delta t \to 0} (16t + 8\Delta t - 2) = 16t - 2 \text{ watts}$$

(*b*) At $t = 2$:

$$P = 16(2) - 2 = 30 \text{ watts}$$

**EXERCISES**

**1.** (*a*) If the charge $q$ at a certain point in a circuit is given by the equation $q = 3t^2 + 2t$ coulombs, find the average current from $t = 0$ to $t = 0.04$ sec.

(*b*) Find the current $I$ at any time $t$.

(*c*) Find the current at $t = 0.01$ sec.

**2.** If the charge $q$ at a certain point in a circuit is given by the equation $q = 2t^2 + t + 0.001$ coulombs, find the current $I$ at any time $t$. Find the current at $t = 0.01$ sec.

**3.** If the charge $q$ at a certain point in a circuit is given by the equation $q = 0.005 + 2t - \frac{1}{2}t^2$ coulombs, find the current $I$ at $t = 0.005$ sec.

**4.** (*a*) If a voltage of $V = 200t^2 + 100t$ volts is applied to a 2-microfarad capacitor, what is the average current in the capacitor from $t = 0$ to $t = 2$ sec?

(*b*) What is the current at any time $t$?

(*c*) What is the current at $t = 0.01$ sec?

**5.** A voltage of $V = 100t^3$ volts is applied to a 1-microfarad capacitor. Find the current when $t = 0.2$ sec.

**6.** At what rate must the applied voltage be changed in a 1-microfarad capacitor if the current is to be 0.002 ampere?

**7.** The energy (work) dissipated by a certain resistor is given by the formula $W = 5t^2 + 2t$ joules. Find the power in the resistor at any time $t$. Find the power at $t = 3$ sec.

**8.** The energy supplied by a battery is given by the equation $W = 2t^3 - t^2 + 2t$. Find the power at $t = 2$ sec.

# 5

# Formulas for Calculating Derivatives

## 5-1 THE DERIVATIVE OF A CONSTANT

**Theorem** *If $f(x) = C$, then $D_x f(x) = 0$.*

*Proof* We must evaluate $\lim_{\Delta x \to 0} \frac{\Delta y}{\Delta x}$. Since $y = f(x) = C$, the function value is always equal to $C$ and $f(x + \Delta x) = C$.

$$\begin{aligned} \Delta y &= f(x + \Delta x) - f(x) \\ &= C - C \\ \frac{\Delta y}{\Delta x} &= \frac{0}{\Delta x} = 0 \\ \lim_{\Delta x \to 0} \frac{\Delta y}{\Delta x} &= 0 \; \blacksquare \end{aligned}$$

Thus the derivative of a constant is always 0. If we think of the geometric interpretation of the derivative this result is quite plausible.

The geometric interpretation of the derivative is the slope of the tangent to the curve at a given point. Since $f(x) = C$ the curve is a horizontal line whose slope is always 0.

## 5-2 THE DERIVATIVE OF THE INDEPENDENT VARIABLE

**Theorem** *If $y = f(x) = x$, then $D_x f(x) = 1$.*

*Proof*

$$\begin{aligned} \Delta y &= f(x + \Delta x) - f(x) \\ &= x + \Delta x - x \\ \frac{\Delta y}{\Delta x} &= \frac{\Delta x}{\Delta x} = 1 \\ D_x f(x) &= \lim_{\Delta x \to 0} \frac{\Delta y}{\Delta x} = 1 \quad \blacksquare \end{aligned}$$

## 5-3 THE POWER RULE FOR POSITIVE INTEGRAL EXPONENTS

**Theorem** *If $y = x^n$, where $n$ is a positive integer, then $D_x y = nx^{n-1}$.*

*Proof*

$$\Delta y = (x + \Delta x)^n - x^n$$

Here we shall apply the binomial theorem.†

$$\begin{aligned} \Delta y = \bigg(x^n &+ \frac{nx^{n-1}\,\Delta x}{1} + \frac{n(n-1)x^{n-2}\,\Delta x^2}{1 \cdot 2} \\ &+ \frac{n(n-1)(n-2)}{1(2)(3)} x^{n-3}\,\Delta x^3 + \cdots + nx\,\Delta x^{n-1} + \Delta x^n\bigg) - x^n \\ = nx^{n-1}\,\Delta x &+ \frac{n(n-1)x^{n-2}\,\Delta x^2}{2} + \frac{n(n-1)(n-2)x^{n-3}\,\Delta x^3}{6} \\ &+ \cdots + nx\,\Delta x^{n-1} + \Delta x^n \end{aligned}$$

$$\begin{aligned} \frac{\Delta y}{\Delta x} = nx^{n-1} &+ \frac{n(n-1)x^{n-2}\,\Delta x}{2} + \frac{n(n-1)(n-2)x^{n-3}\,\Delta x^2}{6} \\ &+ \cdots + nx\,\Delta x^{n-2} + \Delta x^{n-1} \end{aligned}$$

† The binomial theorem tells us that

$$\begin{aligned} (a + b)^n = a^n &+ \frac{na^{n-1}b}{1} + \frac{n(n-1)}{1(2)} a^{n-2}b^2 \\ &+ \frac{n(n-1)(n-2)}{1(2)(3)} a^{n-3}b^3 + \cdots + nab^{n-1} + b^n \end{aligned}$$

Except for the first term, each term has $\Delta x$ as a factor. Therefore,

$$D_x y = \lim_{\Delta x \to 0} \frac{\Delta y}{\Delta x} = nx^{n-1} \blacksquare$$

**Example 1** If $y = x^5$, then

$$y' = 5x^4$$

**Example 2** If $y = x^{10}$, then

$$y' = 10x^9$$

## 5-4 THE DERIVATIVE OF A SUM

**Theorem** *If $y = F(x) = [f(x) + g(x)]$, then $D_x y = D_x f(x) + D_x g(x)$.*

*Proof*

$$\begin{aligned} \Delta y &= F(x + \Delta x) - F(x) \\ &= [f(x + \Delta x) + g(x + \Delta x)] - [f(x) + g(x)] \\ &= f(x + \Delta x) - f(x) + g(x + \Delta x) - g(x) \\ \frac{\Delta y}{\Delta x} &= \frac{f(x + \Delta x) - f(x)}{\Delta x} + \frac{g(x + \Delta x) - g(x)}{\Delta x} \end{aligned}$$

From our theorem on limits we know that the limit of a sum equals the sum of the limits. Therefore,

$$\begin{aligned} \lim_{\Delta x \to 0} \frac{\Delta y}{\Delta x} &= \lim_{\Delta x \to 0} \frac{f(x + \Delta x) - f(x)}{\Delta x} + \lim_{\Delta x \to 0} \frac{g(x + \Delta x) - g(x)}{\Delta x} \\ &= D_x f(x) + D_x g(x) \blacksquare \end{aligned}$$

A similar result follows for a sum of more than two functions and we may generalize the result by saying that the derivative of a sum equals the sum of the derivatives. If we replace the plus sign in our theorem by a minus sign, we get a similar theorem for a difference. Thus, if $y = f(x) - g(x)$, then

$$D_x y = D_x f(x) - D_x g(x)$$

**Example 1** If $y = x^3 + x^2 + x + 7$, find $D_x y$.

*Solution* Since the derivative of a sum equals the sum of the derivatives, we can differentiate term by term:

$$D_x y = D_x x^3 + D_x x^2 + D_x x + D_x 7 = 3x^2 + 2x + 1$$

**Example 2**

$$D_x(x^2 - 3x + 7) = 2x - 3$$

## 5-5 THE DERIVATIVE OF A CONSTANT TIMES A FUNCTION

**Theorem** *If $y = cf(x)$, then $D_xy = c\,D_xf(x)$.*

*Proof*

$$\begin{aligned}\Delta y &= cf(x + \Delta x) - cf(x)\\ &= c[f(x + \Delta x) - f(x)]\\ \frac{\Delta y}{\Delta x} &= c\,\frac{f(x + \Delta x) - f(x)}{\Delta x}\end{aligned}$$

Since the limit of a product equals the product of the limits,

$$\begin{aligned}\lim_{\Delta x \to 0} \frac{\Delta y}{\Delta x} &= \lim_{\Delta x \to 0} c \lim_{\Delta x \to 0} \frac{f(x + \Delta x) - f(x)}{\Delta x}\\ &= c\,D_xf(x) \quad \blacksquare\end{aligned}$$

**Example**

$$D_x(4x^3) = 4(3x^2) = 12x^2$$
$$D_x(5x^2 + 4x - 7) = 10x + 4$$

**EXERCISES**

In Exercises 1 to 10 find $D_xy$.

**1.** $y = x^3 - 4x + 7$

**2.** $y = 6x - 4x^3$

**3.** $y = 4x^4 + \frac{1}{3}x^3 + x^2 - 8$

**4.** $y = x^7 - x^5 + 4x^3 + \frac{1}{2}x^2 + 20$

**5.** $y = 7 - 7x + x^2 + 9x^3$

**6.** $y = 4x^2 - 6x + 8$

**7.** $y = 8x^5 - 4x^3 + x$

**8.** $y = 2x^4 + 3x^2 - 7$

**9.** $y = 2x^8 + 4x^5 - 7x^3$

**10.** $y = 5x^2 + \frac{1}{3}x + 278$

**11.** Find the slope of the tangent line for the following curves.
(*a*) $y = 2x^2 + 6x + 7$, at $x = 2$
(*b*) $y = 3x^3 + 6$, at $x = 3$

**12.** Find the equation of the tangent lines for the following curves at the indicated points.
(*a*) $y = 4x - x^2$, at the point (2,4)
(*b*) $y = x^2 - 8x + 6$, at the point $(1,-1)$

**13.** The line which is perpendicular to the tangent line at the point of tangency is called the *normal* to the curve at that point. For the following curves find the equation of the normal line at the indicated point. (*Hint:* If two lines are perpendicular, their slopes are negative reciprocals.)
(*a*) $y = 4x^2 - 8x + 4$, at the point (2,4)
(*b*) $y = 2x^3 + 4x$, at the point (0,0)
(*c*) $y = (x - 2)^2$, at the point (1,1)

**14.** Find the velocity and acceleration at any time $t$ for a particle in rectilinear motion for each of the given laws of motion.
(*a*) $s = 6t^2 - 8t + 4$
(*b*) $s = 40t - 10t^2$
(*c*) $s = 16t^2 + 80t + 40$

**15.** A ball is thrown vertically upward so that its position at any time $t$ is given by the law of motion $s = 160t - 16t^2$.

(*a*) Find the initial velocity (i.e., the velocity when $t = 0$).

(*b*) How long did it take for the ball to reach its highest point, and how high did it go?

**16.** If the charge $q$ at a certain point in a circuit is given by the equation $q = 0.001 + 4t - \frac{5}{3}t^3$ coulombs, find the current $I$ at $t = 0.05$ sec.

**17.** A voltage of $V = 50t^3 + 40t^2$ is applied to a 2-microfarad capacitor. Find the current when $t = 0.3$ sec.

**18.** The energy supplied by a battery is given by the equation $W = \frac{1}{4}t^4 - \frac{2}{3}t^3 + \frac{3}{2}t^2 - t$. Find the power at $t = 3$ sec.

## 5-6 THE DERIVATIVE OF A PRODUCT OF FUNCTIONS

**Theorem** *If* $y = F(x) = f(x) \cdot g(x)$, *then* $D_x y = f(x) \cdot D_x g(x) + g(x) D_x f(x)$.

*Proof*

$$\Delta y = F(x + \Delta x) - F(x)$$
$$= f(x + \Delta x) \cdot g(x + \Delta x) - f(x) \cdot g(x)$$
$$\frac{\Delta y}{\Delta x} = \frac{f(x + \Delta x) \cdot g(x + \Delta x) - f(x) \cdot g(x)}{\Delta x}$$

The limit as $\Delta x$ approaches 0 of the right-hand side does not give us a familiar expression, in fact, it gives us $\frac{0}{0}$. We would like to somehow rearrange the terms so as to get the familiar expressions

$$\frac{f(x + \Delta x) - f(x)}{\Delta x} \quad \text{and} \quad \frac{g(x + \Delta x) - g(x)}{\Delta x}$$

The limits of the above expressions as $\Delta x \to 0$ are, respectively, $D_x f(x)$ and $D_x g(x)$. To bring about the above expressions, we shall have to employ a trick which is often used in mathematical proofs. We shall subtract and add the expression $f(x + \Delta x) \cdot g(x)$ in the numerator of the right-hand side. (We are thus adding zero to the numerator. This does not change the value of the fraction.) Thus,

$$\frac{\Delta y}{\Delta x} = \frac{f(x + \Delta x) \cdot g(x + \Delta x) - f(x + \Delta x) \cdot g(x) + f(x + \Delta x) \cdot g(x) - f(x) \cdot g(x)}{\Delta x}$$

$$= \frac{f(x + \Delta x) \cdot g(x + \Delta x) - f(x + \Delta x) \cdot g(x)}{\Delta x} + \frac{f(x + \Delta x) \cdot g(x) - f(x) \cdot g(x)}{\Delta x}$$

$$= f(x + \Delta x) \cdot \frac{g(x + \Delta x) - g(x)}{\Delta x} + g(x) \frac{f(x + \Delta x) - f(x)}{\Delta x}$$

$$D_x y = \lim_{\Delta x \to 0} \frac{\Delta y}{\Delta x} = \lim_{\Delta x \to 0} f(x + \Delta x) \cdot \lim_{\Delta x \to 0} \frac{g(x + \Delta x) - g(x)}{\Delta x}$$

$$+ \lim_{\Delta x \to 0} g(x) \cdot \lim_{\Delta x \to 0} \frac{f(x + \Delta x) - f(x)}{\Delta x}$$

$$= f(x) \cdot D_x g(x) + g(x) \cdot D_x f(x)$$

In words, the derivative of a product is the first factor times the derivative of the second plus the second times the derivative of the first.■

**Example 1**

$$\begin{aligned} D_x[(x^2 + 3)(x^3 + x)] &= (x^2 + 3)\, D_x(x^3 + x) + (x^3 + x)\, D_x(x^2 + 3) \\ &= (x^2 + 3)(3x^2 + 1) + (x^3 + x)(2x) \\ &= 3x^4 + 10x^2 + 3 + 2x^4 + 2x^2 \\ &= 5x^4 + 12x^2 + 3 \end{aligned}$$

**Example 2**

$$\begin{aligned} D_x[(2x + 7)(x^3 + 4)] &= (2x + 7)\, D_x(x^3 + 4) + (x^3 + 4)\, D_x(2x + 7) \\ &= (2x + 7)(3x^2) + (x^3 + 4)(2) \\ &= 6x^3 + 21x^2 + 2x^3 + 8 \\ &= 8x^3 + 21x^2 + 8 \end{aligned}$$

**EXERCISES**

In Exercises 1 to 6 find $D_x y$.

**1.** $y = (2x - 7)(4x + 3)$

**2.** $y = (7 - x^2)(3x + 7)$

**3.** $y = (x^2 - 4)(7x + x^3)$

**4.** $y = (x^3 + 1)(4x^2 + x)$

**5.** $y = (x^2 - 8)(3x^2 + 4)$

**6.** $y = (3x + 4)(x^2 - 1)(6x + 7)$

**7.** Find the equation of the tangent line to the curve $y = (x^2 - 7)(3x + 4)$ at $x = 2$.

**8.** Find the equation of the normal line to the curve $y = (3x - 7)(x^2 + 4)$ at $x = 1$.

## 5-7 THE DERIVATIVE OF THE RECIPROCAL OF $f(x)$

**Theorem** *If* $y = \dfrac{1}{f(x)}$, *then* $D_x y = -\dfrac{D_x f(x)}{[f(x)]^2}$.

*Proof*

$$\begin{aligned} \Delta y &= \frac{1}{f(x + \Delta x)} - \frac{1}{f(x)} \\ &= \frac{f(x) - f(x + \Delta x)}{f(x) f(x + \Delta x)} \\ \frac{\Delta y}{\Delta x} &= \frac{f(x) - f(x + \Delta x)}{\Delta x\, f(x) f(x + \Delta x)} \\ &= -\frac{f(x + \Delta x) - f(x)}{\Delta x} \frac{1}{f(x) f(x + \Delta x)} \end{aligned}$$

$$
\begin{aligned}
D_x y &= \lim_{\Delta x \to 0} \frac{\Delta y}{\Delta x} \\
&= \lim_{\Delta x \to 0} - \frac{f(x + \Delta x) - f(x)}{\Delta x} \lim_{\Delta x \to 0} \frac{1}{f(x)f(x + \Delta x)} \\
&= -D_x f(x) \frac{1}{[f(x)]^2} \\
&= -\frac{D_x f(x)}{[f(x)]^2} \blacksquare
\end{aligned}
$$

**Example 1**

$$D_x \frac{1}{x^2 + 2x} = -\frac{D_x(x^2 + 2x)}{(x^2 + 2x)^2} = -\frac{2x + 2}{(x^2 + 2x)^2}$$

**Example 2** If $y = x^{-5}$, find $D_x y$.

*Solution* Though our power rule has been proved only for positive integral exponents, we can rewrite $x^{-5}$ as $\frac{1}{x^5}$ and use the formula for the derivative of the reciprocal of $f(x)$.

$$
\begin{aligned}
y &= x^{-5} = \frac{1}{x^5} \\
D_x y &= -\frac{5x^4}{(x^5)^2} = -\frac{5x^4}{x^{10}} = -\frac{5}{x^6} \\
&= -5x^{-6}
\end{aligned}
$$

This result seems to imply that the power rule for positive integral exponents will also apply to negative integral exponents. We shall now prove this result.

**Theorem** *If $y = x^{-n}$, $n$ a positive integer, then $D_x y = -nx^{-n-1}$.*

*Proof*

$$
\begin{aligned}
y &= x^{-n} = \frac{1}{x^n} \\
D_x y &= -\frac{nx^{n-1}}{(x^n)^2} \\
&= -\frac{nx^{n-1}}{x^{2n}} \\
&= -\frac{n}{x^{n+1}} \\
&= -nx^{-n-1} \blacksquare
\end{aligned}
$$

**Example 3**

$$
\begin{aligned}
D_x(x^{-6}) &= -6x^{-7} \\
D_x(x^{-10}) &= -10x^{-11}
\end{aligned}
$$

## 5-8 THE DERIVATIVE OF A QUOTIENT OF FUNCTIONS

**Theorem** *If* $y = \frac{f(x)}{g(x)}$, $g(x) \neq 0$, *then* $D_x y = \frac{g(x)\, D_x f(x) - f(x)\, D_x g(x)}{[g(x)]^2}$.

*Proof*

$$y = \frac{f(x)}{g(x)} = f(x) \frac{1}{g(x)}$$

We shall use the product rule:

$$\begin{aligned} D_x y &= f(x)\, D_x \frac{1}{g(x)} + \frac{1}{g(x)} D_x f(x) \\ &= f(x) \frac{-D_x g(x)}{[g(x)]^2} + \frac{D_x f(x)}{g(x)} \\ &= \frac{-f(x)\, D_x g(x) + g(x)\, D_x f(x)}{[g(x)]^2} \\ &= \frac{g(x)\, D_x f(x) - f(x)\, D_x g(x)}{[g(x)]^2} \quad \blacksquare \end{aligned}$$

**Example 1** Find $D_x y$ for $y = \frac{2x+1}{x^3 - x}$.

*Solution*

$$\begin{aligned} y &= \frac{2x+1}{x^3-x} \\ D_x y &= \frac{(x^3 - x)\, D_x(2x+1) - (2x+1)\, D_x(x^3 - x)}{(x^3-x)^2} \\ &= \frac{(x^3-x)2 - (2x+1)(3x^2-1)}{(x^3-x)^2} \\ &= \frac{2x^3 - 2x - 6x^3 - 3x^2 + 2x + 1}{(x^3-x)^2} \\ &= \frac{-4x^3 - 3x^2 + 1}{(x^3-x)^2} \end{aligned}$$

**Example 2** Find $D_x y$ for $y = \frac{x^2}{x - x^2}$.

*Solution*

$$\begin{aligned} y &= \frac{x^2}{x - x^2} \\ D_x y &= \frac{(x - x^2)\, D_x x^2 - x^2\, D_x(x - x^2)}{(x-x^2)^2} \\ &= \frac{(x-x^2)2x - x^2(1-2x)}{(x-x^2)^2} \\ &= \frac{2x^2 - 2x^3 - x^2 + 2x^3}{(x-x^2)^2} \\ &= \frac{x^2}{(x-x^2)^2} \end{aligned}$$

**EXERCISES**

In Exercises 1 to 10 find $D_x y$.

**1.** $y = \dfrac{1}{x^5}$

**2.** $y = x^{-6}$

**3.** $y = \dfrac{1}{x}$

**4.** $y = \dfrac{x^2 + 2}{x - x^2}$

**5.** $y = 3x^{-5} + 4x^{-2}$

**6.** $y = \dfrac{x^2 - 1}{x + x^2}$

**7.** $y = \dfrac{2x - 1}{x^2 + 1}$

**8.** $y = \dfrac{4}{x} - \dfrac{3}{x^2}$

**9.** $y = \dfrac{2x + 7}{x - 1}$

**10.** $y = \dfrac{1 - x^2}{4x + 3}$

**11.** Find the slope of the tangent line for the curve $y = \dfrac{6x + 4}{2x - x^2}$ at the point (1,10).

**12.** Find the equation of the tangent line for the curve $y = \dfrac{1}{x^2}$ at the point (1,1).

**13.** Find the equation of the normal line to the curve $y = \dfrac{x - 2}{x^2 + 1}$ at the point (2,0).

**14.** Find the equation of the normal line to the curve $y = \dfrac{1}{x + 1}$ at the point (0,1).

**15.** Find the velocity at any time $t$ for a particle in rectilinear motion for each of the given laws of motion.

(*a*) $s = \dfrac{t + 2}{t^2 - 1}$

(*b*) $s = 2t + \dfrac{1}{t}$

(*c*) $s = \dfrac{t + 1}{t^2}$

## 5-9 THE CHAIN RULE

Let us consider the pair of functions

$$y = 3u^2 + 1 \qquad u = 2x - 3$$

If we choose a value for $x$, it will produce a corresponding value for $u$ which in turn will produce a corresponding value for $y$. Thus, we can say that $y$ is a function of $x$ using $u$ as a link in the chain. For example, if $x = 2$, $u = 1$, and $y = 4$; if $x = 3$, $u = 3$ and $y = 28$. For this particular example we could use the method of substitution to determine a rule of correspondence which would give us $y$ as a function of $x$.

$$\begin{aligned} y &= 3(2x - 3)^2 + 1 \\ &= 3(4x^2 - 12x + 9) + 1 \\ &= 12x^2 - 36x + 28 \end{aligned}$$

On the other hand, let us consider the following pair of functions

$$y = \sqrt{\tfrac{1}{2} - u} \qquad u = \sqrt{1 - x}$$

When $x = \frac{3}{4}$, $u = \frac{1}{2}$ and $y = 0$. However, when $x = 0$, $u = 1$ and $y$ is not defined. Therefore, we must be careful when setting up a chain, where $y = f(u)$ and $u = g(x)$, to make sure that the values of $u$ corresponding to the values of $x$ for the function $g(x)$ be in the domain of the function $f(u)$. When this is done, we sometimes use the notation $y = f[g(x)]$ and we call the function a composite function.

In our first example, $y = 3u^2 + 1$ and $u = 2x - 3$, we pointed out that $y$ is a function of $x$, and by using the method of substitution we found that $y = 12x^2 - 36x + 28$. In this case, $D_x y = 24x - 36$.

In many cases, however, the method of substitution gives rise to complicated expressions which may be very difficult to differentiate. We shall develop a formula for differentiation which will not require substitution before differentiation. The formula is generally referred to as the *chain rule.*

**Theorem** *If $y = f(u)$ and $u = g(x)$, where for every value of $x$ there exists a corresponding value of $y$, and if $D_u f(u)$ and $D_x g(x)$ both exist, then*

$$D_x y = D_u f(u) \cdot D_x g(x)$$

*Proof*

$$y = f(u) \qquad u = g(x)$$

For every value of $x$ we get a corresponding value for $u$ and this in turn gives us a corresponding value for $y$. Thus, using $u$ as a link we can say that $y$ is a function of $x$. Now, if we introduce an increment $\Delta x$ this will give rise to an increment $\Delta u$ which in turn will bring about an increment $\Delta y$. We would like to find $\lim_{\Delta x \to 0} \frac{\Delta y}{\Delta x}$. We note the following:

$$\Delta u = g(x + \Delta x) - g(x)$$
$$\Delta y = f(u + \Delta u) - f(u)$$
$$D_x y = \lim_{\Delta x \to 0} \frac{\Delta y}{\Delta x}$$
$$D_u y = \lim_{\Delta u \to 0} \frac{\Delta y}{\Delta u}$$

and

$$D_x u = \lim_{\Delta x \to 0} \frac{\Delta u}{\Delta x}$$

Since we are interested in finding $\lim_{\Delta x \to 0} \frac{\Delta y}{\Delta x}$, we shall form the difference quotient $\frac{\Delta y}{\Delta x}$.

$$\frac{\Delta y}{\Delta x} = \frac{f(u + \Delta u) - f(u)}{\Delta x}$$

If $\Delta u \neq 0$†, we can multiply by $\frac{\Delta u}{\Delta u}$, and, rearranging the terms, we have

$$\frac{\Delta y}{\Delta x} = \frac{f(u + \Delta u) - f(u)}{\Delta u} \frac{\Delta u}{\Delta x}$$

But,

$$\Delta u = g(x + \Delta x) - g(x)$$

Therefore,

$$\frac{\Delta y}{\Delta x} = \frac{f(u + \Delta u) - f(u)}{\Delta u} \frac{g(x + \Delta x) - g(x)}{\Delta x}$$

We have already pointed out that if a function is differentiable in an interval, then it is also continuous in that interval. Therefore, the function $u = g(x)$, which is differentiable, must also be continuous, and from the properties of continuous functions we can say that as $\Delta x \to 0$ then $\Delta u \to 0$. Thus, when $\Delta x \to 0$, then $\Delta u \to 0$ and

$$\frac{f(u + \Delta u) - f(u)}{\Delta u} \to D_u f(u)$$

$$\frac{g(x + \Delta x) - g(x)}{\Delta x} \to D_x g(x)$$

Hence,

$$\frac{\Delta y}{\Delta x} = \frac{f(u + \Delta u) - f(u)}{\Delta u} \frac{g(x + \Delta x) - g(x)}{\Delta x}$$

$$D_x y = \lim_{\Delta x \to 0} \frac{\Delta y}{\Delta x} = D_u f(u)\, D_x g(x)$$

$$D_x y = D_u f(u) \cdot D_x g(x) \blacksquare$$

† It can be proved that the results of this theorem hold also for the case where $\Delta u = 0$. A proof of this can be found in more advanced texts on the calculus.

**Example 1**

$$\begin{aligned} y &= 3u^2 + 1 \qquad & u &= 2x - 3 \\ D_u y &= 6u & D_x u &= 2 \\ D_x y &= D_u y D_x u \\ &= 6u(2) \\ &= 12u \\ &= 12(2x - 3) \\ &= 24x - 36 \end{aligned}$$

This result checks with our previous result using the method of substitution.

**Example 2**

$$\begin{aligned} y &= u - 2u^2 \qquad u = 3x^2 - 7 \\ D_x y &= D_u y \, D_x u \\ &= (1 - 4u)6x \\ &= 6x - 24xu \\ &= 6x - 24x(3x^2 - 7) \\ &= -72x^3 + 174x \end{aligned}$$

**Example 3** If $y = t^2 + 2t$ and $x = 1 - 2t$, find $D_x y$.

*Solution* Unlike our previous examples, in this example both variables, $x$ and $y$, are given in terms of a third variable $t$. The equations given in the form $x = f(t)$ and $y = g(t)$ are referred to as *parametric equations* and the variable $t$ is referred to as a *parameter*.

From the chain rule we know that if

$$y = F(x) \qquad \text{and} \qquad x = G(t)$$
$$D_t y = D_x y \, D_t x$$

Then if $D_t x \neq 0$, we can divide both sides by $D_t x$ and thus

$$D_x y = \frac{D_t y}{D_t x}$$

This relationship will be particularly useful in our example. Hence,

$$\begin{aligned} y &= t^2 + 2t \\ D_t y &= 2t + 2 \\ x &= 1 - 2t \\ D_t x &= -2 \\ D_x y &= \frac{D_t y}{D_t x} = \frac{2t + 2}{-2} = -t - 1 \end{aligned}$$

*Remark* For some parametric equations it is possible to eliminate the parameter $t$ and write $y$ as a function of $x$. In our example

$$\begin{aligned} y &= t^2 + 2t \\ x &= 1 - 2t \\ 2t &= 1 - x \\ t &= \frac{1 - x}{2} \end{aligned}$$

Hence

$$\begin{aligned} y &= \left(\frac{1 - x}{2}\right)^2 + 2\frac{1 - x}{2} \\ &= \tfrac{1}{4}(1 - 2x + x^2) + 1 - x \\ &= \tfrac{1}{4}x^2 - \tfrac{3}{2}x + \tfrac{5}{4} \\ D_xy &= \tfrac{1}{2}x - \tfrac{3}{2} \end{aligned}$$

Using the formula $D_xy = \dfrac{D_ty}{D_tx}$, we found

$$D_xy = -t - 1$$

Since $t = \dfrac{1 - x}{2}$,

$$\begin{aligned} D_xy &= -\frac{1 - x}{2} - 1 \\ &= \tfrac{1}{2}x - \tfrac{3}{2} \end{aligned}$$

**EXERCISES**

Find $D_xy$.

**1.** $y = t^2 - 2t,\ t = 4 + x^2$

**2.** $y = 4 - t + t^2,\ t = \dfrac{1}{x}$

**3.** $y = \dfrac{1}{t^2},\ t = 1 + \dfrac{1}{x}$

**4.** $y = 3u^2 + 4,\ u = 3x + 2$

**5.** $y = \dfrac{u}{u + 1},\ u = x^2 - 3$

**6.** $y = u^3 - 2,\ u = 8x + 2$

**7.** $y = t^2 - t,\ x = t^2 + 1$

**8.** $y = \tfrac{1}{2}t^2,\ x = \tfrac{1}{3}t^3$

**9.** $y = t^2 + 1,\ x = 2t - 4$

**10.** $y = \dfrac{t}{t + 1},\ x = \dfrac{1}{t + 1}$

## 5-10 THE DERIVATIVE OF A FUNCTION TO A POWER

We have found that if $y = x^n$, then $D_xy = nx^{n-1}$. Will the same formula work if we try to find the derivative of $y = (x^2 + 2x)^2$? If the formula were to be applied to the above function, we would conclude that $D_xy = 2(x^2 + 2x) = 2x^2 + 4x$. If we multiply out $(x^2 + 2x)^2$, we get

$$\begin{aligned} y &= x^4 + 4x^3 + 4x^2 \\ D_xy &= 4x^3 + 12x^2 + 8x \end{aligned}$$

Clearly the power rule for $x^n$ does not work for a function of $x$ to a power $[f(x)]^n$. We can, however, approach the above problem in the following way: Let $u = (x^2 + 2x)$. Then we have

$$y = u^2 \quad \text{and} \quad u = x^2 + 2x$$

By the chain rule

$$\begin{aligned} D_x y &= D_u y \, D_x u \\ &= 2u(2x + 2) \\ &= 4xu + 4u \\ &= 4x(x^2 + 2x) + 4(x^2 + 2x) \\ &= 4x^3 + 8x^2 + 4x^2 + 8x \\ &= 4x^3 + 12x^2 + 8x \end{aligned}$$

This leads us to the following theorem.

**Theorem** *If $y = [f(x)]^n$, then $D_x y = n[f(x)]^{n-1} D_x f(x)$.*

*Proof*

$$y = [f(x)]^n$$

Let $u = f(x)$. Then

$$\begin{aligned} y &= u^n \quad \text{and} \quad u = f(x) \\ D_x y &= D_u y \, D_x u \\ &= nu^{n-1} D_x u \\ &= n[f(x)]^{n-1} D_x f(x) \quad \blacksquare \end{aligned}$$

**Example 1** If $y = (x^3 + 2x)^7$, find $D_x y$.

*Solution* Let $u = x^3 + 2x$. Then

$$\begin{aligned} y &= u^7 \qquad u = x^3 + 2x \\ D_x y &= (7u^6)(3x^2 + 2) \\ &= 7(x^3 + 2x)^6(3x^2 + 2) \end{aligned}$$

**Example 2** Find the derivative of $y = (x - 3x^2)^8$.

*Solution* We shall not go through the process of substitution, but we shall apply the formula directly. If $y = [f(x)]^n$, then $D_x y = n[f(x)]^{n-1} D_x f(x)$. Therefore, if $y = (x - 3x^2)^8$, then

$$D_x y = 8(x - 3x^2)^7(1 - 6x)$$

**EXERCISES**

Find $D_x y$.

**1.** $y = (x - 2x^3)^5$

**2.** $y = (3x^2 + 4)^8$

**3.** $y = \left(1 + \dfrac{1}{x}\right)^6$

**4.** $y = (1 - 3x^2)^6$

**5.** $y = (3x^2 + 4)^3$

**6.** $y = \left(x^2 - \dfrac{1}{x^2}\right)^2$

## 5-11 IMPLICIT DIFFERENTIATION

Until now we have been considering functions where the rule of correspondence for a function expresses the dependent variable directly in terms of the independent variable, for example, $y = x^3 + 2x^2 - x + 3$ or $y = x - 3x^2$. A function such as this is called an *explicit function.* If, however, we must solve for the dependent variable in order to express it in terms of the independent variable, the function is called an *implicit function.*

**Example 1** $x^2 + 2xy = 4$ is a rule of correspondence for an implicit function. Here

$$2xy = 4 - x^2$$

$$y = \frac{4 - x^2}{2x} = \frac{2}{x} - \frac{x}{2}$$

**Example 2** $y^4 - xy^3 + x^2 - 7 = 0$ is a rule of correspondence for an implicit function. However, here we would have considerable difficulty in trying to solve for $y$ in terms of $x$. Nevertheless, we can still think of $y$ as a function of $x$, where the function is defined implicitly.

We shall now illustrate a method for finding $D_xy$ for functions defined implicitly.

**Example 3** Find $D_xy$ for $x^2 + 2xy = 4$.

*Solution* We have already shown that when we solve for $y$ in terms of $x$,

$$y = \frac{2}{x} - \frac{x}{2}$$

$$D_xy = -\frac{2}{x^2} - \frac{1}{2}$$

We can also find $D_xy$ without actually solving for $y$ in terms of $x$. Since we know that $y = f(x)$, we can think of our implicit function as

$$x^2 + 2x[f(x)] = 4$$

We shall differentiate term by term:

$$2x + 2x\, D_xf(x) + 2f(x) = 0$$

[where we have used the product rule to differentiate $2xf(x)$].

$$2x\, D_xf(x) = -2x - 2f(x)$$

$$D_xf(x) = -\frac{x + f(x)}{x}$$

$$= -1 - \frac{f(x)}{x}$$

$$= -1 - \frac{y}{x}$$

To check this result, we shall substitute the value of $y$.

$$y = \frac{2}{x} - \frac{x}{2}$$

$$\begin{aligned} D_x f(x) &= -1 - \frac{\frac{2}{x} - \frac{x}{2}}{x} \\ &= -1 - \frac{2}{x^2} + \frac{1}{2} \\ &= -\frac{1}{2} - \frac{2}{x^2} \end{aligned}$$

**Example 4** Find $D_x y$ for $y^4 - xy^3 + x^2 - 7 = 0$.

*Solution* Here again $y = f(x)$, and our equation is equivalent to

$$[f(x)]^4 - x[f(x)]^3 + x^2 - 7 = 0$$

Differentiating, we get

$$\begin{aligned} 4[f(x)]^3\, D_x f(x) - x3[f(x)]^2\, D_x f(x) - [f(x)]^3 + 2x - 0 &= 0 \\ D_x f(x)\,\{4[f(x)]^3 - 3x[f(x)]^2\} &= [f(x)]^3 - 2x \\ D_x f(x) &= \frac{[f(x)]^3 - 2x}{4[f(x)]^3 - 3x[f(x)]^2} \\ D_x y &= \frac{y^3 - 2x}{4y^3 - 3xy^2} \end{aligned}$$

**Example 5** Find $D_x y$ for $xy^2 + y^2 - 2x = 5$.

*Solution* Again we shall think of $y = f(x)$, but we shall not substitute. We have pointed out that $D_x[f(x)]^n = n[f(x)]^{n-1}\, D_x f(x)$. Thus,

$$D_x(y^n) = ny^{n-1}\, D_x y$$

or, using a shorter notation,

$$D_x(y^n) = ny^{n-1}y'$$

Therefore, differentiating $xy^2 + y^2 - 2x = 5$ term by term, we get

$$\begin{aligned} x(2yy') + y^2(1) + 2yy' - 2 &= 0 \\ y'(2xy + 2y) &= 2 - y^2 \\ y' &= \frac{2 - y^2}{2xy + 2y} \end{aligned}$$

**Example 6** Find the equation of the tangent line to the curve $2x^2 + y^2 + 2xy = 5$ at the point (1,1).

*Solution* The equation of the tangent line will have the form $y - y_1 = m(x - x_1)$. We know that $x_1 = 1$ and $y_1 = 1$. We must find $m$.

$$2x^2 + y^2 + 2xy = 4$$
$$4x + 2yy' + 2xy' + 2y = 0$$
$$y'(2y + 2x) = -4x - 2y$$
$$y' = \frac{-4x - 2y}{2x + 2y} = -\frac{2x + y}{x + y}$$

At the point (1,1):

$$m = y' = -\frac{2 + 1}{1 + 1} = -\frac{3}{2}$$

The equation of the tangent line is then

$$y - 1 = -\tfrac{3}{2}(x - 1)$$

It is important to remember that the derivative of $y^n$ is *not* $ny^{n-1}$ but $ny^{n-1}y'$.

**EXERCISES**

Find $D_xy$ in terms of $x$ and $y$.

**1.** $3x - 2y = 7$

**2.** $x^2 + 4y = 8$

**3.** $x^2y = 5$

**4.** $x^2 - 3xy + y = 7$

**5.** $3xy + y^2 - x = 3$

**6.** $xy^2 + 2x^2y = 5$

**7.** $x^2 + y^2 + 2x - 4y = 0$

**8.** $y^3 + 3xy - 4 = 0$

**9.** $xy^2 + 4y = 3$

**10.** $x^2y^2 + 2x - y = 8$

**11.** Find the equation of the tangent line to the curve $x^2 + y^2 = 25$ at the point (4,3).

**12.** Find the equation of the normal line to the curve $x^2y + 2x - y = 10$ at the point (2,2).

**13.** Find the equation of the normal line to the curve $xy = 4$ at the point $(8,\frac{1}{2})$.

**14.** The angle between two curves is defined as the angle between the tangent lines to the curves at the point of intersection. Two curves that intersect at right angles are said to intersect orthogonally. Prove that the hyperbolas $x^2 - y^2 = 9$ and $xy = 4$ intersect orthogonally.

**15.** Show that the parabolas $y = \frac{1}{2}x^2 - 2$ and $y = -\frac{1}{8}x^2 + \frac{1}{2}$ intersect orthogonally.

## 5-12 FRACTIONAL EXPONENTS

We shall now consider the derivative of $y = x^{p/q}$ where $p$ and $q$ are positive integers. If the power rule were to work in the case of fractional exponents, we would have

$$D_xy = \frac{p}{q}x^{p/q-1}$$

We shall now prove that this is actually the case.

**Theorem** *If* $y = x^{p/q}$, *then* $y' = \dfrac{p}{q} x^{p/q-1}$.

*Proof*

$$y = x^{p/q}$$

We shall raise both sides of the equation to the $q$th power and then differentiate implicitly.

$$\begin{aligned} y^q &= x^p \\ qy^{q-1}y' &= px^{p-1} \\ y' &= \frac{p}{q}\frac{x^{p-1}}{y^{q-1}} \end{aligned}$$

Since $y = x^{p/q}$,

$$\begin{aligned} y' &= \frac{p}{q}\frac{x^{p-1}}{(x^{p/q})^{q-1}} \\ &= \frac{p}{q}\frac{x^{p-1}}{x^{p-p/q}} \\ &= \frac{p}{q} x^{p/q-1} \quad \blacksquare \end{aligned}$$

If either $p$ or $q$ is negative, we can apply our methods for negative exponents to show that the power rule applies in that case too.

**Example 1**

$$\begin{aligned} y &= x^{\frac{1}{3}} \\ y' &= \tfrac{1}{3}x^{-\frac{2}{3}} \end{aligned}$$

**Example 2**

$$\begin{aligned} y &= \sqrt{x^2 + 4} \\ &= (x^2 + 4)^{\frac{1}{2}} \\ y' &= \tfrac{1}{2}(x^2 + 4)^{-\frac{1}{2}}2x \\ &= \frac{x}{\sqrt{x^2 + 4}} \end{aligned}$$

**EXERCISES**

Find $y'$.

**1.** $y = \sqrt{x} + \dfrac{1}{\sqrt{x}}$

**2.** $y = \sqrt[3]{x}$

**3.** $y = \sqrt{2x - 1}$

**4.** $y = \dfrac{1}{\sqrt{x^2 - 4}}$

**5.** $y = \dfrac{x + 2}{\sqrt{x}}$

**6.** $y = (x^3 - 1)^{\frac{2}{3}}$

**7.** $y = (4 - x^2)^{\frac{1}{3}}$

**8.** $y = (2 + 3x)^{-\frac{1}{2}}$

**9.** $y = \dfrac{1}{x} + \sqrt{5}$

**10.** $y = \sqrt{\dfrac{1}{x}}$

# 6
# Applications of Derivatives

## 6-1 GEOMETRIC SIGNIFICANCE OF THE SIGN OF THE DERIVATIVE

Let us consider a function $y = f(x)$. If at a point $(x_1,y_1)$, $f'(x_1)$ is positive, then the slope is positive and $\lim_{\Delta x \to 0} \frac{\Delta y}{\Delta x}$ is positive. Thus, when $\Delta x$ is positive, $\Delta y$ is positive, and when $\Delta x$ is negative, $\Delta y$ is negative. In either case, as we move from left to right the curve is rising. Similarly, if $f'(x_1)$ is negative, then when $\Delta x$ is positive $\Delta y$ is negative and when $\Delta x$ is negative $\Delta y$ is positive. Therefore, as we move from left to right the curve is falling (Fig. 6-1).

We can summarize this discussion with the following rule: *At any point $(x,y)$ if $f'(x)$ is positive, the curve is rising, and if $f'(x)$ is negative, the curve is falling.* If $f'(x) = 0$ at a given point, then the curve is neither rising nor falling, and the given point is said to be a stationary point.

**Example 1** Determine where the curve $y = \frac{1}{3}x^3 - 3x^2 + 8x$ is rising and where it is falling (Fig. 6-2*b*).

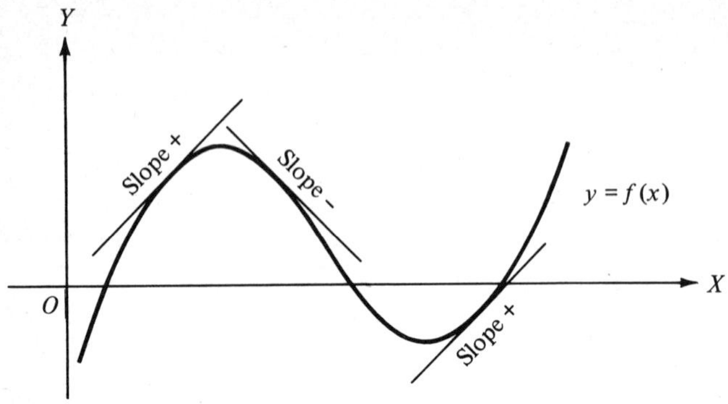


Fig. 6-1

*Solution* We know that where $f'(x)$ is positive the curve is rising and where $f'(x)$ is negative the curve is falling.

$$\begin{aligned} y &= \tfrac{1}{3}x^3 - 3x^2 + 8x \\ y' &= x^2 - 6x + 8 \\ &= (x - 2)(x - 4) \end{aligned}$$

We note first that at $x = 2$ and at $x = 4$ the curve is neither rising nor falling.

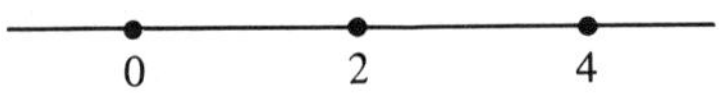


Fig. 6-2a

When $x < 2$ (Fig. 6-2*a*), $(x - 2)$ is negative and $(x - 4)$ is negative and their product is positive.
When $2 < x < 4$, $(x - 2)$ is positive but $(x - 4)$ is negative and their product is negative.
When $x > 4$, $(x - 2)$ is positive and $(x - 4)$ is positive and their product is positive.

Thus, the curve is rising when $x < 2$ and also when $x > 4$, and falling between $x = 2$ and $x = 4$.

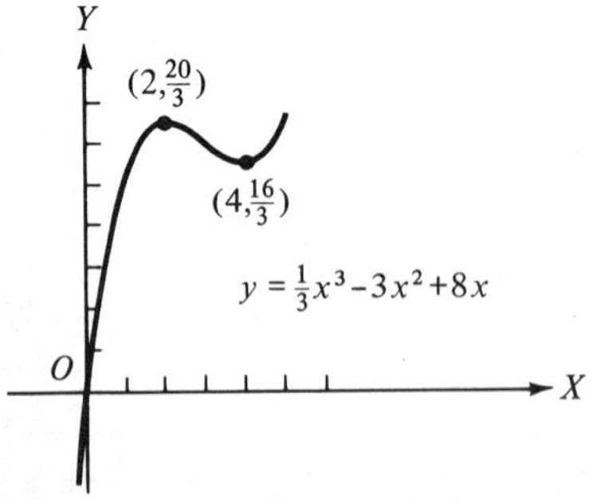


Fig. 6-2b

**Example 2** Determine where the curve $y = x^3$ is rising and where it is falling (Fig. 6-3).

*Solution*

$$y = x^3$$
$$y' = 3x^2$$

When $x = 0$, the curve is neither rising nor falling.
When $x \neq 0$, $y'$ is always positive.
Thus, the curve is rising at all points except at the point (0,0).

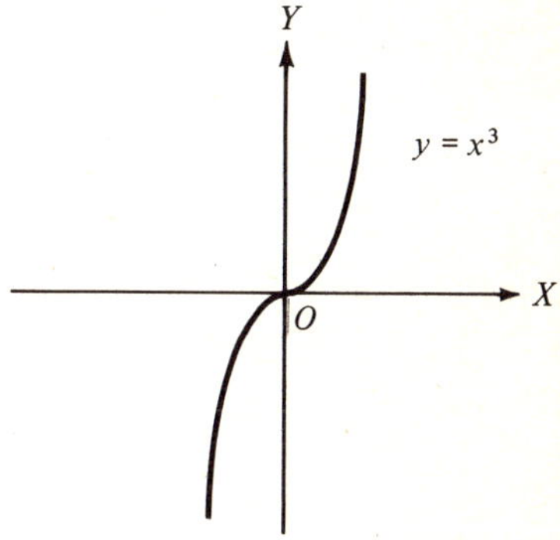


Fig. 6-3

**EXERCISES**

In Exercises 1 to 4 determine whether the curve is rising, falling, or stationary for the given value of $x$.

1. $y = 2x^2 - 3x + 2, x = 5$
2. $y = 4 - x^2, x = 3$
3. $y = 7 - 12x + x^3, x = -2$
4. $y = x^3 + \frac{1}{2}x^2 - 3x + 4, x = 0$

In Exercises 5 to 10 determine where the given curves are rising and where they are falling.

5. $y = 2x^2 - 4x + 6$
6. $y = 1 - x^3$
7. $y = \dfrac{1}{x}$
8. $y = x^2 - 6x + 8$
9. $y = 2x^3 - 6x + 4$
10. $y = 4 - 2x$
11. For what values of $x$ is the curve $y = \dfrac{x - 1}{x}$ rising?

## 6-2 MAXIMA AND MINIMA

**Definition** *A relative maximum point on a curve $y = f(x)$ is a point on the curve for which the $y$ value is greater than the $y$ value of any other point immediately preceding or following it. Similarly, a relative minimum point on a curve $y = f(x)$ is a point on the curve for which the $y$ value* *is less than the $y$ value of any other point immediately preceding or following it.*

In Fig. 6-4, points $B$ and $D$ are relative maximum points, and points $C$ and $E$ are relative minimum points. We note that the relative minimum point $E$ has a greater $y$ value than the relative maximum point $B$. It is thus essential that we distinguish between the absolute minimum point and a relative minimum point and between the absolute maximum and a relative maximum. In Fig. 6-4 the absolute maximum is the point $F$ and the absolute minimum is the point $A$.

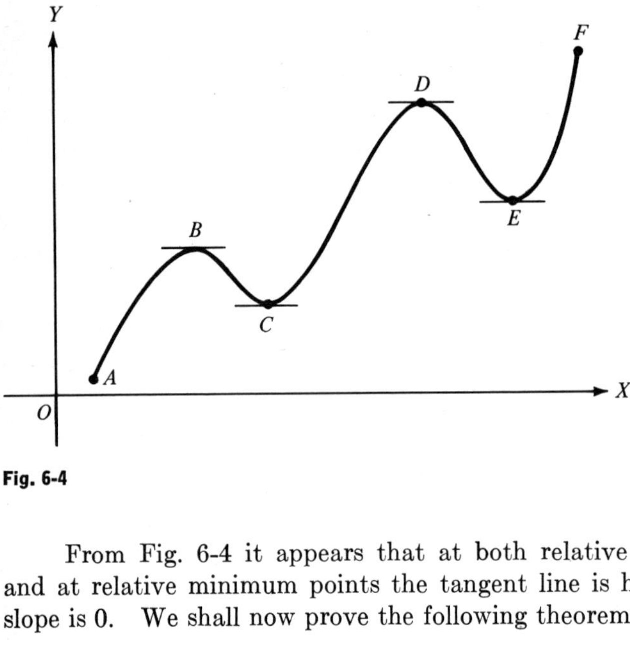


**Fig. 6-4**

From Fig. 6-4 it appears that at both relative maximum points and at relative minimum points the tangent line is horizontal and the slope is 0. We shall now prove the following theorem.

**Theorem** *If the function $f(x)$ is defined and differentiable in the interval between $x = a$ and $x = b$ $(a \leq x \leq b)$ and if the function has a relative maximum or a relative minimum point at $x = c$ then $f'(c) = 0$.*

*Proof* We shall prove the theorem for the case when $x = c$ is a minimum point. A similar proof can be used for the case when $x = c$ is a maximum point. Now,

$$f'(c) = \lim_{\Delta x \to 0} \frac{f(c + \Delta x) - f(c)}{\Delta x}$$

Since the function has a minimum at $x = c$, $f(c) < f(c + \Delta x)$ for all $c + \Delta x$ close to $c$ (that is, $\Delta x$ close to 0), and

$$f(c + \Delta x) - f(c) > 0$$

1. $\dfrac{f(c + \Delta x) - f(c)}{\Delta x} > 0 \qquad \text{when } \Delta x > 0 \qquad \text{(Why?)}$

2. $\dfrac{f(c + \Delta x) - f(c)}{\Delta x} < 0 \qquad \text{when } \Delta x < 0$

In (1):

$$\lim_{\Delta x \to 0} \frac{f(c + \Delta x) - f(c)}{\Delta x} \geq 0$$

and in (2):

$$\lim_{\Delta x \to 0} \frac{f(c + \Delta x) - f(c)}{\Delta x} \leq 0$$

Since the function is differentiable at $x = c$, both limits must be equal. Thus,

$$f'(c) = \lim_{\Delta x \to 0} \frac{f(c + \Delta x) - f(c)}{\Delta x} = 0 \; \blacksquare$$

The theorem that we have now proved tells us that if we have a relative maximum or a relative minimum at $x = c$ and if the function has a derivative at $x = c$, then $f'(c) = 0$. If the derivative does not exist at $x = c$, then we cannot say anything about $f'(c)$ since it does not exist. The theorem does not claim that the converse is true. Thus, if $f'(c) = 0$ we cannot say that there is a relative maximum or a relative minimum at $x = c$. There may be a relative maximum at $x = c$, or there may be a relative minimum at $x = c$, or the point $[c,f(c)]$ may be neither a relative maximum nor a relative minimum point.

## 6-3 FIRST DERIVATIVE TEST FOR MAXIMA AND MINIMA

We have already established that if $f(x)$ is differentiable and if there is a maximum point at $x = a$, then $f'(a) = 0$. Geometrically this says that at $x = a$ the slope of the tangent line is 0. From our definition of a relative maximum point the curve must be rising to the left of $x = a$ and falling to the right of $x = a$. When the curve is rising, the slope is positive and $f'(x)$ is positive. When the curve is falling, the slope is negative and $f'(x)$ is negative (Fig. 6-5).

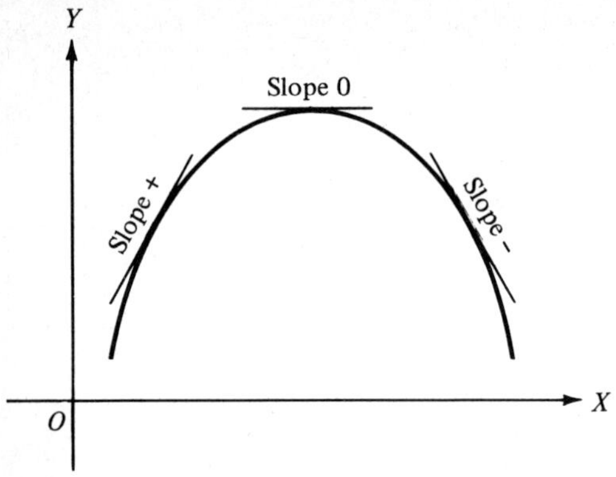


Fig. 6-5

We can summarize our discussion with the following test for finding a relative maximum point.

1. Find all values of $x$ for which $f'(x) = 0$.
2. If $f'(a) = 0$, then test the values of $x$ just less than $x = a$ (being careful not to go beyond any of the other $x$ values being tested) and note the corresponding sign of $f'(x)$ for these values.
3. Now test values of $x$ just greater than $x = a$ and note the corresponding sign of $f'(x)$ for these values.
4. If as $x$ increases through $x = a$ the sign of $f'(x)$ changes from $+$ to $-$, then there is a relative maximum at $x = a$.

If there is a relative minimum at $x = b$, then $f'(b) = 0$. From our definition of a relative minimum the curve must be falling to the left of

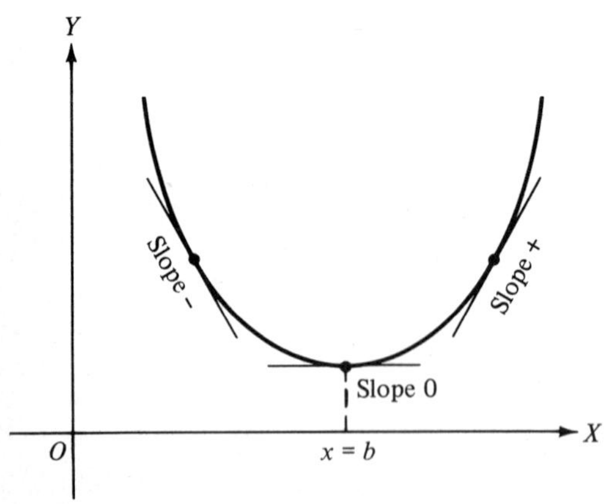


Fig. 6-6

$x = b$ and rising to the right of $x = b$. Thus, if as $x$ increases through $x = b$, the sign of $f(x)$ changes from $-$ to $+$, then there is a relative minimum at $x = b$ (Fig. 6-6).

*First derivative test for maxima and minima*

| Maximum at $x = a$ if $f'(a) = 0$ and $f'(x)$ changes from $+$ to $-$ |
|---|
| Minimum at $x = b$ if $f'(b) = 0$ and $f'(x)$ changes from $-$ to $+$ |

We note that if $f'(c) = 0$ and $f'(x)$ does not change its sign as $x$ increases through $x = c$, then there is neither a relative maximum nor a relative minimum at $x = c$.

**Example 1** Find the maxima and minima for the function $f(x) = 4 - x^2$ (Fig. 6-7).

*Solution* First we must find $f'(x)$.

$$f'(x) = -2x$$

and

$$f'(x) = 0 \qquad \text{when } x = 0$$

Thus, if there is a relative maximum or a relative minimum it must occur at $x = 0$. (We shall only consider functions that are differentiable and therefore a relative maximum or a relative minimum can occur only where the first derivative equals zero.)

We shall now test values of $x$ less than 0 and values of $x$ greater than 0.

When $x < 0$, $f'(x) = -2x > 0$.
When $x > 0$, $f'(x) = -2x < 0$.

Thus, $f'(x)$ changes from $+$ to $-$ and we have a relative maximum at $x = 0$. There is no relative minimum.

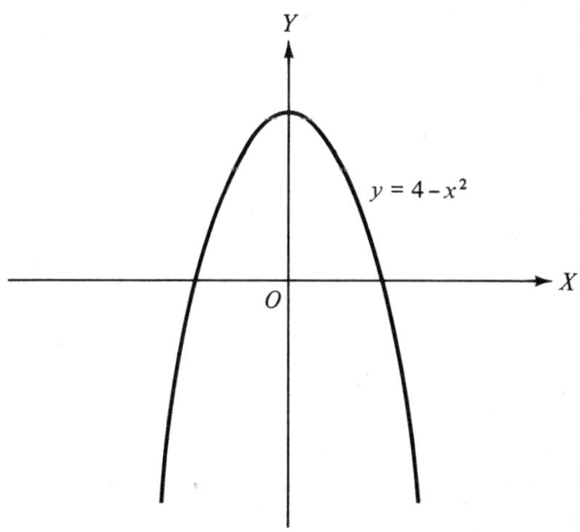


Fig. 6-7

**Example 2** Find the maxima and minima for the function $y = \frac{1}{3}x^3 - 3x^2 + 8x$ (Fig. 6-8*b*).

*Solution*

$$\begin{aligned} y &= \tfrac{1}{3}x^3 - 3x^2 + 8x \\ y' &= x^2 - 6x + 8 \\ &= (x - 2)(x - 4) \end{aligned}$$

We now have two values to test: $x = 2$ and $x = 4$. The values for which $y'$ equals 0 are sometimes referred to as critical values. Thus, the critical values are $x = 2$ and $x = 4$.

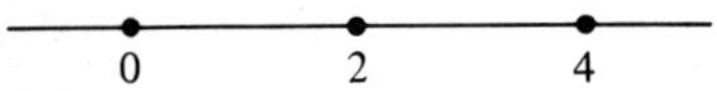


**Fig. 6-8a**

When $x < 2$ (Fig. 6-8*a*), $x - 2 < 0$, $x - 4 < 0$, and $y' = (x - 2)(x - 4) > 0$.
When $2 < x < 4$, $x - 2 > 0$, $x - 4 < 0$, and $y' = (x - 2)(x - 4) < 0$.

Since $y'$ changes from $+$ to $-$, there is a relative maximum point at $x = 2$.

When $2 < x < 4$, $x - 2 > 0$, $x - 4 < 0$, and $y' < 0$.
When $x > 4$, $x - 2 > 0$, $x - 4 > 0$, and $y' > 0$.

Since $y'$ changes from $-$ to $+$, there is a relative minimum at $x = 4$.

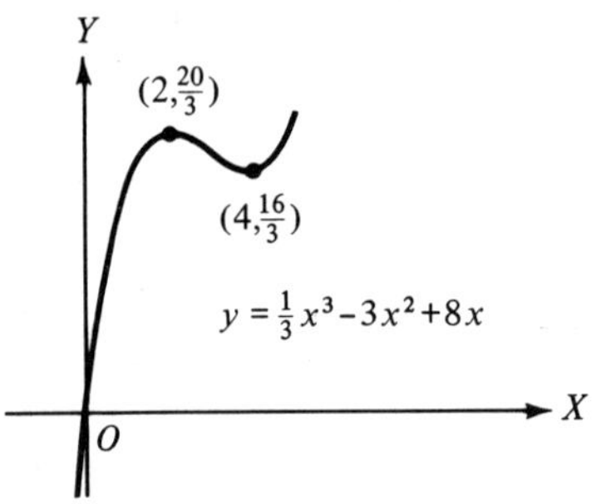


**Fig. 6-8b**

**Example 3** Find the maxima and minima for the function $y = x^3$ (Fig. 6-9).

*Solution*

$$\begin{aligned} y &= x^3 \\ y' &= 3x^2 \end{aligned}$$

The critical value is thus $x = 0$.

When $x < 0$, $y' > 0$.
When $x > 0$, $y' > 0$.

$y'$ does not change its sign, and at $x = 0$ there is neither a relative maximum nor a relative minimum point.

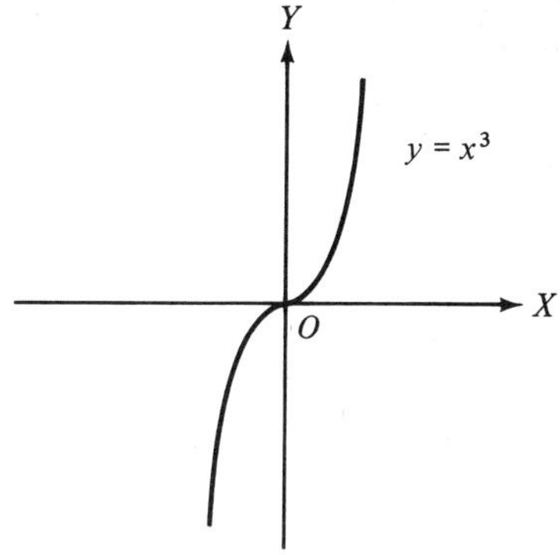


Fig. 6-9

In finding maxima and minima the term "relative" is often omitted. It is, however, to be understood that we are referring to relative maximum or minimum points, and not to the absolute maximum or minimum points.

**EXERCISES**

For the following curves find the maxima and minima.

**1.** $y = x^2 - 4x + 3$

**2.** $y = 4 + 8x - x^2$

**3.** $y = 2x^3 + 3x^2 - 12x + 1$

**4.** $y = \dfrac{1}{x}$

**5.** $y = x^3 - 3x^2 + 4$

**6.** $y = x + \dfrac{1}{x}$

**7.** $x^2y + 4y - 1 = 0$

**8.** $y = 2x^3 - 6x + 1$

## 6-4 SIGNIFICANCE OF THE SIGN OF THE SECOND DERIVATIVE

We have found that the sign of the first derivative tells us whether the function is increasing or decreasing. If the first derivative is positive at a point, then the curve is rising at that point. If the first derivative is negative at a point, the curve is falling at that point. We shall now consider the significance of the sign of the second derivative (i.e., the derivative of the first derivative).

In Fig. 6-10 the curve is concave downward along $ABC$ and concave upward along $CDE$. Let us examine the slope of the tangent line along $ABC$. As a point moves from $A$ to $B$, the slope of the tangent line is always positive but it is decreasing. At point $B$ the slope is 0, and as we move from $B$ to $C$ the slope is negative and continues to decrease. Thus, as we move along the curve where it is concave downward, the slope is always decreasing. Since $y'$ represents the slope at any point and $y'$ is a function of $x$, we are saying that when the curve is concave downward, the function of $x$ (that is, $y'$) must be decreasing. To test when a function is decreasing we need merely to find where its derivative is negative. In this case we want to know where $y''$ is negative.

If at a given point $y''$ is negative, then the slope of the tangent line is decreasing at that point and the curve is concave downward.

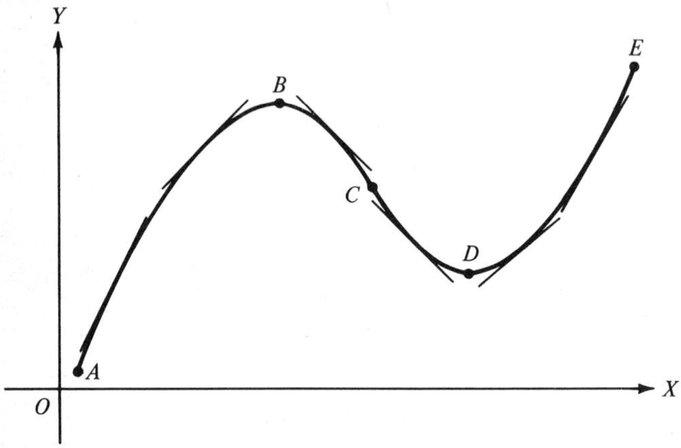


Fig. 6-10

Similarly, when the curve is concave upward, the slope of the tangent line increases as $x$ increases. Thus, when $y''$ is positive, the slope is increasing and the curve is concave upward.

*Concavity*

| | |
|---|---|
| $y''$ + | Concave upward |
| $y''$ − | Concave downward |

**Example 1** Determine where the curve $y = x^3 + 3x^2 + 1$ is concave upward and where it is concave downward (Fig. 6-11).

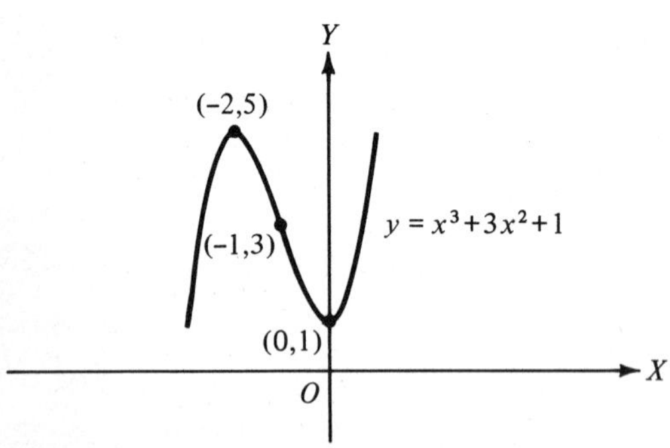


Fig. 6-11

*Solution*

$$\begin{aligned} y &= x^3 + 3x^2 + 1 \\ y' &= 3x^2 + 6x \\ y'' &= 6x + 6 \\ &= 6(x + 1) \end{aligned}$$

When $x < -1$, $y'' < 0$ and the curve is concave downward.
When $x > -1$, $y'' > 0$ and the curve is concave upward.

A point on a curve where the curve changes from concave downward to concave upward or vice versa is called an *inflection point.* In the above problem the inflection point is at $x = -1$. Since at the inflection point the curve is neither concave downward nor concave upward, the second derivative $y''$ cannot be negative or positive at that point and must equal 0. The converse of this statement is not true. It is possible for the second derivative to equal 0 at a point without the point being an inflection point. If $f''(x_1) = 0$, and the curve changes concavity at $x = x_1$, then there is an inflection point at $x = x_1$. However, if $f''(x_1) = 0$, and if $x_1$ is an inner point of an interval for which $f''(x)$ is positive, then the curve is concave upward at $x = x_1$. If $f''(x_1) = 0$, and $x_1$ is an inner point of an interval for which $f''(x)$ is negative, then the curve is concave downward at $x = x_1$.

**Example 2** For the curve $y = x^3$ determine the concavity and inflection points (Fig. 6-12).

*Solution*

$$\begin{aligned} y &= x^3 \\ y' &= 3x^2 \\ y'' &= 6x \end{aligned}$$

When $x < 0$, $y'' < 0$ and therefore the curve is concave downward.
When $x > 0$, $y'' > 0$ and therefore the curve is concave upward.
When $x = 0$, $y'' = 0$, and we have an inflection point at $x = 0$.

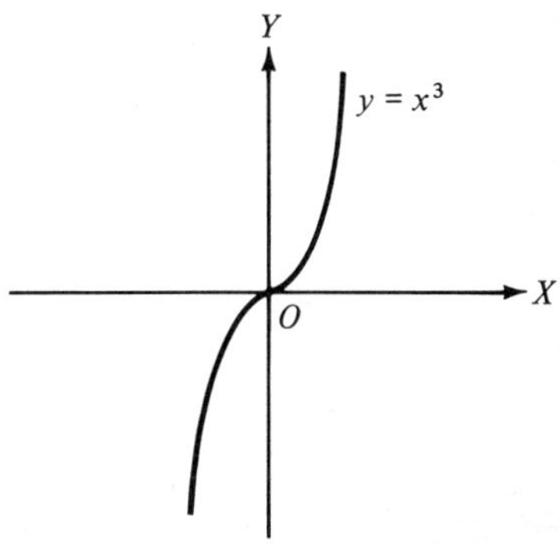


**Fig. 6-12**

**Example 3** For the curve $y = x^4$ determine the concavity and inflection points (Fig. 6-13).

*Solution*

$$y = x^4$$
$$y' = 4x^3$$
$$y'' = 12x^2$$

When $x < 0$, $y'' > 0$.
When $x > 0$, $y'' > 0$.

Thus, the curve is concave upward when $x$ is positive or negative. Though $y''(0) = 0$ the curve is concave upward at $x = 0$. There are no inflection points.

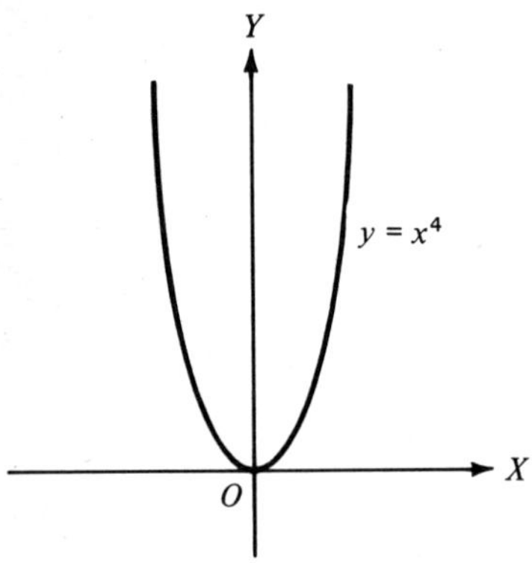


**Fig. 6-13**

### EXERCISES

Determine the direction of concavity and inflection points for the following curves.

**1.** $y = 2x^2 - 8x + 4$

**2.** $y = 5 - 2x - x^2$

**3.** $y = \dfrac{1}{x}$

**4.** $y = 2x^3 - 6x^2 + 8$

**5.** $y = 2x^4 - 4x^2$

**6.** $y = 1 - x^3$

**7.** $y = 4x^3 + 6x^2 - 24x + 4$

**8.** $y = \dfrac{x}{x + 1}$

## 6-5 THE SECOND DERIVATIVE TEST

We have found that if there is a maximum or a minimum point at $x = a$ then $f'(a)$ must equal 0. If we look at the concavity of a curve at a maximum point, we see that the curve is concave downward and $f''(x) < 0$. If $x = b$ is a minimum point, then $f'(b) = 0$, the curve is concave upward at $x = b$, and $f''(b) > 0$ (Fig. 6-14$a$ and $b$).

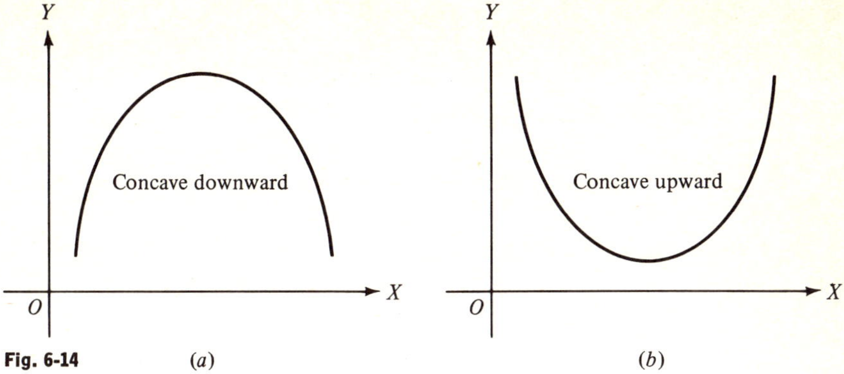


**Fig. 6-14**

*Second derivative test*

| Maximum at $x = a$ if $f'(a) = 0$ and $f''(a) < 0$ |
|---|
| Minimum at $x = b$ if $f'(b) = 0$ and $f''(b) > 0$ |

**Example 1** Find the maxima and minima for the curve $y = x^3 + 3x^2 - 9x$ (Fig. 6-15).

*Solution* We shall use the second derivative test.

$$\begin{aligned} y &= x^3 + 3x^2 - 9x \\ y' &= 3x^2 + 6x - 9 \\ &= 3(x^2 + 2x - 3) \\ &= 3(x + 3)(x - 1) \end{aligned}$$

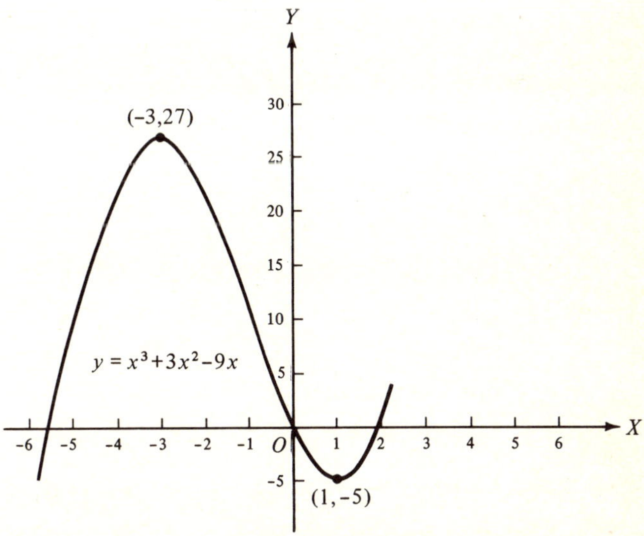


**Fig. 6-15**

The critical values are $x = -3$ and $x = 1$.

$$y'' = 6x + 6$$
$$y''(-3) = -12$$

Since $y'' < 0$, $x = -3$ is a maximum.

$$y''(1) = 12$$

$x = 1$ is a minimum.

**Example 2** Find the maxima and minima for the curve $y = x^3$.

*Solution* We have already solved this problem using the first derivative test. We shall now try the second derivative test.

$$\begin{aligned} y &= x^3 \\ y' &= 3x^2 \qquad x = 0 \text{ a critical value} \\ y'' &= 6x \\ y''(0) &= 0 \end{aligned}$$

From the fact that $y''(0) = 0$ we can only conclude that the second derivative test fails and that at $x = 0$ there may be a maximum point, a minimum point, or an inflection point. To make a final determinaton we must go back to the first derivative test:

When $x < 0$, $y' > 0$.
When $x > 0$, $y' > 0$.

$y'$ does not change its sign and therefore it is neither a relative maximum nor a relative minimum. To verify that it is in fact an inflection point we note:

When $x < 0$, $y'' < 0$.
When $x > 0$, $y'' > 0$.

The curve changes from concave downward to concave upward at $x = 0$.

**Example 3** Find the maxima and minima for the curve $y = x^4$.

*Solution*

$$\begin{aligned} y &= x^4 \\ y' &= 4x^3 \qquad x = 0 \text{ a critical value} \\ y'' &= 12x^2 \\ y''(0) &= 0 \end{aligned}$$

Again the second derivative test fails and we must go back to the first derivative.

When $x < 0$, $y' < 0$.
When $x > 0$, $y' > 0$.

Since $y'$ changes from $-$ to $+$, there is a minimum point at $x = 0$.

**EXERCISES**

Wherever possible use the second derivative test to find the maxima and minima for the following curves.

**1.** $y = 8 + 16x + 4x^2$

**2.** $y = 7 + 4x - 2x^2$

**3.** $y = 1 - \frac{1}{x}$

**4.** $y = x^4 - 4x^2$

**5.** $y = 4x^3 - 12x + 2$

**6.** $x^2y = -1$

**7.** $y = 2x^3 - 24x + 24$

**8.** $y = 2x^3 - 9x^2 - 60x + 10$

## 6-6 USING DERIVATIVES FOR CURVE TRACING

We will now illustrate the usefulness of derivatives in curve sketching.

**Example 1** Give a complete discussion of the curve $y = \dfrac{1}{x^2 + 1}$ and sketch.

*Solution* (See Fig. 6-16*b*.)

(*a*) *Intercepts* When $x = 0$, $y = 1$.

$$y \neq 0$$

(*b*) *Symmetry* If we replace $x$ by $-x$ the equation remains unchanged. If we replace $y$ by $-y$ the equation changes. There is symmetry with respect to the $Y$ axis only.

(*c*) *Extent* $y = \dfrac{1}{x^2 + 1}$ and all values of $x$ may be used.

$$x^2y + y = 1$$
$$x^2y = 1 - y$$
$$x^2 = \frac{1 - y}{y}$$
$$x = \pm\sqrt{\frac{1 - y}{y}}$$

**Fig. 6-16a**

When $y < 0$ (Fig. 6-16*a*), the expression inside the square root is negative, and thus there are no real values for $x$ when $y < 0$.
When $y = 0$, the expression is not defined.
When $0 < y \le 1$, the expression inside the square root is positive.
When $y > 1$, we again get imaginary values.

*Extent*

| All values of $x$ |
|---|
| $0 < y \le 1$ |

(*d*) *Asymptotes*

$$y = \frac{1}{x^2 + 1} \qquad \text{no vertical asymptote}$$

$$x = \pm\sqrt{\frac{1 - y}{y}} \qquad y = 0 \text{ is a horizontal asymptote}$$

(*e*) *Maxima and minima*

$$y = \frac{1}{x^2 + 1}$$

$$y' = \frac{-2x}{(x^2 + 1)^2} \qquad x = 0 \text{ is a critical value}$$

$$y'' = \frac{-2(x^2 + 1)^2 + (2x)2(x^2 + 1)2x}{(x^2 + 1)^4}$$

$$= \frac{-2(x^2 + 1) + 8x^2}{(x^2 + 1)^3}$$

$$= \frac{6x^2 - 2}{(x^2 + 1)^3}$$

$$y''(0) = \frac{-2}{1} < 0$$

There is a maximum point at $x = 0$.
(*f*) *Concavity and inflection points*

$$y'' = \frac{2(3x^2 - 1)}{(x^2 + 1)^3}$$

The denominator of $y''$ is always positive. The numerator will be positive if

$$3x^2 - 1 > 0$$
$$3x^2 > 1$$
$$x^2 > \tfrac{1}{3}$$
$$x > \frac{1}{\sqrt{3}} \qquad \text{or} \qquad x < -\frac{1}{\sqrt{3}}$$

Thus:

When $x < -\dfrac{1}{\sqrt{3}}$, the curve is concave upward.

When $-\dfrac{1}{\sqrt{3}} < x < \dfrac{1}{\sqrt{3}}$, the curve is concave downward.

When $x > \dfrac{1}{\sqrt{3}}$, the curve is concave upward.

There are inflection points at $x = -\dfrac{1}{\sqrt{3}}$ and $x = +\dfrac{1}{\sqrt{3}}$.

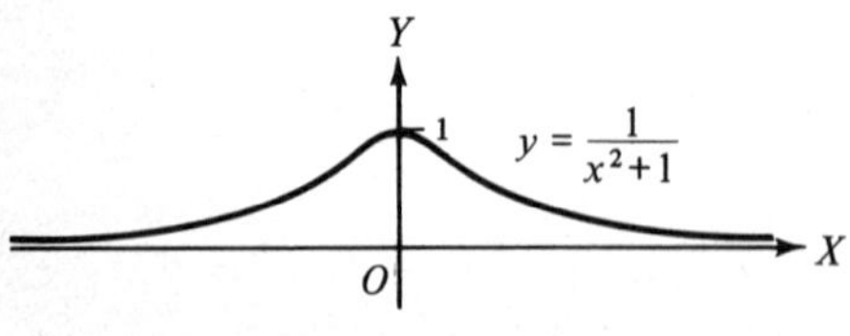


Fig. 6-16b

**Example 2** Give a complete discussion of the curve $y = \dfrac{x^2}{x-2}$. Sketch the curve.

*Solution* (See Fig. 6-17*b*.)

$$y = \frac{x^2}{x-2}$$

(*a*) *Intercepts* When $x = 0$, $y = 0$.

(*b*) *Symmetry* The equation changes when we replace $x$ by $-x$ or $y$ by $-y$. There is no symmetry with respect to the $X$ axis, the $Y$ axis, or the origin.

(*c*) *Extent*

$$y = \frac{x^2}{x-2} \qquad x \neq 2$$

$$xy - 2y = x^2$$

$$x^2 - xy + 2y = 0$$

Using the quadratic formula, $x = \dfrac{-b \pm \sqrt{b^2 - 4ac}}{2a}$, with $a = 1$, $b = -y$, and $c = 2y$,

$$x = \frac{y \pm \sqrt{y^2 - 8y}}{2} = \frac{y \pm \sqrt{y(y-8)}}{2}$$

**Fig. 6-17a**

When $y < 0$ (Fig. 6-17*a*), the expression in the radical is positive.
When $y = 0$, $x = 0$.
When $0 < y < 8$, the expression in the radical is negative and there are no real values for $x$.
When $y = 8$, $x = 4$.
When $y > 8$, the expression in the radical is positive.

*Extent*

| All $x \neq 2$ | |
|---|---|
| $y \leq 0$ | $y \geq 8$ |

(*d*) *Asymptotes*

$$y = \frac{x^2}{x-2} \qquad x = 2 \text{ a vertical asymptote}$$

$$x = \frac{y \pm \sqrt{y(y-8)}}{2}$$

There is no horizontal asymptote.

(*e*) *Maxima and minima*

$$y = \frac{x^2}{x-2}$$

$$y' = \frac{2x(x-2) - x^2}{(x-2)^2} = \frac{x^2 - 4x}{(x-2)^2} = \frac{x(x-4)}{(x-2)^2}$$

$x = 0$ and $x = 4$ are the critical values.

$$y'' = \frac{(x-2)^2(2x-4) - (x^2-4x)2(x-2)}{(x-2)^4}$$
$$= \frac{(x-2)(2x-4) - 2(x^2-4x)}{(x-2)^3} = \frac{8}{(x-2)^3}$$
$$y''(0) = -1 < 0$$
$$y''(4) = 1 > 0$$

There is a maximum point at $x = 0$ and a minimum point at $x = 4$.

(*f*) *Concavity and inflection points*

$$y'' = \frac{8}{(x-2)^3}$$

When $x > 2$, $y'' > 0$ and the curve is concave upward.
When $x < 2$, $y'' < 0$ and the curve is concave downward.

The curve is concave downward when $x < 2$ and concave upward when $x > 2$. Since $y''$ is never equal to 0, there are no inflection points.

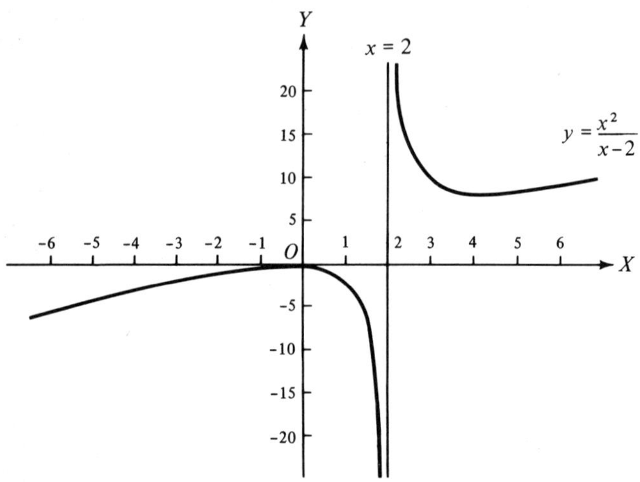


**Fig. 6-17b**

**EXERCISES**

Discuss and sketch the following curves.

**1.** $y = 8 - 2x^2$

**2.** $y = 2x^2 - 12x + 16$

**3.** $y = 2x^3 - 3x^2 - 12x - 5$

**4.** $y = \dfrac{x}{x+2}$

**5.** $x^2y - 4y - 2 = 0$

**6.** $y = \dfrac{1}{x^4}$

**7.** $y = x^3 - 3x$

**8.** $y = \dfrac{x+1}{x-1}$

## 6-7 VERBAL PROBLEMS INVOLVING MAXIMA AND MINIMA

In practice, there are many problems which involve finding a maximum value or a minimum value. In these problems, however, we are concerned with the absolute maximum or the absolute minimum and not the relative maximum or minimum. For a curve which is defined over an interval $a \leq x \leq b$ and which is differentiable for all values of the domain, the absolute maximum value must occur either at one of the endpoints or it must be one of the relative maximum points. Thus, to find a maximum value we must first find all the relative maximum values and compare them with the values at the endpoints. The same procedure applies for finding minimum values. In many cases, a knowledge of the curve involved will eliminate consideration of the endpoints and we will only have to find the relative maximum or minimum points.

**Example 1** A fence is needed to enclose a rectangular area consisting of 200 yd² in a beach area. Only three sides of fence are necessary since no fence is needed at the water's edge. What dimensions should we choose for the rectangle so that the least amount of fence is needed?

*Solution* This is a minimization problem, and we wish to find the dimensions of the rectangle so that the amount of fence used is a minimum. We shall denote the sides of the rectangle by $x$ and $y$ (Fig. 6-18).

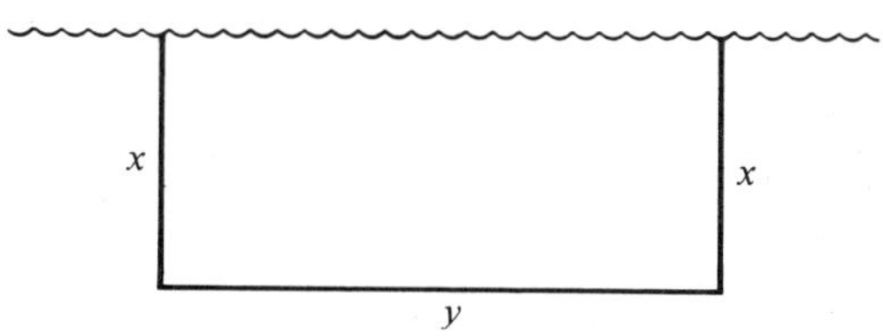


**Fig. 6-18**

Since the area is 200 yd²,

$$xy = 200$$

$$y = \frac{200}{x}$$

The amount of fence $A$ will actually be the perimeter.

$$A = 2x + y$$

$$= 2x + \frac{200}{x}$$

To find the relative minimum we must find $A'$ and set it equal to 0.

$$A' = 2 - \frac{200}{x^2} = 0$$

$$\frac{200}{x^2} = 2$$

$$200 = 2x^2$$

$$100 = x^2$$

$$x = 10$$

$$y = \tfrac{200}{10} = 20$$

To check that this is actually a relative minimum, we note that

$$A'' = \frac{400}{x^3}$$

$$A''(10) > 0$$

In this problem $A = f(x)$ and the domain of $f(x)$ is all positive values of $x$. There are no endpoints to test, but a quick glance at the graph of the curve corresponding to the equation $A = 2x + \dfrac{200}{x}$ shows that the relative minimum is also the absolute minimum (Fig. 6-19).

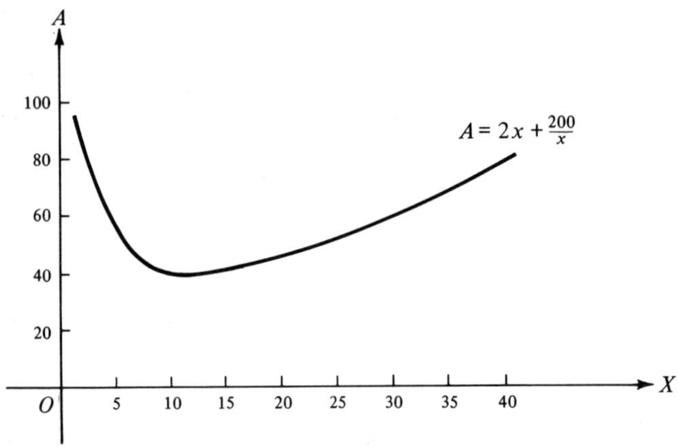


**Fig. 6-19**

**Example 2** An open box is to be made from a piece of cardboard 20 in. long and 20 in. wide by cutting out a square from each corner and turning up the sides. What dimensions for the cut-out squares will result in the greatest volume for the box?

*Solution* This is a maximum problem. We shall denote a side of the cut-out square by $x$ (Fig. 6-20).

$$\begin{aligned} V &= (\text{length})(\text{width})(\text{height}) \\ &= (20 - 2x)(20 - 2x)x \\ &= 4x^3 - 80x^2 + 400x \end{aligned}$$

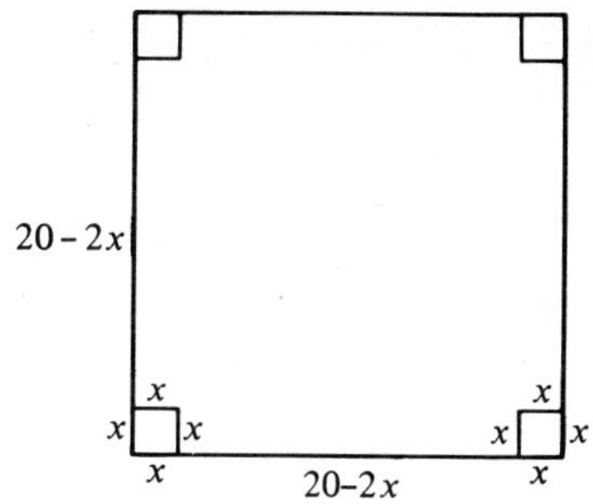


Fig. 6-20

We also note that since $20 - 2x$ must be positive, $x$ must be between 0 and 10 $(0 < x < 10)$.

$$\begin{aligned} V' &= 12x^2 - 160x + 400 \\ &= 4(x - 10)(3x - 10) \end{aligned}$$

The critical values are

$$x = \tfrac{10}{3} \qquad x = 10$$

Since both $x = 0$ and $x = 10$ will give us a 0 volume, we need only to consider $x = \frac{10}{3}$.

$$\begin{aligned} V'' &= 24x - 160 \\ V''(\tfrac{10}{3}) &= -80 \end{aligned}$$

At $x = \frac{10}{3}$ there is a relative maximum.

We have already tested the endpoints $x = 0$ and $x = 10$ which both give us $V = 0$. The maximum volume occurs when $x = \frac{10}{3}$ and $V = 16{,}000/27 = 592\frac{16}{27}$ in.$^3$ (Fig. 6-21).

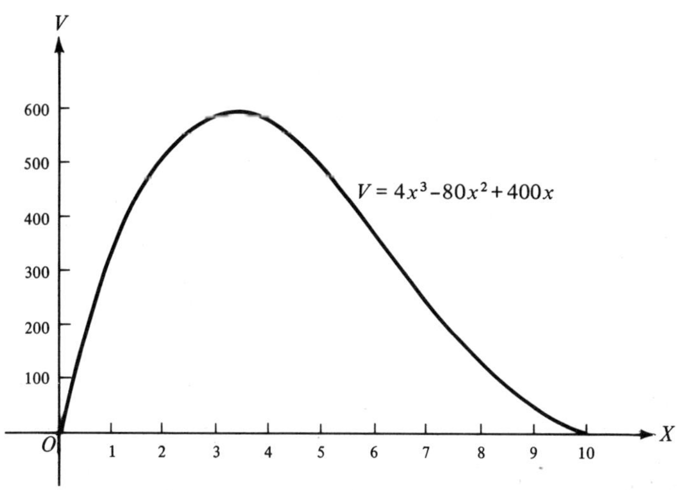


Fig. 6-21

**Example 3** A manufacturer wishes to construct a cylindrical tin can to hold 1 quart of orange juice. What dimensions for the can will require the least amount of material?

*Solution: Method I* Let $k$ represent the number of cubic inches in a quart.† For a right circular cylinder (Fig. 6-22)

$$V = \pi r^2 h$$

The material necessary for the can will be the top, the bottom, and the lateral surface area.

$$\begin{aligned} A &= \pi r^2 + \pi r^2 + 2\pi rh \\ &= 2\pi r^2 + 2\pi rh \end{aligned}$$

Since $V = k$,

$$\begin{aligned} \pi r^2 h &= k \\ h &= \frac{k}{\pi r^2} \\ A &= 2\pi r^2 + 2\pi r \frac{k}{\pi r^2} \\ &= 2\pi r^2 + \frac{2k}{r} \\ A' &= 4\pi r - \frac{2k}{r^2} \end{aligned}$$

To minimize $A$, we set $A'$ equal to 0:

$$\begin{aligned} A' &= 4\pi r - \frac{2k}{r^2} = 0 \\ 4\pi r &= \frac{2k}{r^2} \\ r^3 &= \frac{k}{2\pi} \\ r &= \sqrt[3]{\frac{k}{2\pi}} \\ h &= \frac{k}{\pi r^2} \\ &= \frac{k}{\pi \left(\sqrt[3]{\frac{k}{2\pi}}\right)^2} \\ &= 2\sqrt[3]{\frac{k}{2\pi}} \end{aligned}$$

The second derivative test will show that we have actually found a relative minimum, and an examination of the curve will show that it is the absolute minimum.

*Solution: Method II* As before,

$$V = \pi r^2 h \qquad A = 2\pi r^2 + 2\pi rh$$

† $k = 57.75$.

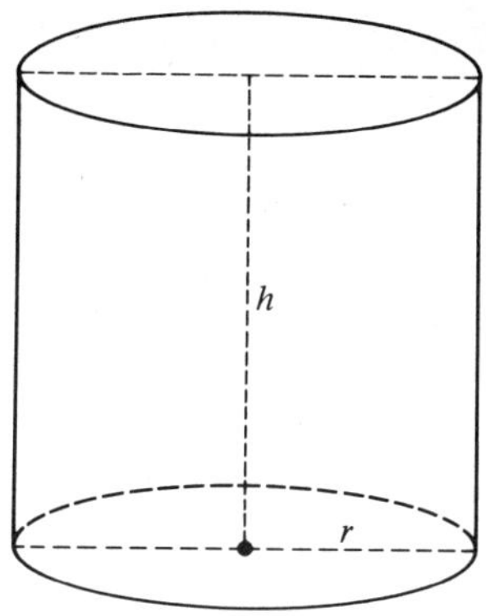


Fig. 6-22

We shall not substitute for $h$ in terms of $r$. Since $h$ is a function of $r$, we shall differentiate implicitly.

$$\begin{aligned} V &= \pi r^2 h = k \\ D_r V &= 2\pi rh + \pi r^2 \, D_r h = 0 \end{aligned}$$

Using the prime notation,

$$\begin{aligned} 2\pi rh + \pi r^2 h' &= 0 \\ 2h + rh' &= 0 \\ rh' &= -2h \\ A &= 2\pi r^2 + 2\pi rh \\ A' &= 4\pi r + 2\pi h + 2\pi rh' \end{aligned}$$

For a minimum, $A' = 0$.

$$\begin{aligned} 4\pi r + 2\pi h + 2\pi rh' &= 0 \\ 2r + h + rh' &= 0 \end{aligned}$$

Now, $rh' = -2h$,

$$\begin{aligned} 2r + h - 2h &= 0 \\ 2r - h &= 0 \\ h &= 2r \end{aligned}$$

We now have the relative dimensions and we must find the actual values of $r$ and $h$.

$$\begin{aligned} V &= \pi r^2 h = k \\ \pi r^2 (2r) &= k \\ 2\pi r^3 &= k \\ r^3 &= \frac{k}{2\pi} \\ r &= \sqrt[3]{\frac{k}{2\pi}} \\ h &= 2\sqrt[3]{\frac{k}{2\pi}} \end{aligned}$$

The second method is very useful in problems where relative dimensions are necessary.

In solving maximum and minimum problems, an analysis of the nature of the problem will often indicate that the relative maximum or minimum must also be the absolute maximum or minimum. In these cases it is only necessary to find the relative maximum or relative minimum values.

### EXERCISES

**1.** There are 40 ft of wire available for the construction of a rectangular fence. What dimensions for the fence will enclose the largest area?

**2.** For an open cylindrical can with a fixed volume determine the relative dimensions which will require the least amount of material.

**3.** A rectangular box with a square base that is closed on top is to be made from 24 $\text{ft}^2$ of cardboard. Determine the dimensions of the box with maximum volume.

**4.** What number exceeds its square by the greatest amount?

**5.** A merchant estimates that he can sell every one of his 60 tomatoes at the current price of 10 cents per tomato. For every day that he waits, the price of tomatoes will increase by 1 cent each, but the tomatoes will spoil at a rate of 2 per day. If all tomatoes are sold on the same day, determine how many days he should wait before selling so as to maximize the income from these tomatoes.

**6.** Determine the least amount of fence necessary to enclose a rectangular area of 400 $\text{yd}^2$.

**7.** An open box is to be made from a square of tin 12 in. on each side by cutting out equal squares from the four corners and folding up the sides. Determine the size of the cut-out square which gives the box a maximum volume.

**8.** An open box having a volume of 32 $\text{ft}^3$ is to have a square base. Determine the dimensions of the box which will require the least amount of material.

**9.** A magazine publisher finds that he can get 10,000 subscribers at a yearly rate of $4.00 each. He will get 200 more subscribers for each 5-cent decrease in the rate. What rate will yield the maximum gross income?

**10.** Determine the relative dimensions of the largest rectangle that can be inscribed in a circle of radius 2.

**11.** For a fixed amount of material (i.e., fixed total surface area) determine the relative dimensions of the right circular cylinder of maximum volume.

**12.** A Norman window has the form of a rectangle surmounted by a semicircle whose diameter is the width of the rectangle. Find the relative dimensions which will admit the most light if the perimeter of the window is 40 ft.

## 6-8 RELATED RATES

If a variable $y$ is a differentiable function of a variable $x$, then the derivative of $y$ with respect to $x$ will give us the rate of change of $y$ per unit of change in $x$. If both $x$ and $y$ vary with respect to time, then the rate of change per unit of time of one variable can be expressed in terms of the rate of change per unit of time of the second variable. Problems in which the differentiation of the variables is taken with respect to time are often referred to as *related rate problems.*

**Example 1** A boy makes a spherical snowball (Fig. 6-23) of radius 6 in. and leaves it in the sun to melt. If the radius of the snowball decreases at the rate of $\frac{1}{2}$ in./ min, how fast was the volume decreasing when the radius was 3 in. long?

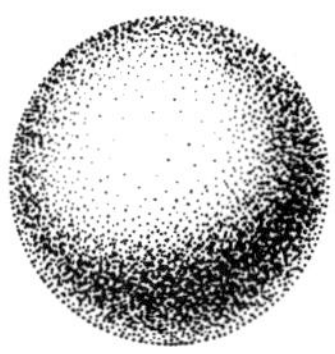

Fig. 6-23

*Solution* The volume of a sphere is given as $V = \frac{4}{3}\pi r^3$. We wish to find $D_tV$ (the rate of change of $V$ with respect to $t$) when the radius is equal to 3 in. We are also given the fact that $D_tr = -\frac{1}{2}$. (Why the minus sign?) Differentiating our equation with respect to time, we get

$$\begin{aligned} D_tV &= \tfrac{4}{3}\pi 3r^2\, D_tr \\ &= 4\pi r^2\, D_tr \\ &= 4\pi(3)^2(-\tfrac{1}{2}) \\ &= -18\pi \end{aligned}$$

Thus, when $r = 3$, the volume is decreasing at the rate of $18\pi$ in.$^3$/min.

**Example 2** Sand is being poured onto the ground, forming a conical pile with its altitude equal to the radius of the base. If the sand is being poured at the rate of 10 ft$^3$/min, how fast is the radius increasing when it is 5 ft long (Fig. 6-24)?

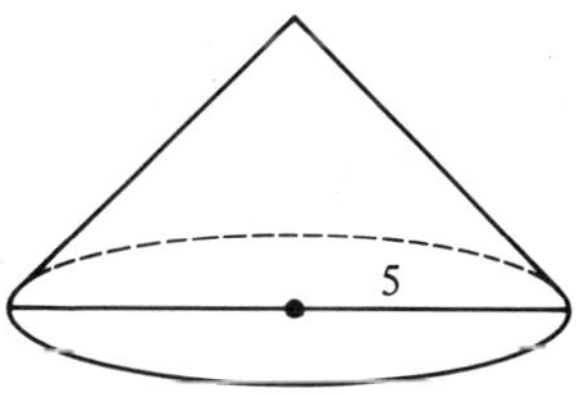


Fig. 6-24

*Solution* Let $V$ = volume, $h$ = altitude, and $r$ = radius. Then for the cone,

$$V = \tfrac{1}{3}\pi r^2 h$$

We notice that $V$ is a function of two variables, $r$ and $h$. If possible, we would like to eliminate one of the variables and follow our standard procedure of differentiation. Since we are given the fact that the altitude always equals the radius, in this problem we can set

$$h = r \qquad V = \tfrac{1}{3}\pi r^3$$

Differentiating with respect to time,

$$\begin{aligned} D_tV &= \tfrac{1}{3}\pi 3r^2\, D_tr \\ &= \pi r^2\, D_tr \end{aligned}$$

We would like to know the value of $D_t r$ when $r = 5$. Since the sand is falling at the rate of 10 ft³/min,

$$D_t V = 10$$

Thus,

$$10 = \pi(25) D_t r$$

$$D_t r = \frac{10}{25\pi} = \frac{2}{5\pi} \text{ ft/min}$$

**Example 3** A boy flies a kite which is 120 ft above his hand. If the wind carries the kite horizontally at the rate of 30 ft/min, at what rate is the string being paid out when there is 150 ft of string out?

*Solution* From the given information we can construct a right triangle (Fig. 6-25). Using the pythagorean theorem, we find that when the hypotenuse $y = 150$, then

$$\begin{aligned} x^2 + (120)^2 &= (150)^2 \\ x^2 &= (150)^2 - (120)^2 \\ &= 22{,}500 - 14{,}400 \\ &= 8{,}100 \\ x &= 90 \end{aligned}$$

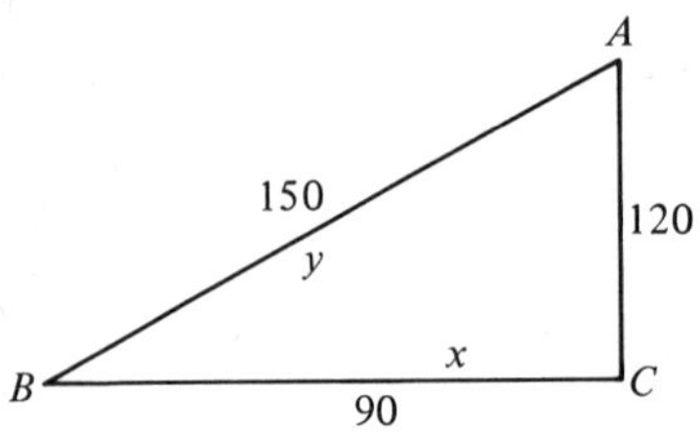


**Fig. 6-25**

We point out that since the kite is moving horizontally, sides $\overline{AB}$ and $\overline{BC}$ are variables, but $\overline{AC}$ is constant. Side $\overline{BC}$ is changing at the rate of 30 ft/sec and thus $D_t x = 30$. We wish to find $D_t y$ when $y = 150$. Referring back to the triangle,

$$\begin{aligned} x^2 + (120)^2 &= y^2 \\ 2x\, D_t x + 0 &= 2y\, D_t y \\ D_t y &= \frac{x\, D_t x}{y} \\ &= \frac{90(30)}{150} = 18 \text{ ft/min} \end{aligned}$$

**EXERCISES**

**1.** The radius of an expanding circular plate is increasing at the rate of 2 in./min. How fast is the area increasing when the radius is 8 in. long?

**2.** A ship leaves port at 10 A.M. and sails eastward at the rate of 30 mph. Another ship leaves the same port at noon and sails southward at the rate of 45 mph. How fast are the two ships separating at 2 P.M.?

**3.** Water is leaking out of the bottom of a cylindrical tank of radius 8 ft at the rate of 4 ft³/min. How fast is the surface falling?

**4.** A 17-ft ladder leaning against a wall starts to slide down. How fast is the distance from the top of the ladder to the base of the wall decreasing when the distance from the bottom of the ladder to the wall is 15 ft and the bottom of the ladder is sliding away from the wall at the rate of $\frac{1}{2}$ ft/min?

**5.** Liquid is being poured into an inverted cone whose altitude is 18 in. and whose diameter is 12 in. at the rate of $3\pi$ in.$^3$/min. How fast is the height of the liquid in the cone rising when the water level in the cone is at the half-way mark (that is, $h = 9$)?

**6.** A spherical ball of ice is melting with the radius decreasing at the rate of $\frac{1}{2}$ in./min. How fast is the surface area decreasing when the radius is 2 in. long?

**7.** A dock stands 7 ft above the deck of a boat. A rope is attached to the front of the deck and the boat is pulled into the dock at the rate of 4 ft/min. At what rate is the boat moving toward the dock when it is 24 ft away?

**8.** A balloon is blown up at the rate of 32 in.$^3$/min. How fast is the radius increasing when it is equal to 8 in.?

**9.** Water is flowing out of a conical reservoir 14 ft in diameter and 21 ft deep (vertex down) at the constant rate of 10 ft$^3$/min. How fast is the water level falling when the depth of the water is 5 ft?

**10.** Two ships $A$ and $B$ are approaching the same port. Ship $A$ is following a course due north at a rate of 20 mph. Ship $B$ is following a course due west at 15 mph. At a certain time it is found that ship $A$ is 30 miles away from the port and ship $B$ is 40 miles away from the port. How fast is the distance between the two ships decreasing at that time?

## 6-9 DIFFERENTIALS

Looking back at the definition of the derivative, we recall that if $y = f(x)$, then $f'(x) = \lim_{\Delta x \to 0} \frac{\Delta y}{\Delta x}$. Thus, it would not be correct to say $\frac{\Delta y}{\Delta x} = f'(x)$. We can, however, say that

$$\frac{\Delta y}{\Delta x} = f'(x) + \epsilon \qquad \text{where } \epsilon \to 0 \text{ as } \Delta x \to 0$$

That is to say that the difference between $\frac{\Delta y}{\Delta x}$ and $f'(x)$ is some number $\epsilon$ which approaches 0 when $\Delta x$ approaches 0. Considering $\Delta y$, we can say

$$\Delta y = f'(x)\,\Delta x + \epsilon\,\Delta x$$

Let us consider the geometric significance of the above result (Fig. 6-26). Let $P$ be a fixed point on the curve of the function $y = f(x)$ and let $Q$ be a neighboring point on the curve. $\overline{QS}$ represents $\Delta y$ and $\overline{PS}$ represents $\Delta x$. The slope of the tangent line $\overline{PR}$ is $\frac{\overline{RS}}{\overline{PS}}$, and since $f'(x)$ is also the slope, we can conclude that

$$\frac{\overline{RS}}{\overline{PS}} = f'(x)$$

$$\begin{aligned}\overline{RS} &= f'(x)\overline{PS} \\ &= f'(x)\,\Delta x\end{aligned}$$

From our diagram, $\Delta y = \overline{RS} + \overline{QR}$.

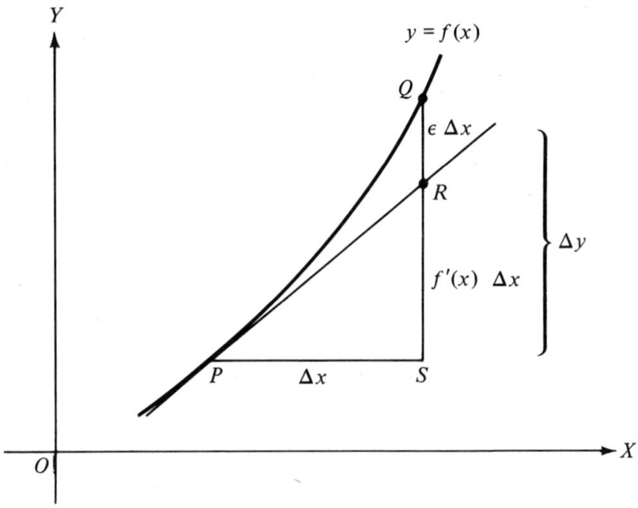


Fig. 6-26

Using the fact that $\Delta y = f'(x)\,\Delta x + \epsilon\,\Delta x$, we find that $\overline{QR} = \epsilon\,\Delta x$. $\overline{RS} = f'(x)\,\Delta x$ is sometimes referred to as the principal part of $\Delta y$ and it is often used as an approximation for $\Delta y$. If $\Delta x$ is large, then $\overline{RS}$ will be a very poor approximation for $\Delta y$. However, if $\Delta x$ is very small, then $\epsilon\,\Delta x$ is even smaller, and $\overline{RS}$, the principal value of $\Delta y$, is a good approximation for $\Delta y$.

The terminology that is most often used for $\overline{RS} = f'(x)\,\Delta x$ is the term *differential* of $y$, and that is denoted by $dy$. Thus,

$$dy = f'(x)\,\Delta x$$

When the differential terminology is used, $\Delta x$ is replaced by $dx$ and we write

$$dy = f'(x)\,dx$$

**Example 1** Find $dy$ for

(a) $y = x^2$
(b) $y = 3x^2 + 4x$
(c) $y = (x^2 + 1)^3$

*Solution* Since $dy = f'(x)\,dx$, we have

(a) $y = x^2$
$dy = 2x\,dx$

(b) $y = 3x^2 + 4x$
$dy = (6x + 4)\,dx$

(c) $y = (x^2 + 1)^3$
$dy = [3(x^2 + 1)^2 2x]\,dx$
$= 6x(x^2 + 1)^2\,dx$

**Example 2** (a) Find the exact change in the value of $y = x^2$ when $x$ changes from 3 to 3.1.

(b) Find the approximate change by using differentials.

*Solution*

(a) The exact change is $\Delta y$.

$$\begin{aligned} x &= 3 \\ \Delta x &= 0.1 \\ y &= x^2 \\ \Delta y &= (x + \Delta x)^2 - x^2 \\ &= x^2 + 2x\,\Delta x + \Delta x^2 - x^2 \\ &= 2x\,\Delta x + \Delta x^2 \\ &= 2(3)(0.1) + (0.1)^2 \\ &= 0.6 + 0.01 \\ &= 0.61 \end{aligned}$$

The exact change is 0.61.

(b) $y = x^2$
$dy = 2x\,dx = 2(3)(0.01) = 0.6$
The approximate change $dy = 0.6$, and the error due to approximation is 0.01.

**Example 3** Use differentials to approximate the value of $\sqrt{26}$.

*Solution* We know that $\sqrt{25} = 5$. We shall consider the function $y = \sqrt{x} = x^{\frac{1}{2}}$. Let $x = 25$ and $\Delta x = 1$. Then

$$\begin{aligned} y &= x^{\frac{1}{2}} \\ dy &= \tfrac{1}{2}x^{-\frac{1}{2}}\,dx \\ &= \frac{1}{2\sqrt{x}}\,dx \\ &= \frac{1}{2(5)}(1) = \tfrac{1}{10} = 0.1 \end{aligned}$$

Thus,

$$\sqrt{26} \approx 5 + 0.1 = 5.1$$

**Example 4** Approximate the value of $\sqrt[3]{26.5}$.

*Solution* Let $y = x^{\frac{1}{3}}$. We must choose a value of $x$ for which we can easily find the cube root of $x$. Since we want $\sqrt[3]{26.5}$, we will let $x = 27$ and $\Delta x = -0.5$.

$$\begin{aligned} y &= x^{\frac{1}{3}} \\ dy &= \tfrac{1}{3}x^{-\frac{2}{3}}\,dx \\ &= \frac{1}{3(\sqrt[3]{x})^2}\,dx \\ &= \frac{1}{3(\sqrt[3]{27})^2}(-0.5) \\ &= \frac{1}{3(3)^2}(-0.5) \\ &= \frac{-0.5}{27} \\ &= -0.019 \end{aligned}$$

Thus,

$$\sqrt[3]{26.5} \approx 3 - 0.019 = 2.981$$

**EXERCISES**

Find the differential $dy$.

**1.** $y = 3x^2 + 4$

**2.** $y = (2x^2 - 1)^2$

**3.** $y = \dfrac{1}{x}$

**4.** $y = \sqrt{x^2 + 4}$

**5.** $y = \dfrac{x}{x + 1}$

**6.** $y = (x^3 + 4x)^{\frac{4}{3}}$

**7.** $y = x\sqrt{1 + x}$

**8.** $y = \sqrt{1 - x^2}$

**9.** Find the exact change in the value of the function $y = x^2$ when $x$ changes from 4 to 4.01. Also, find the approximate change by using differentials.

**10.** Approximate the value of $\sqrt{35}$.

**11.** Approximate the value of $\sqrt[3]{65}$.

**12.** Approximate the value of $\sqrt{15.9}$.

## ▶ 6-10 ROLLE'S THEOREM AND THE MEAN VALUE THEOREM

A very important and useful theorem is the *mean value theorem* for derivatives. Before considering this theorem we shall look at a special case known as Rolle's theorem.

**Rolle's theorem** *If $y = f(x)$ is a function which is continuous in the closed interval $a \leq x \leq b$ and differentiable in the open interval $a < x < b$, and if*

$$f(a) = f(b) = 0$$

*then there is at least one number c between a and b* ($a < c < b$) *where the derivative* $f'(c) = 0$ (*Fig.* 6-27).

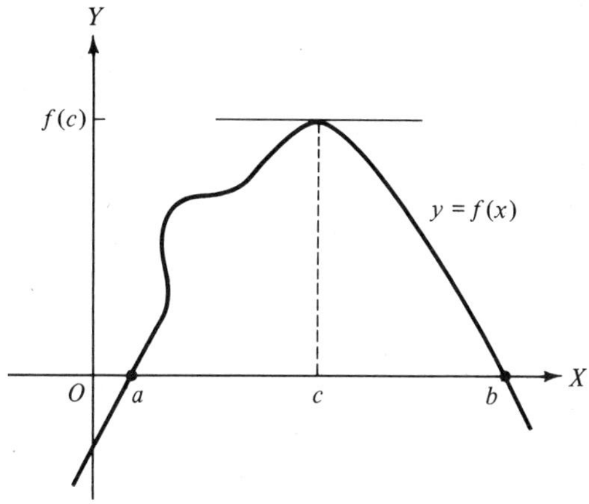


**Fig. 6-27**

This theorem tells us that if the function is continuous in the closed interval and differentiable in the open interval, and if the curve crosses the $X$ axis at $x = a$ and $x = b$, then there is at least one number $c$ between $a$ and $b$ where the tangent line is horizontal. The theorem does not preclude the existence of two or more such numbers between $a$ and $b$. We can see from Fig. 6-28 that if the function is continuous but is not differentiable, then a horizontal tangent line may not exist between $x = a$ and $x = b$.

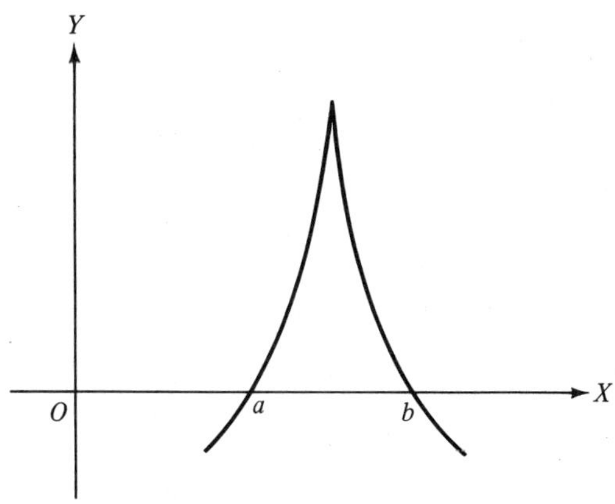


**Fig. 6-28**

Certainly, if the function is not continuous between $x = a$ and $x = b$, there may not exist a horizontal tangent line (Fig. 6-29).

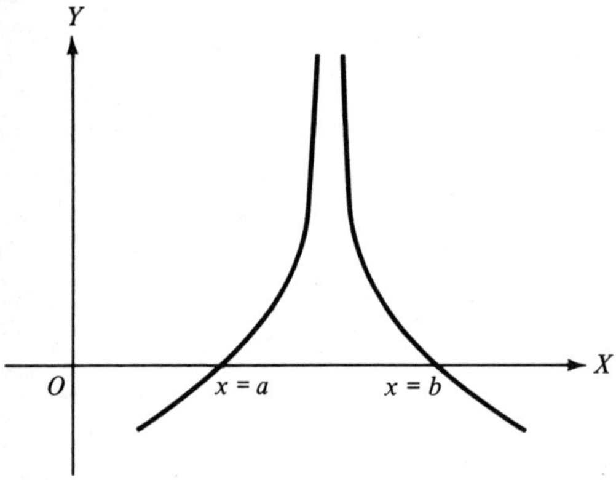


**Fig. 6-29**

*Proof* To prove this theorem we consider two cases.

*Case* 1 If $f(x) = 0$ for all values of $x$, then certainly the theorem is satisfied.

*Case* 2 If $f(x)$ is not 0 for all values of $x$ between $x = a$ and $x = b$, then there are values of $x$ between $x = a$ and $x = b$ for which the function is either positive or negative.

By the properties of continuous functions, the function must, therefore, have a greatest value (maximum) or a least value (minimum) or both, between $x = a$ and $x = b$. Let $x = c$, $a < c < b$, denote the $x$ value of such a maximum or minimum point. Since the derivative at such a turning point (maximum or minimum) must be 0, we can conclude that

$$f'(c) = 0 \blacksquare$$

In the hypothesis of Rolle's theorem, $f(a) = f(b) = 0$ and the slope of the chord connecting the points $[a, f(a)]$ and $[b, f(b)]$ is $\dfrac{f(b) - f(a)}{b - a} = 0$. Rolle's theorem tells us that between $x = a$ and $x = b$ we can find an $x$ value $x = c$ such that the slope of the curve at $x = c$ is equal to the slope of the chord between $[a, f(a)]$ and $[b, f(b)]$ [that is, $f'(c) = 0$]. This result is generalized by the mean value theorem for derivatives which does not require $f(a) = f(b) = 0$.

**Mean value theorem** *If $y = f(x)$ is a function which is continuous in the closed interval $a \leq x \leq b$ and differentiable in the open interval $a < x < b$, then there is at least one number $c$ between $a$ and $b$ $(a < c < b)$ such that*

$$f'(c) = \frac{f(b) - f(a)}{b - a}$$

*or, equivalently,*

$$\boxed{f(b) - f(a) = f'(c)(b - a)}$$

Geometrically, the mean value theorem tells us that if the function is continuous in the closed interval $a \leq x \leq b$ and differentiable in the open interval $a < x < b$, then there exists at least one $x$ value $x = c$, between $a$ and $b$, where the slope of the tangent line is the same as the slope of the chord connecting the points $[a, f(a)]$ and $[b, f(b)]$,

$$f'(c) = \frac{f(b) - f(a)}{b - a}$$

and the two lines are parallel (Fig. 6-30). In the special case where $f(a) = f(b) = 0$, $\dfrac{f(b) - f(a)}{b - a} = 0$ and we have Rolle's theorem. We now proceed to the proof of the mean value theorem.

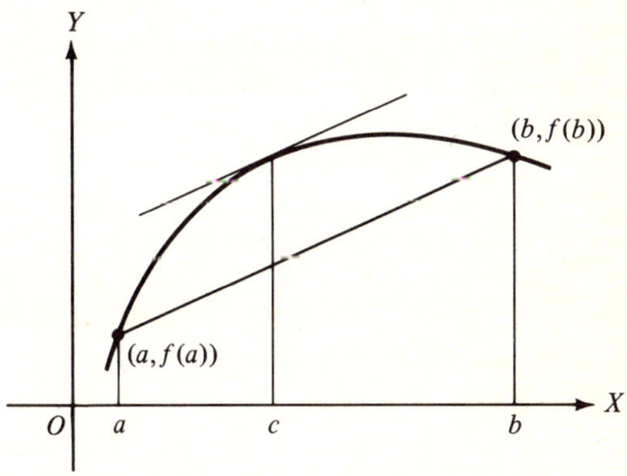


**Fig. 6-30**

*Proof* We shall form a new function $F(x)$ which geometrically will give us the distance between the chord and the curve $y = f(x)$. We recall that the equation of a straight line is given by the equation

$y - y_1 = m(x - x_1)$. For the given chord,

$$y - f(a) = \frac{f(b) - f(a)}{b - a}(x - a)$$

$$y = f(a) + \frac{f(b) - f(a)}{b - a}(x - a)$$

Thus, the vertical distance between the curve and the chord can be given by

$$F(x) = f(x) - \left[f(a) + \frac{f(b) - f(a)}{b - a}(x - a)\right]$$

$$= f(x) - f(a) - \frac{f(b) - f(a)}{b - a}(x - a)$$

We note that since $f(x)$ is continuous in the closed interval $a \leq x \leq b$ and differentiable in the open interval $a < x < b$, and the same is true for the function $m(x - a)$, then $F(x)$ is a function which is continuous for $a \leq x \leq b$ and differentiable for $a < x < b$. Moreover,

$$F(a) = f(a) - f(a) - \frac{f(b) - f(a)}{b - a}(a - a) = 0$$

$$F(b) = f(b) - f(a) - \frac{f(b) - f(a)}{b - a}(b - a) = 0$$

Thus, $F(a) = F(b) = 0$ and the hypotheses of Rolle's theorem are satisfied. Hence, there exists at least one number $c$, $a < c < b$, for which $F'(c) = 0$. Now,

$$F'(x) = f'(x) - \frac{f(b) - f(a)}{b - a}$$

$$F'(c) = f'(c) - \frac{f(b) - f(a)}{b - a} = 0$$

$$f'(c) = \frac{f(b) - f(a)}{b - a}$$

$$f(b) - f(a) = f'(c)(b - a) \blacksquare$$

**Example 1** For the function $y = x^2 + 1$, $a = 0$, and $b = 2$, we shall apply the mean value theorem.

$$f(x) = x^2 + 1$$

$$f'(x) = 2x$$

$$\frac{f(b) - f(a)}{b - a} = \frac{5 - 1}{2} = \frac{4}{2} = 2$$

$$2x = 2$$

$$x = 1$$

Thus, at $x = 1$ the tangent line is parallel to the chord connecting the points (0,1) and (2,5) (Fig. 6-31).

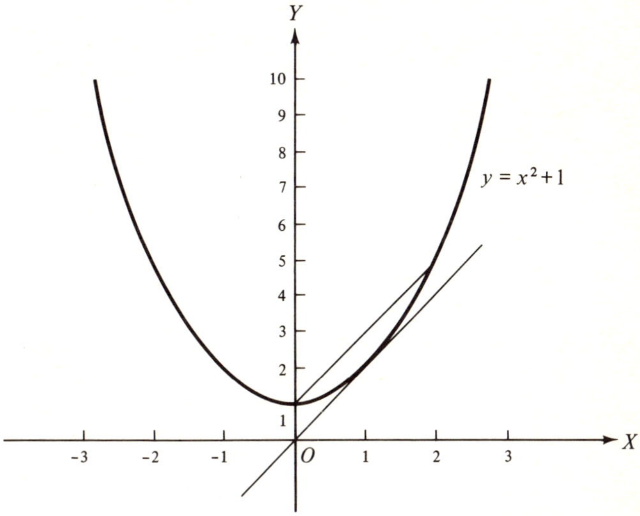


**Fig. 6-31**

## EXERCISES

Using the mean value theorem, find the appropriate value or values of $c$.

**1.** $f(x) = x^3;\ a = 0,\ b = 3$

**2.** $f(x) = x^2 + 4x + 4;\ a = 1,\ b = 3$

**3.** $f(x) = 4 - x^2;\ a = 0,\ b = 1$

**4.** $f(x) = \dfrac{1}{x};\ a = 2,\ b = 4$

**5.** $f(x) = \sqrt{x};\ a = 4,\ b = 9$

# 7
# Antidifferentiation

## 7-1 REVERSING THE PROCESS OF DIFFERENTIATION

We have considered the question of finding the derivative of a function. We shall now consider the problem of reversing the process of differentiation and finding an *antiderivative.* Suppose we are given an expression such as $3x^2$. Can we find a function whose derivative is $3x^2$?

One function that satisfies the condition is

$$y = x^3$$

Are there any other functions? If we think a moment we find that the condition is satisfied by the functions

$$y = x^3 + 1$$
$$y = x^3 + 2$$
and
$$y = x^3 - 3$$

This indicates that any function in the form $y = x^3 + C$ will satisfy the condition. Thus, if $F(x)$ is a function such that $F'(x) = f(x)$, then

$F(x) + C$ will be an antiderivative of $f(x)$. Though we shall not prove it here, it can be proved as a theorem that if $F(x)$ and $G(x)$ are functions defined over an interval $a \leq x \leq b$, and $F'(x) = G'(x)$ over the interval, then $F(x) = G(x) + C$, where $C$ is some constant. This theorem tells us that two antiderivatives of a given function can differ at most by a constant.

An antiderivative is sometimes referred to as an indefinite integral and the process of finding the antiderivative is sometimes referred to as *integration*. When using the integration terminology, the symbol used to denote the antiderivative of $f(x)$ is

$$\int f(x)\,dx$$

Thus,
$$\int 3x^2\,dx = x^3 + C$$

The $C$ is referred to as the *constant of integration.*

Just as we had rules for differentiation, we can develop rules for antidifferentiation. In fact, since antidifferentiation reverses the process of differentiation, our rules for antidifferentiation will merely reverse the process of differentiation. We shall state some fundamental theorems of antidifferentiation using the integration notation.

1. $\int kf(x)\,dx = k\int f(x)\,dx \qquad$ where $k$ is a constant
2. $\int [f(x) \pm g(x)]\,dx = \int f(x)\,dx \pm \int g(x)\,dx$
3. $\int x^n\,dx = \dfrac{x^{n+1}}{n+1} + C \qquad$ if $n \neq -1$
4. $\int k\,dx = kx + C \qquad$ where $k$ is a constant

Since the variable used does not alter the integration, property 3 is often stated as

$$\int u^n\,du = \frac{u^{n+1}}{n+1} + C \qquad \text{if } n \neq -1$$

This last form will be very useful when we wish to integrate a function to a power.

**Example 1**

(a) $\int x^6\,dx = \dfrac{x^7}{7} + C$

(b) $\int x^{-3}\,dx = \dfrac{x^{-2}}{-2} + C = -\dfrac{1}{2x^2} + C$

(c) $\int \sqrt{x}\,dx = \int x^{\frac{1}{2}}\,dx = \dfrac{x^{\frac{3}{2}}}{\frac{3}{2}} + C = \frac{2}{3}x^{\frac{3}{2}} + C$

**Example 2**

$$(a)\ \int (x^3 - x)\,dx = \int x^3\,dx - \int x\,dx = \frac{x^4}{4} - \frac{x^2}{2} + C$$

$$(b)\ \int (4x^2 - 2x + 4)\,dx = \int 4x^2\,dx - \int 2x\,dx + \int 4dx = 4\int x^2\,dx - 2\int x\,dx + \int 4dx = 4\frac{x^3}{3} - \frac{2x^2}{2} + 4x + C = \frac{4x^3}{3} - x^2 + 4x + C$$

$$(c)\ \int \left(\frac{2}{x^2} - \frac{3}{x^3}\right) dx = \int \frac{2}{x^2}\,dx - \int \frac{3}{x^3}\,dx = 2\int x^{-2}\,dx - 3\int x^{-3}\,dx = 2\frac{x^{-1}}{-1} - \frac{3x^{-2}}{-2} + C = -\frac{2}{x} + \frac{3}{2x^2} + C$$

**EXERCISES**

Evaluate the following integrals (i.e., find the antiderivatives). Check each problem by differentiating your result.

1. $\int x^2\,dx$
2. $\int x^{-5}\,dx$
3. $\int x^{\frac{2}{3}}\,dx$
4. $\int x^{-\frac{1}{2}}\,dx$
5. $\int \sqrt[3]{x}\,dx$
6. $\int \frac{1}{\sqrt{x}}\,dx$
7. $\int (4x^3 - 5x^2 + x)\,dx$
8. $\int (3x^2 - 4x + 7)\,dx$
9. $\int (2x^2 + 3x^3)\,dx$
10. $\int (7x^7 - 5x^5 + 3x^3)\,dx$
11. $\int x^2(x + 3)\,dx$
12. $\int \left(x^4 - \frac{1}{x^2} + 3x^2\right) dx$
13. $\int x(x^2 + 2)\,dx$
14. $\int \frac{x^2 + 2x}{x^4}\,dx$
15. $\int x^2 \sqrt{x}\,dx$
16. $\int \sqrt{x}\,(x^2 - 2)\,dx$
17. $\int (x + 1)(x - 1)\,dx$
18. $\int \frac{2 - x}{x^3}\,dx$
19. $\int (x + 3)^2\,dx$
20. $\int \frac{1 - x}{\sqrt{x}}\,dx$

## 7-2 THE POWER FORMULA

We recall that the derivative of $y = x^n$ is $y' = nx^{n-1}$, but the derivative of $y = [f(x)]^n$ is $y' = n[f(x)]^{n-1} \cdot f'(x)$ (making use of the chain rule). In reversing the process of differentiation we shall have to reverse the effect of the chain rule.

**Example 1** Find $\int(x^2 + 4)^5 \cdot 2x\,dx$.

*Solution* Since the derivative of $(x^2 + 4)^6$ is $6(x^2 + 4)^5 \cdot 2x$, then all we have to do is divide $(x^2 + 4)^6$ by 6. Thus,

$$\int(x^2 + 4)^5 \cdot 2x\,dx = \frac{(x^2 + 4)^6}{6} + C$$

**Example 2** Find $\int(x^3 - 7)^8 \cdot 3x^2\,dx$.

*Solution* The results of the previous example indicate that we must examine the function $y = (x^3 - 7)^9$. If $y = (x^3 - 7)^9$, then

$$y' = 9(x^3 - 7)^8 \cdot 3x^2$$

Here again, if we divide by the constant 9 and take $y = \dfrac{(x^3 - 7)^9}{9} + C$, then

$$y' = (x^3 - 7)^8 \cdot 3x^2$$

Thus,

$$\int(x^3 - 7)^8 \cdot 3x^2\,dx = \frac{(x^3 - 7)^9}{9} + C$$

In analyzing the two previous examples, we note first that in each case we were given an expression in the form

$$\int[f(x)]^n \cdot f'(x)\,dx$$

In the first example $f(x)$ was $x^2 + 4$ and $f'(x)$ was $2x$. In the second example $f(x)$ was $x^3 - 7$ and $f'(x)$ was $3x^2$. We also notice that in each example

$$\int[f(x)]^n \cdot f'(x)\,dx = \frac{f(x)^{n+1}}{n + 1} + C$$

To verify this result in general, we merely point out that the derivative with respect to $x$ of $\dfrac{[f(x)]^{n+1}}{n + 1} + C$ is

$$(n + 1)\,\frac{[f(x)]^n}{n + 1} \cdot f'(x) = [f(x)]^n \cdot f'(x)$$

**Example 3** Find $\int(x^3 - 2x)^5(3x^2 - 2)\,dx$.

*Solution* Here

$$f(x) = x^3 - 2x \qquad n = 5$$
$$f'(x) = 3x^2 - 2$$

Hence

$$\int(x^3 - 2x)^5(3x^2 - 2)\,dx = \frac{(x^3 - 2x)^6}{6} + C$$

**Example 4** Evaluate $\int (x^3 + 9)^2 \cdot x^2\,dx$.

*Solution* We can let $f(x) = x^3 + 9$ and $n = 2$. Then, $f'(x) = 3x^2$. The expression is almost in the form $\int [f(x)]^n \cdot f'(x)\,dx$. The only thing that is missing is the factor 3 in $f'(x)$. This can easily be remedied in the following way:

$$\int (x^3 + 9)^2 \cdot x^2\,dx = \int (x^3 + 9)^2(\tfrac{1}{3})(3)x^2\,dx$$

Since multiplying by $\frac{1}{3}(3)$ is equivalent to multiplying by 1, this does not change the original integral. Continuing,

$$\begin{aligned}\int (x^3 + 9)^2(\tfrac{1}{3})3x^2\,dx &= \int \tfrac{1}{3}(x^3 + 9)^2 \cdot 3x^2\,dx\\ &= \tfrac{1}{3}\int (x^3 + 9)^2 \cdot 3x^2\,dx \qquad \text{(Why?)}\\ &= \frac{1}{3}\left[\frac{(x^3 + 9)^3}{3} + C\right]\end{aligned}$$

**Example 5** Evaluate $\int (x^2 - 2x)^5(x - 1)\,dx$.

*Solution* Let $f(x) = x^2 - 2x$, and $n = 5$. Then,

$$f'(x) = 2x - 2 = 2(x - 1)$$

We shall use the method of the previous example:

$$\begin{aligned}\int (x^2 - 2x)^5(x - 1)\,dx &= \int \tfrac{1}{2}(x^2 - 2x)^5 \cdot 2(x - 1)\,dx\\ &= \tfrac{1}{2}\int (x^2 - 2x)^5 \cdot (2x - 2)\,dx\\ &= \frac{1}{2}\left[\frac{(x^2 - 2x)^6}{6} + C\right]\end{aligned}$$

**Example 6** Evaluate $\int \sqrt{x^2 + 1}\,x\,dx$.

*Solution*

$$\sqrt{x^2 + 1} = (x^2 + 1)^{\frac{1}{2}}$$

Let $f(x) = x^2 + 1$ and $n = \frac{1}{2}$. Then

$$\begin{aligned}f'(x) &= 2x\\ \int (x^2 + 1)^{\frac{1}{2}}x\,dx &= \int \tfrac{1}{2}(x^2 + 1)^{\frac{1}{2}}2x\,dx\\ &= \tfrac{1}{2}\int (x^2 + 1)^{\frac{1}{2}}2x\,dx\\ &= \frac{1}{2}\left[\frac{(x^2 + 1)^{\frac{3}{2}}}{\frac{3}{2}} + C\right]\\ &= \frac{(x^2 + 1)^{\frac{3}{2}}}{3} + C'\end{aligned}$$

## EXERCISES

Evaluate the following integrals:

1. $\int (1 + x^2)^3 2x\,dx$
2. $\int (x^3 + 9)^5 3x^2\,dx$
3. $\int (7 - x^3)^4 3x^2\,dx$
4. $\int (x^3 - 3x^2)^2(3x^2 - 6x)\,dx$
5. $\int (x^4 - x^2)^7(4x^3 - 2x)\,dx$
6. $\int (x - x^3)^3(3x^2 - 1)\,dx$
7. $\int (x^5 - 4x)^{\frac{3}{2}}(5x^4 - 4)\,dx$
8. $\int \left(x + \frac{1}{x}\right)^2\left(1 - \frac{1}{x^2}\right)dx$
9. $\int 12(4x^3 - 12x)^2(x^2 - 1)\,dx$
10. $\int 2(x^2 + 4)^{\frac{1}{2}}x\,dx$

**11.** $\int(4 - x^4)x^3\,dx$

**12.** $\int(x^4 - 2x^2)^3(x^3 - x)\,dx$

**13.** $\int(3x^4 - 4x^3)^{-\frac{1}{2}}(x^3 - x^2)\,dx$

**14.** $\int(7 - 4x^3)^4 3x^2\,dx$

**15.** $\int \sqrt{x^2 - 4}\,x\,dx$

**16.** $\displaystyle\int \frac{3x^2}{\sqrt{x^3 - 4}}\,dx$

**17.** $\displaystyle\int \frac{x}{\sqrt{4 - 6x^2}}\,dx$

**18.** $\displaystyle\int \frac{1}{(7 - x)^{\frac{3}{2}}}\,dx$

▶ **19.** $\int(x^2 + 1)^2\,dx$

▶ **20.** $\int(2 - x^3)^2\,dx$

## 7-3 USING DIFFERENTIALS FOR THE POWER FORMULA

In working with the integral symbol $\int f(x)\,dx$ it is often convenient to think of the $dx$ as the differential of $x$. Thus, in the formula

$$\int u^n\,du = \frac{u^{n+1}}{n+1} + C$$

the $du$ can be thought of as the differential of $u$.

**Example 1** Evaluate $\int(x^3 - 3x^2)^5(3x^2 - 6x)\,dx$.

*Solution* We can think of this as being in the form $\int[f(x)]^n \cdot f'(x)\,dx$. However, we can also think of this as being in the form $\int u^n\,du$, where

$$u = f(x) \qquad n = 5$$
$$du = f'(x)\,dx$$

Thus

$$\int(x^3 - 3x^2)^5(3x^2 - 6x)\,dx = \int u^5\,du = \frac{u^6}{6} + C = \frac{(x^3 - 3x^2)^6}{6} + C$$

**Example 2** Evaluate $\int(x^3 - 3x)^{\frac{1}{2}}(x^2 - 1)\,dx$.

*Solution* Let $u = x^3 - 3x$. Then

$$\begin{aligned} du &= (3x^2 - 3)\,dx \\ &= 3(x^2 - 1)\,dx \\ \tfrac{1}{3}du &= (x^2 - 1)\,dx \end{aligned}$$

Thus,

$$\begin{aligned} \int(x^3 - 3x)^{\frac{1}{2}}(x^2 - 1)\,dx &= \int u^{\frac{1}{2}}(\tfrac{1}{3})\,du \\ &= \tfrac{1}{3}\int u^{\frac{1}{2}}\,du \\ &= \frac{1}{3}\left(\frac{u^{\frac{3}{2}}}{\frac{3}{2}} + C\right) \\ &= \tfrac{2}{9}u^{\frac{3}{2}} + \tfrac{1}{3}C \\ &= \tfrac{2}{9}(x^3 - 3x)^{\frac{3}{2}} + C' \end{aligned}$$

**Example 3** Evaluate $\displaystyle\int \frac{x}{\sqrt{x^2 - 4}}\,dx$.

*Solution*

$$\int \frac{x}{\sqrt{x^2 - 4}}\,dx = \int (x^2 - 4)^{-\frac{1}{2}}x\,dx$$

Let $u = x^2 - 4$ and $n = -\frac{1}{2}$. Then

$$\begin{aligned} du &= 2x\,dx \\ \tfrac{1}{2}du &= x\,dx \\ \int (x^2-4)^{-\frac{1}{2}} x\,dx &= \int u^{-\frac{1}{2}}(\tfrac{1}{2})\,du \\ &= \tfrac{1}{2}\int u^{-\frac{1}{2}}\,du \\ &= \frac{1}{2}\left(\frac{u^{\frac{1}{2}}}{\frac{1}{2}} + C\right) \\ &= u^{\frac{1}{2}} + C' \\ &= \sqrt{x^2-4} + C' \end{aligned}$$

**EXERCISES**

Evaluate the following integrals.

**1.** $\int (7-4x^3)^4 x^2\,dx$

**2.** $\int (5x^3-15x)^{\frac{3}{5}}(x^2-1)\,dx$

**3.** $\int \sqrt{1-4x^2}\,x\,dx$

**4.** $\int \frac{x^2}{\sqrt{x^3-4}}\,dx$

**5.** $\int \sqrt[3]{4-2x^3}\,x^2\,dx$

**6.** $\int \frac{3x^3}{\sqrt{x^4-5}}\,dx$

## 7-4 GEOMETRIC APPLICATIONS OF THE INDEFINITE INTEGRAL

The geometric interpretation of the derivative was the slope of the tangent line to the curve at any point. We can now consider the question of determining an equation of a curve whose slope is known.

**Example 1** Find an equation of a curve whose slope is equal to $2x$ at any point on the curve.

*Solution* We know that the slope $m$ is equal to $y'$. Thus $y' = 2x$. Integrating, we get

$$y = x^2 + C$$

If we choose $C = 0$, we get the curve

$$y = x^2$$

If we choose $C = 1$, we get the curve

$$y = x^2 + 1$$

which has the same slope as the curve $y = x^2$ but has an ordinate which is always one greater than the ordinate for the corresponding $x$ values of the curve $y = x^2$. Thus, we have found a *family of curves* all of which have a slope of $2x$ (Fig. 7-1).

**Example 2** Find the equation of the curve whose slope is $4 - 2x$ and which passes through the point (2,6).

*Solution* Since $y' = 4 - 2x$,

$$y = 4x - x^2 + C$$

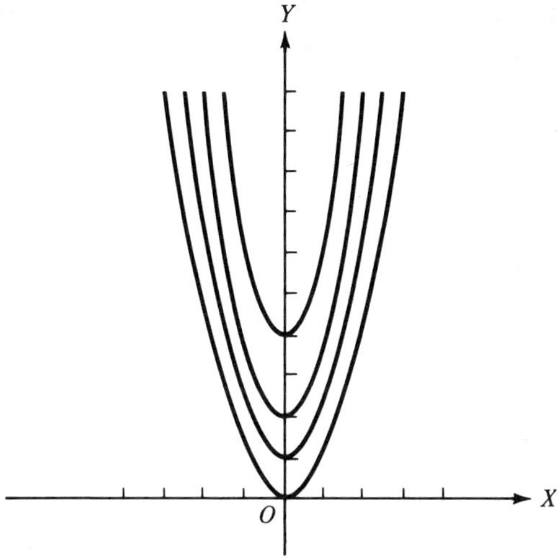


**Fig. 7-1**

The curve passes through the point (2,6), and if we substitute these values into the equation, we get

$$6 = 4(2) - (2)^2 + C$$
$$6 = 4 + C$$
$$2 = C$$

Therefore, $y = 4x - x^2 + 2$ is the equation of the curve we are looking for.

In this example the additional information telling us a point on the curve enabled us to choose the one curve out of the family of curves given by the equation $y = 4x - x^2 + C$.

**EXERCISES**

**1.** (*a*) Find the equation of the family of curves which have a slope of $4x - 2$ at any point $(x,y)$.

(*b*) From this family of curves, find the equation of the curve whose $y$ intercept is 5.

**2.** (*a*) Find the equation of the family of curves which have a slope of $3x^2 - 1$ at any point $(x,y)$.

(*b*) From this family of curves, find the equation of the curve which passes through the point (1,4).

**3.** Find the equation of the curve whose slope at any point $(x,y)$ is $4x - 3x^2$ and which passes through the point $(3,-2)$.

**4.** Find the equation of the curve whose slope at any point $(x,y)$ is $\frac{1}{x^2}$ and which passes through the point $(-1,-3)$.

**5.** Find the equation of the curve for which $y'' = 12x$ and which passes through the point (1,1) with an angle of inclination of 45°. (*Hint:* The tangent of the angle of inclination is the slope.)

**6.** Find the equation of the curve for which $y'' = 6x + 2$ and which passes through the points (0,3) and (1,1).

## 7-5 PHYSICAL APPLICATIONS OF THE INDEFINITE INTEGRAL

When working with the derivative, we found that when we were given the law of motion of an object in rectilinear motion, we were able to determine the velocity and acceleration. We shall now use integration to reverse the problem by starting with the acceleration and velocity and determining the law of motion $s = f(t)$.

**Example** A man throws an object out of a window 160 ft above the ground. If the object is thrown up with an initial velocity of 48 ft/sec, how high will the object go, and when will it reach the ground?

*Solution* We shall neglect air resistance and assume that the only force acting upon the object is that of gravity, which we shall take as $-32$ ft/sec². Thus, $a = -32$. Since $a = \frac{dv}{dt}$, the velocity $v$ at any time $t$ is the antiderivative of $a$ with respect to $t$. Therefore,

$$v = -32t + C$$

We know that the initial velocity is 48 ft/sec. Therefore, $v = 48$ when $t = 0$.

$$\begin{aligned} 48 &= -32(0) + C \\ C &= 48 \\ v &= -32t + 48 \\ v &= \frac{ds}{dt} \\ s &= -16t^2 + 48t + k \end{aligned}$$

Since the object was thrown from a height of 160 ft above the ground, $s = 160$ when $t = 0$, and

$$\begin{aligned} 160 &= -16(0) + 48(0) + k \\ s &= -16t^2 + 48t + 160 \end{aligned}$$

where $s$ represents the height of the object above the ground. We now proceed to answer given questions. To find how high the object will go, we point out that at the highest point the velocity is 0 since the object will stop at that second.

$$\begin{aligned} v &= -32t + 48 \\ 0 &= -32t + 48 \\ 32t &= 48 \\ t &= \tfrac{3}{2} \\ s &= -16t^2 + 48t + 160 \\ &= -16(\tfrac{3}{2})^2 + 48(\tfrac{3}{2}) + 160 \\ &= -36 + 72 + 160 \\ s &= 196 \end{aligned}$$

The object will reach a maximum height of 196 ft, and this will occur $1\frac{1}{2}$ sec after the object is thrown. To find when the object will reach the ground, we turn to the equation for $s$ to find when $s = 0$.

$$\begin{aligned} s &= -16t^2 + 48t + 160 \\ 0 &= -16t^2 + 48t + 160 \\ t^2 - 3t - 10 &= 0 \\ (t + 2)(t - 5) &= 0 \\ t &= -2 \qquad \text{Reject (Why?)} \\ t &= 5 \end{aligned}$$

Thus, 5 sec after the object is thrown up, it will reach the ground.

## EXERCISES

**1.** A particle moving in a straight line has a velocity of $v = 2t + 4$. If $s = 13$ when $t = 2$ find $s$ at any time $t$.

**2.** A ball falls from rest from a balloon 1,600 ft above the ground. When will the ball hit the ground and with what velocity?

**3.** (*a*) If a ball is thrown up from the ground with an initial velocity of $v_0$, what is its height after $t$ sec?

(*b*) If a ball is thrown up from a height of $s_0$, with an initial velocity of $v_0$, what is its height after $t$ sec?

**4.** If a car decelerates at a constant rate of 20 ft/sec when the brakes are applied, how long after the brakes are applied will it take a car traveling at 30 mph to stop, and how far will it go before stopping? (*Hint:* 1 mph $= \frac{88}{60}$ ft/sec and when $t = 0$, $v = 30$ mph and $s = 0$.)

**5.** If an object falls from the ledge of a window 800 ft above the ground, with what velocity will it hit the ground?

**6.** If a ball is thrown down from a roof 192 ft above the ground with an initial velocity of $-16$ ft/sec, how long will it take to reach the ground, and with what velocity will it strike the ground?

**7.** A coin is dropped into a well 64 ft deep. How long will it take to hit the bottom of the well and with what velocity will it strike?

**8.** A man standing on the ground throws an object upward to a man on a roof top 96 ft above the ground. If it takes 2 sec for the object to reach the hand of the man on the roof, with what velocity was it thrown?

# 8
# Trigonometric Functions

## 8-1 BRIEF REVIEW OF TRIGONOMETRY

We recall from trigonometry that an angle in standard position has its initial side on the positive $X$ axis and its vertex at the origin. The positive direction of the angle is taken in the counterclockwise direction (Fig. 8-1).

The measurement of angles can be given either in terms of degrees or radians. A degree is $\frac{1}{360}$ of a complete revolution, so that a complete revolution measures 360°. Radian measure is given in terms of the radius and the arc $s$ subtended by the central angle $\theta$ (Fig. 8-2).

If $r$ is the radius of a sector of a circle and $s$ is the arc subtended by the central angle $\theta$, then the number of radians in the angle $\theta$ is defined as the quotient $\frac{s}{r}$.

$$\theta = \frac{s}{r} \qquad \text{radians}$$

Thus, if arc $s$ contains (has a length of) 1 radius unit, then the number

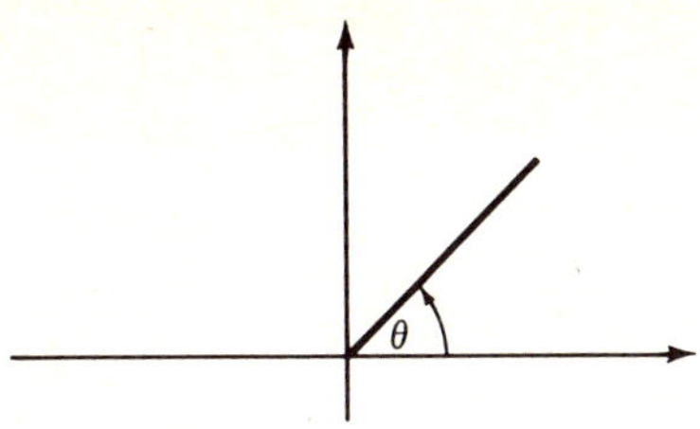


Fig. 8-1

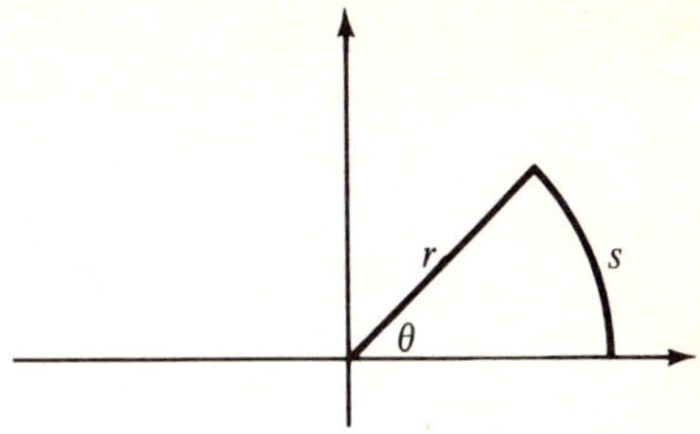


Fig. 8-2

of radians in angle $\theta$ is 1, since $\theta = \frac{s}{r} = \frac{r}{r} = 1$. If arc $s$ contains 2 radius units, then the number of radians in angle $\theta$ is 2 $\left(\theta = \frac{s}{r} = \frac{2r}{r} = 2\right)$. Because $s$ and $r$ are in the same units, radian measure gives us a real number as a result.

To see the relationship between degree and radian measure we can consider an angle $\theta$ produced by a complete revolution. In degree measure $\theta = 360°$. For radian measure we note that $s$ is the circumference of a circle and thus $s = 2\pi r$.

$$\theta = \frac{s}{r} = \frac{2\pi r}{r} = 2\pi \text{ radians}$$

Since we are working with the same angle $\theta$ we can conclude that

$$360° = 2\pi \text{ radians}$$

or

$$180° = \pi \text{ radians}$$

Another result which will prove to be useful is a formula for the area of a sector of a circle (Fig. 8-3). We shall set up a proportion between any angle $\theta$ and the angle $2\pi$ and compare the corresponding areas. If $A$ is the area of the sector of the circle, we can establish the following proportion

$$\frac{\theta}{2\pi} = \frac{A}{\pi r^2}$$

$$A = \tfrac{1}{2} r^2 \theta$$

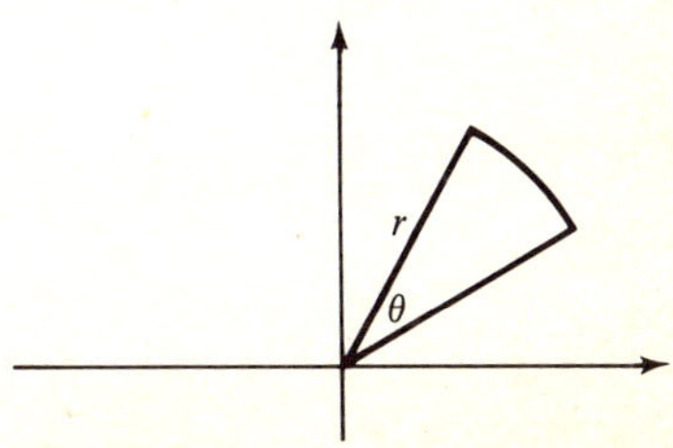


Fig. 8-3

Here we have used the fact that corresponding to an angle of $2\pi$ we have the area of the entire circle which equals $\pi r^2$.

We now recall the definitions of the trigonometric functions. Let $\theta$ be an angle in standard position, $P(x,y)$ a point on the terminal side of angle $\theta$, and $r$ the length of the segment from the vertex to the point $P$ (Fig. 8-4). The trigonometric functions are defined as follows:

$$\sin\theta = \frac{y}{r} \qquad \csc\theta = \frac{r}{y}$$
$$\cos\theta = \frac{x}{r} \qquad \sec\theta = \frac{r}{x}$$
$$\tan\theta = \frac{y}{x} \qquad \cot\theta = \frac{x}{y}$$

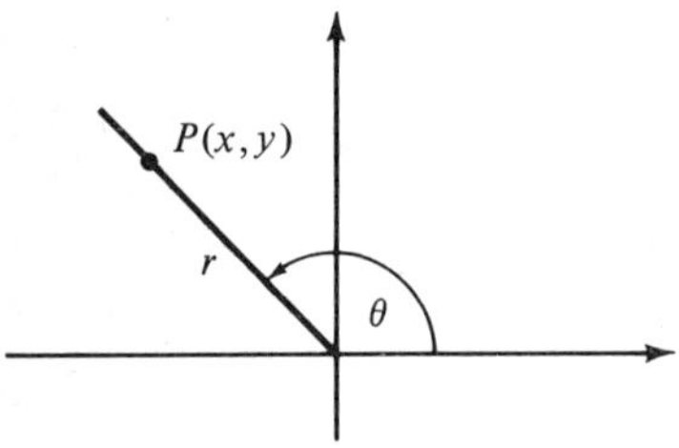


Fig. 8-4

From the above definitions we can readily show the following:

$$x = r\cos\theta \qquad y = r\sin\theta$$
$$\sin\theta = \frac{1}{\csc\theta} \qquad \tan\theta = \frac{1}{\cot\theta}$$
$$\cos\theta = \frac{1}{\sec\theta} \qquad \tan\theta = \frac{\sin\theta}{\cos\theta}$$
$$\sin^2\theta + \cos^2\theta = 1 \qquad \tan^2\theta + 1 = \sec^2\theta$$

Additional trigonometric formulas which will be useful are the following:

$$\sin(A \pm B) = \sin A\cos B \pm \cos A\sin B$$
$$\cos(A \pm B) = \cos A\cos B \mp \sin A\sin B$$
$$\sin A - \sin B = 2\cos\frac{A+B}{2}\sin\frac{A-B}{2}$$
$$\cos A - \cos B = -2\sin\frac{A+B}{2}\sin\frac{A-B}{2}$$
$$\sin 2A = 2\sin A\cos A$$
$$\cos 2A = \cos^2 A - \sin^2 A = 2\cos^2 A - 1 = 1 - 2\sin^2 A$$
$$\tan 2A = \frac{2\tan A}{1 - \tan^2 A}$$

$$\sin \tfrac{1}{2}A = \pm \sqrt{\frac{1 - \cos A}{2}}$$

$$\cos \tfrac{1}{2}A = \pm \sqrt{\frac{1 + \cos A}{2}}$$

$$\tan \tfrac{1}{2}A = \pm \sqrt{\frac{1 - \cos A}{1 + \cos A}} = \frac{\sin A}{1 + \cos A} = \frac{1 - \cos A}{\sin A}$$

$$\sin \left(\frac{\pi}{2} - A\right) = \cos A$$

$$\cos \left(\frac{\pi}{2} - A\right) = \sin A$$

$$\sin (-\theta) = -\sin \theta$$

$$\cos (-\theta) = \cos \theta$$

## 8-2 Limit $\lim_{\theta \to 0} \frac{\sin \theta}{\theta}$

Before we proceed to the differentiation of the trigonometric functions, we will consider $\lim_{\theta \to 0} \frac{\sin \theta}{\theta}$. If we apply the quotient rule, we obtain the indeterminate form $\frac{0}{0}$. We shall resort to a geometric argument to determine this limit. Let $\theta$ be a small positive angle at the center of a circle of radius 1, $\theta = \measuredangle POQ$. Let $\overline{RQ}$ be tangent to the circle at $Q$ and intersect side $\overline{OP}$ at $R$. From Fig. 8-5 we find that

$$\text{Area } \triangle POQ < \text{area sector } POQ < \text{area } \triangle ROQ$$

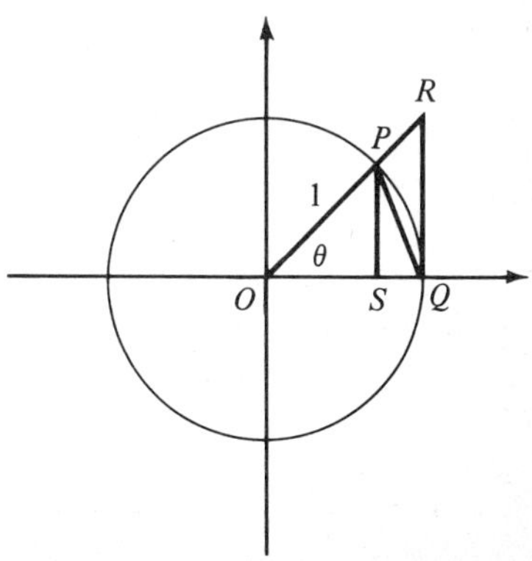


Fig. 8-5

We now make the following observations.

In $\triangle OPS$: $$\sin\theta = \frac{\overline{PS}}{\overline{OP}} = \frac{\overline{PS}}{1} = \overline{PS}$$

In $\triangle ORQ$: $$\tan\theta = \frac{\overline{RQ}}{\overline{OQ}} = \frac{\overline{RQ}}{1} = \overline{RQ}$$

$$\begin{aligned}
\text{Area } \triangle OPQ &= \tfrac{1}{2}\cdot\overline{OQ}\cdot\overline{PS} \\
&= \tfrac{1}{2}(1)\sin\theta \\
&= \tfrac{1}{2}\sin\theta \\
\text{Area } \triangle ORQ &= \tfrac{1}{2}\cdot\overline{OQ}\cdot\overline{RQ} \\
&= \tfrac{1}{2}(1)\tan\theta \\
&= \tfrac{1}{2}\tan\theta \\
\text{Area sector } POQ &= \tfrac{1}{2}r^2\theta \\
&= \tfrac{1}{2}(1)\theta \\
&= \tfrac{1}{2}\theta
\end{aligned}$$

Substituting these results into our inequality, we get

$$\begin{gathered}
\tfrac{1}{2}\sin\theta < \tfrac{1}{2}\theta < \tfrac{1}{2}\tan\theta \\
\sin\theta < \theta < \tan\theta
\end{gathered}$$

Dividing by $\sin\theta$,†

$$1 < \frac{\theta}{\sin\theta} < \frac{1}{\cos\theta}$$

If we now take reciprocals, this will reverse the inequality, giving us

$$1 > \frac{\sin\theta}{\theta} > \cos\theta$$

Thus, $\dfrac{\sin\theta}{\theta}$ is between $\cos\theta$ and 1. Considering the limit as $\theta$ approaches 0, we find that $\cos\theta$ approaches 1, and $\dfrac{\sin\theta}{\theta}$, which is between 1 and $\cos\theta$, must approach 1. In our discussion we have only considered a small positive angle $\theta$. If $\theta$ is negative, we can express $\theta = -\beta$ where $\beta$ is positive. Then,

$$\lim_{\theta\to 0-}\frac{\sin\theta}{\theta} = \lim_{\beta\to 0+}\frac{\sin(-\beta)}{-\beta} = \lim_{\beta\to 0+}\frac{-\sin\beta}{-\beta} = \lim_{\beta\to 0+}\frac{\sin\beta}{\beta} = 1$$

† Note: Since we have considered a small positive angle $\theta$, $\sin\theta$ is positive, and the sense of the inequality remains the same when we divide by $\sin\theta$.

Therefore, whether $\theta$ is positive or negative, we have proved

$$\boxed{\lim_{\theta \to 0} \frac{\sin \theta}{\theta} = 1}$$

**Example** Find $\lim_{\theta \to 0} \dfrac{\tan \theta}{\theta}$.

*Solution* We shall express $\tan \theta$ as $\dfrac{\sin \theta}{\cos \theta}$.

$$\lim_{\theta \to 0} \frac{\tan \theta}{\theta} = \lim_{\theta \to 0} \frac{\sin \theta}{\theta \cos \theta} = \lim_{\theta \to 0} \frac{\sin \theta}{\theta} \cdot \lim_{\theta \to 0} \frac{1}{\cos \theta} = 1(1) = 1$$

**EXERCISES**

Evaluate the following limits.

**1.** $\lim_{\theta \to 0} \dfrac{\sin 2\theta}{\theta}$

**2.** $\lim_{\theta \to 0} 2 \cos \theta$

**3.** $\lim_{\theta \to 0} \dfrac{\sin \theta}{5\theta}$

**4.** $\lim_{\theta \to 0} \dfrac{2 \tan \theta}{\theta}$

**5.** $\lim_{\theta \to 0} \dfrac{\sin \frac{1}{2}\theta}{\theta}$

**6.** $\lim_{\theta \to 0} \dfrac{\theta}{\sin \theta}$

**7.** $\lim_{\theta \to 0} \tan \theta \csc \theta$

**8.** $\lim_{\theta \to 0} \dfrac{\sin 2\theta}{\sin 5\theta}$

**9.** $\lim_{\theta \to 0} \dfrac{\theta}{\sin 2\theta}$

**10.** $\lim_{\theta \to 0} \dfrac{\sin (-\theta)}{\theta}$

### 8-3 THE DERIVATIVE OF sin $x$ AND cos $x$

We are now ready to consider the derivative of $y = \sin x$. The definition of the derivative requires us to find $\Delta y$, $\dfrac{\Delta y}{\Delta x}$, and $\lim_{\Delta x \to 0} \dfrac{\Delta y}{\Delta x}$.

$$y = \sin x$$
$$\Delta y = \sin (x + \Delta x) - \sin x$$

At this point we need some way to combine $\sin (x + \Delta x) - \sin x$. We recall from our trigonometric formulas that

$$\sin A - \sin B = 2 \cos \frac{A + B}{2} \sin \frac{A - B}{2}$$

We shall let $A = x + \Delta x$, $B = x$. Then,

$$\sin (x + \Delta x) - \sin x = 2 \cos \frac{2x + \Delta x}{2} \sin \frac{\Delta x}{2}$$

$$\frac{\Delta y}{\Delta x} = \frac{2 \cos \left(x + \frac{\Delta x}{2}\right) \sin \frac{\Delta x}{2}}{\Delta x}$$

We shall now try to arrange the terms so that we can apply our result $\lim_{\theta \to 0} \frac{\sin \theta}{\theta} = 1$. Since we have $\sin \frac{\Delta x}{2}$, we shall think of $\theta = \frac{\Delta x}{2}$. Dividing the numerator and denominator of our quotient by 2, we obtain

$$\frac{\Delta y}{\Delta x} = \cos \left(x + \frac{\Delta x}{2}\right) \frac{\sin \frac{\Delta x}{2}}{\frac{\Delta x}{2}}$$

$$\lim_{\Delta x \to 0} \frac{\Delta y}{\Delta x} = \lim_{\Delta x \to 0} \cos \left(x + \frac{\Delta x}{2}\right) \lim_{\Delta x \to 0} \frac{\sin \frac{\Delta x}{2}}{\frac{\Delta x}{2}}$$

$$= (\cos x)(1) = \cos x$$

$$\boxed{\text{If } y = \sin x \qquad D_x y = \cos x}$$

**Example 1** Find the derivative of $y = \sin 2x$.

*Solution* We have a formula for $\sin x$ and we shall have to adjust it for the derivative of the sine of a function of $x$. Let $u = 2x$. Then,

$$y = \sin u \qquad u = 2x$$

This suggests the use of the chain rule:

$$\begin{aligned} D_x y &= D_u y \cdot D_x u \\ &= (\cos u)(2) \\ &= 2 \cos u \\ &= 2 \cos 2x \end{aligned}$$

We can generalize this result and say if $y = \sin u$, where $u$ is a function of $x$, then $D_x y = \cos u \, D_x u$

$$\boxed{D_x \sin u = \cos u \, D_x u}$$

To derive a formula for the derivative of $y = \cos x$, we recall the trigonometric identities

$$\sin\left(\frac{\pi}{2} - x\right) = \cos x$$

$$\cos\left(\frac{\pi}{2} - x\right) = \sin x$$

$$\begin{aligned} y &= \cos x \\ &= \sin\left(\frac{\pi}{2} - x\right) \\ D_x y &= \left[\cos\left(\frac{\pi}{2} - x\right)\right](-1) \\ &= -\sin x \end{aligned}$$

$$\boxed{D_x \cos x = -\sin x}$$

As before, if instead of $x$ we are given a function of $x$, we can use the chain rule to obtain the formula:

$$\boxed{D_x \cos u = -\sin u \cdot D_x u}$$

**Example 2** Find the derivative of $y = \cos 5x$.

*Solution*

$$\begin{aligned} y &= \cos 5x \\ y' &= -(\sin 5x)(5) \\ &= -5 \sin 5x \end{aligned}$$

**Example 3** Find the derivative of $y = \cos x^2$.

*Solution*

$$\begin{aligned} y &= \cos x^2 \\ y' &= -(\sin x^2)(2x) \\ &= -2x \sin x^2 \end{aligned}$$

**Example 4** Find the derivative of $y = \dfrac{\sin x^2}{x}$.

*Solution*

$$\begin{aligned} y &= \frac{\sin x^2}{x} \\ y' &= \frac{x \cdot (\cos x^2)2x - \sin x^2}{x^2} \\ &= \frac{2x^2 \cos x^2 - \sin x^2}{x^2} \end{aligned}$$

**Example 5** Find the derivative of $y = \sin^2 x$.

*Solution*

$$y = \sin^2 x$$

This is a function to a power and thus

$$\begin{aligned} y' &= 2(\sin x)\, D_x \sin x \\ &= 2 \sin x \cos x \end{aligned}$$

**Example 6** A boy flies a kite which is 100 ft above the ground. If the string is being let out at a rate of 10 ft/sec, and the wind carries the kite horizontally directly away from the boy, find the rate of change of the angle of elevation of the kite when the angle is 30° (Fig. 8-6).

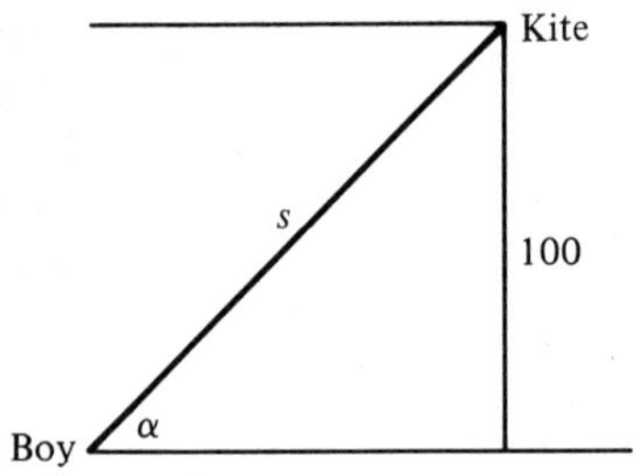


**Fig. 8-6**

*Solution* Let $\alpha$ be the angle of elevation and let $s$ be the amount of string. Then $\dfrac{ds}{dt} = 10$ ft/sec.

$$\sin \alpha = \frac{100}{s}$$

We want to find $\dfrac{d\alpha}{dt}$ when $\alpha = 30°$. Since $30° = \dfrac{\pi}{6}$ radians, when $\alpha = \dfrac{\pi}{6}$ radians,

$$\sin \alpha = \tfrac{1}{2}$$

Therefore, $\qquad \dfrac{100}{s} = \dfrac{1}{2} \qquad s = 200$

Now, $\qquad \sin \alpha = \dfrac{100}{s}$

Differentiating with respect to $t$,

$$\cos \alpha \frac{d\alpha}{dt} = -\frac{100}{s^2}\frac{ds}{dt}$$

Substituting the given values,

$$\frac{\sqrt{3}}{2}\frac{d\alpha}{dt} = \frac{-100}{200(200)}(10)$$

$$\frac{\sqrt{3}}{2}\frac{d\alpha}{dt} = -\frac{1}{40}$$

$$\frac{d\alpha}{dt} = -\frac{1}{20\sqrt{3}}$$

$$= \frac{-\sqrt{3}}{60}\text{ radians/sec}$$

**EXERCISES**

Find $D_xy$ for each of the following functions.

**1.** $y = \sin 3x$

**2.** $y = \sin 2x^2$

**3.** $y = \cos\frac{1}{x}$

**4.** $y = \cos^2 x$

**5.** $y = \cos^2 3x$

**6.** $y = \dfrac{x}{\sin x}$

**7.** $y = x\cos x$

**8.** $y = x\sin 2x^3$

**9.** $y = \dfrac{x^2}{\cos x}$

**10.** $y = \dfrac{\cos x}{x^2}$

**11.** $y = \sin^2\frac{2}{x}$

**12.** $y = \sin^2 x + \cos^2 2x$

**13.** Find the maximum and minimum points and points of inflection for the curve $y = \sin x + \cos x$ for $0 \le x \le 2\pi$.

## 8-4 THE DERIVATIVE OF tan $x$, cot $x$, sec $x$, AND csc $x$

The remaining trigonometric functions can all be expressed in terms of the sine and cosine functions.

**Example 1** Find the derivative of $y = \tan x$.

*Solution*

$$y = \tan x$$

$$y = \frac{\sin x}{\cos x}$$

$$y' = \frac{\cos x\, D_x \sin x - \sin x\, D_x \cos x}{\cos^2 x}$$

$$= \frac{\cos^2 x + \sin^2 x}{\cos^2 x}$$

$$= \frac{1}{\cos^2 x} = \sec^2 x$$

$$\boxed{D_x \tan x = \sec^2 x}$$

Using the chain rule, one obtains the more general result

$$\boxed{D_x \tan u = \sec^2 u \cdot D_x u}$$

In a similar way we can show that $D_x \cot x = -\csc^2 x$, and $D_x \cot u = -\csc^2 u\, D_x u$. This will be left as an exercise for the student.

**Example 2** Find the derivative of $y = \sec x$.

*Solution*

$$\begin{aligned} y &= \sec x \\ &= \frac{1}{\cos x} \\ y' &= \frac{-(-\sin x)}{\cos^2 x} \\ &= \frac{\sin x}{\cos^2 x} = \frac{\sin x}{\cos x}\frac{1}{\cos x} \\ &= \tan x \sec x \end{aligned}$$

$$\boxed{D_x \sec x = \tan x \cdot \sec x}$$

More generally,

$$\boxed{D_x \sec u = \tan u \sec u \cdot D_x u}$$

It will be left for the student to show that $D_x \csc x = -\csc x \cot x$, and $D_x \csc u = -\csc u \cot u \cdot D_x u$.

**Example 3** Find the derivative of $y = \tan^2 x$.

*Solution*

$$\begin{aligned} y &= \tan^2 x \\ y' &= 2 \tan x\, D_x \tan x \\ &= 2 \tan x \sec^2 x \end{aligned}$$

**EXERCISES**

Find $D_x y$ for the following functions.

**1.** $y = \cot 2x$

**2.** $y = \csc 3x$

**3.** $y = \tan 3x$

**4.** $y = \sec x^2$

**5.** $y = x \tan x$

**6.** $y = \dfrac{\tan x}{x}$

**7.** $y = \cot^2 x$

**8.** $y = \csc \dfrac{1}{x}$

**9.** $y = \tan^2 3x$

**10.** $y = \sin^2 3x - \cot^2 3x$

Determine the maxima and minima for the following curves over the interval $0 \leq x < 2\pi$.

**11.** $y = x - \tan x$

**12.** $y = \tan x + \cot x$

## 8-5 INTEGRATION OF THE TRIGONOMETRIC FUNCTIONS

Since integration (antidifferentiation) merely reverses the process of differentiation, we immediately obtain the following formulas:

$$\begin{aligned} \int \cos u\, du &= \sin u + C \\ \int \sin u\, du &= -\cos u + C \\ \int \sec^2 u\, du &= \tan u + C \\ \int \csc^2 u\, du &= -\cot u + C \\ \int \tan u \sec u\, du &= \sec u + C \\ \int \cot u \csc u\, du &= -\csc u + C \end{aligned}$$

**Example 1**

$$\begin{aligned} \int \cos 2x\, dx &= \tfrac{1}{2}\int \cos 2x \cdot 2dx \qquad \text{(Why is there a } \tfrac{1}{2} \text{ in front of the integral sign?)} \\ &= \tfrac{1}{2} \sin 2x + C \end{aligned}$$

As a check we point out that if $y = \frac{1}{2} \sin 2x + C$,

$$\begin{aligned} y' &= \tfrac{1}{2}(\cos 2x)(2) \\ &= \cos 2x \end{aligned}$$

**Example 2** Evaluate $\int x \sin x^2\, dx$.

*Solution* We shall try to apply the formula for $\int \sin u\, du$. To do this we will let $u = x^2$. Then,

$$du = 2x\, dx$$

Thus, we must introduce a factor of 2 within the integral in order to have the form $\sin u\, du$. To compensate for the factor 2 we must also introduce a factor of $\frac{1}{2}$. In doing this we are actually multiplying by 1 and thus not changing the value of the integral. Since a constant can be brought outside the integral we have the following:

$$\begin{aligned} \int x \sin x^2\, dx &= \int \tfrac{1}{2} \sin x^2 \cdot 2x\, dx \\ &= \tfrac{1}{2}\int \sin x^2 \cdot 2x\, dx \end{aligned}$$

This is in the form

$$\tfrac{1}{2}\int \sin u\, du$$

and since $\frac{1}{2}\int \sin u\, du = -\frac{1}{2} \cos u + C$,

$$\tfrac{1}{2}\int \sin x^2 \cdot 2x\, dx = -\tfrac{1}{2} \cos x^2 + C$$

**Example 3** Evaluate $\int \sin^3 2x \cos 2x\, dx$.

*Solution* The $\sin^3 2x$ suggests the use of the general power formula

$$\int u^n\, du = \frac{u^{n+1}}{n+1} + C$$

Let $u = \sin 2x$ and $n = 3$. Then

$$du = (\cos 2x)\, 2dx$$

In order to have the exact expression for $du$ we must introduce a factor of 2. Thus

$$\begin{aligned}\int \sin^3 2x \cos 2x\, dx &= \int \tfrac{1}{2} \sin^3 2x(\cos 2x)\, 2dx \\ &= \tfrac{1}{2}\int \sin^3 2x(\cos 2x)\, 2dx\end{aligned}$$

This is now in the form $\frac{1}{2}\int u^3\, du$. Integrating,

$$\begin{aligned}\tfrac{1}{2}\int u^3\, du &= \frac{1}{2}\frac{u^4}{4} + C \\ \tfrac{1}{2}\int \sin^3 2x(2\cos 2x)\, dx &= \frac{1}{2}\,\frac{\sin^4 2x}{4} + C \\ &= \frac{\sin^4 2x}{8} + C\end{aligned}$$

It is important to realize that the $2 \cos 2x\, dx$ is the $du$ term in the form $\int u^n\, du$ and that this factor will not be present after the integration has been completed.

### EXERCISES

Evaluate the following integrals:

**1.** $\int \cos 5x\, dx$

**2.** $\int \sin 8t\, dt$

**3.** $\int \sec^2 3x\, dx$

**4.** $\int \csc^2 2x\, dx$

**5.** $\int 3x(\sin x^2)\, dx$

**6.** $\int (\cos x^3)x^2\, dx$

**7.** $\int (\sec^2 x^2)x\, dx$

**8.** $\int \left(\sin \frac{1}{x}\right) \frac{1}{x^2}\, dx$

**9.** $\int \sin x \cos x\, dx$

**10.** $\int \cos^2 x \sin x\, dx$

**11.** $\int \tan^5 t \sec^2 t\, dt$

**12.** $\int \sqrt{\sin^3 x} \cos x\, dx$

**13.** $\int \frac{\sin x}{\cos^2 x}\, dx$

**14.** $\int x \sin^3 x^2 \cos x^2\, dx$

**15.** $\int \sin^2 \frac{t}{2} \cos \frac{t}{2}\, dt$

**16.** $\int \sqrt{1 + \cos 2t}\, \sin 2t\, dt$

**17.** $\int \frac{\sin x}{\sqrt{1 + \cos x}}\, dx$

**18.** $\int \cos (\pi - x)\, dx$

**19.** $\int \csc^2 x \cot x\, dx$

**20.** $\int \sec^3 x \tan x\, dx$

# 9 The Definite Integral

## 9-1 THE SIGMA NOTATION

It is often necessary to consider sums such as $1 + 2 + 3 + 4 + \cdots + n$ or $1^2 + 2^2 + 3^2 + 4^2 + \cdots + n^2$, where the three dots indicate that the middle terms have not been written out. In considering sums such as these, we shall find it convenient to use the sigma notation $\Sigma$ to indicate the sum. The sum $1 + 2 + 3 + 4 + \cdots + n$ can be written as

$$\sum_{k=1}^{n} k$$

The above symbol is to be interpreted as follows: First write whatever is to the right of the sigma sign, replacing the letter $k$, wherever it appears, by the number 1. Then, add to this the same expression, replacing the letter $k$ by the number 2. Then, add to this sum the same expression, replacing the letter $k$ by the number 3. We continue adding in this fashion until the letter $k$ has been replaced by the number $n$. The letter $k$ at the bottom of the sigma is called the *index*, the number next to $k$ is called the *lower limit,* and the number above the sigma sign is called the *upper limit.* The lower limit tells us where to start and the upper limit tells us where to stop. Other letters, such as $i$ and $j$, are often used for the index instead of the letter $k$.

**Example 1**

$$\sum_{k=2}^{5} k^2$$

Here our lower limit is 2 and our upper limit is 5.

$$\sum_{k=2}^{5} k^2 = 2^2 + 3^2 + 4^2 + 5^2$$

**Example 2**

$$\sum_{y=3}^{5} (1 + y^3) = (1 + 3^3) + (1 + 4^3) + (1 + 5^3)$$

**Example 3**

$$\sum_{k=1}^{6} x_k = x_1 + x_2 + x_3 + x_4 + x_5 + x_6$$

We shall find the following formulas useful in our discussion of areas.

$$\sum_{k=1}^{n} k = 1 + 2 + 3 + 4 + \cdots + n = \frac{n(n+1)}{2}$$

$$\sum_{k=1}^{n} k^2 = 1^2 + 2^2 + 3^2 + 4^2 + \cdots + n^2 = \frac{n(n+1)(2n+1)}{6}$$

$$\sum_{k=1}^{n} k^3 = 1^3 + 2^3 + 3^3 + 4^3 + \cdots + n^3 = \left[\frac{n(n+1)}{2}\right]^2$$

The method of mathematical induction can be used to prove the above formulas.

**EXERCISES**

In Exercises 1 to 6 write out the sums.

**1.** $\displaystyle\sum_{k=1}^{5} (2 + k)$

**2.** $\displaystyle\sum_{j=1}^{4} x^j$

**3.** $\displaystyle\sum_{k=1}^{7} 2x_k$

**4.** $\displaystyle\sum_{i=1}^{6} \frac{2}{i+1}$

**5.** $\displaystyle\sum_{k=3}^{8} \frac{k+1}{k+2}$

**6.** $\displaystyle\sum_{j=1}^{8} j^3$

In Exercises 7 to 10 write the sums using the Σ notation.

**7.** $f(x_1) + f(x_2) + f(x_3) + f(x_4) + f(x_5)$

**8.** $2x_1 + 2x_2 + 2x_3 + 2x_4$

**9.** $1^2 + 2^2 + 3^2 + 4^2 + 5^2 + 6^2 + 7^2 + 8^2$

**10.** $f(x_1)\,\Delta x + f(x_2)\,\Delta x + f(x_3)\,\Delta x + f(x_4)\,\Delta x$

Evaluate the following sums using the given formulas.

**11.** $\sum_{k=1}^{10} k$

**12.** $\sum_{k=1}^{10} k^2$

**13.** $\sum_{k=1}^{100} k$

**14.** $\sum_{k=1}^{8} k^3$

**15.** $\sum_{k=1}^{1,000} k$

**16.** $\sum_{k=1}^{100} k^3$

## 9-2 AREAS

We now turn to the question of finding the area of a region in a plane. Let us consider the region bounded by the $X$ axis, the curve $y = 6$, and the lines $x = 2$ and $x = 6$ (Fig. 9-1). This region is completely bounded by straight lines and is in fact a rectangle. Thus, the area $A =$ (length) (width).

$$A = 6(4) = 24$$

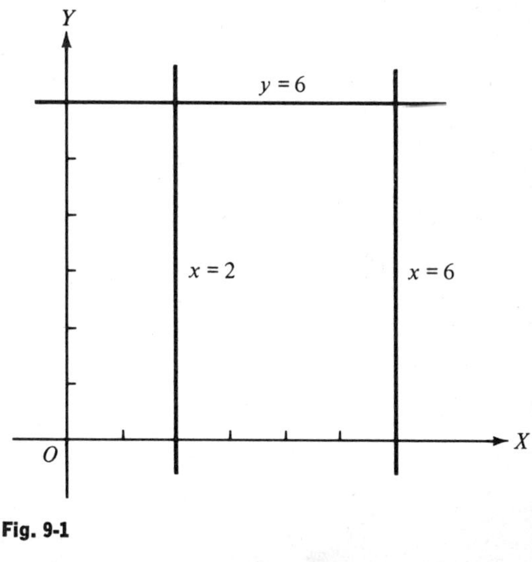


**Fig. 9-1**

If our region is bounded entirely by straight lines, we can always divide the region up into rectangles and triangles and evaluate these areas using the familiar formulas for the areas of triangles and rectangles. On the other hand, if our region is bounded by curves other than straight lines, new procedures will be necessary.

We now consider a region bounded by the curve $y = f(x)$,† the $X$ axis, and the lines $x = a$ and $x = b$ (Fig. 9-2).

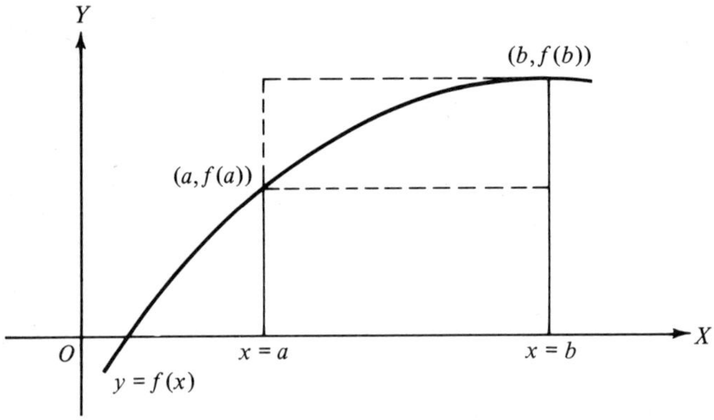


**Fig. 9-2**

Though we cannot yet find the area of the given region, we can try to approximate the area by using triangles, rectangles, or trapezoids. At present we shall restrict ourselves to using rectangles. If we wish we can draw a horizontal line through the point $[a,f(a)]$ to form a rectangle, and we can use this rectangle to approximate the area. A glance at Fig. 9-2 shows that this rectangle greatly underestimates the area of the region. We can also draw a horizontal line through the point $[b,f(b)]$ and form the larger rectangle. This time our rectangle overestimates the area of the region. We could also choose a horizontal line somewhere between $f(a)$ and $f(b)$ and use the corresponding rectangle as an approximation for the area.

We shall attempt to work with the larger rectangle. To improve our approximation we can choose a point $c$ midway between $a$ and $b$ and construct two rectangles rather than one (Fig. 9-3).

We can see from Fig. 9-3 that a large portion of the excess region has now been eliminated by using two rectangles rather than one. There is no reason for us to stop here. We can improve our approximation by

† Note: We shall first consider the case where $f(x) > 0$. The other cases will be considered later in the chapter.

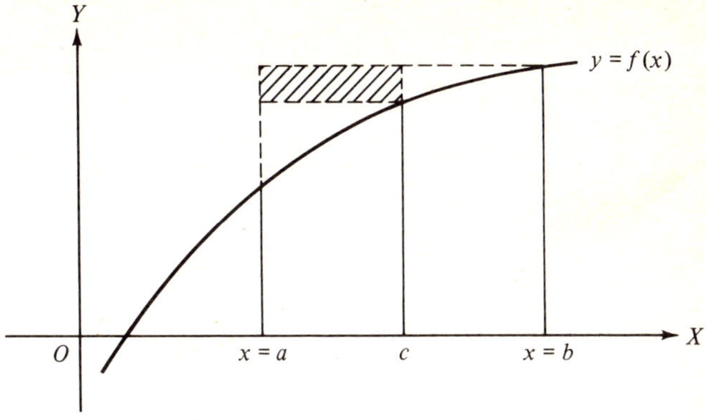


Fig. 9-3

considering four rectangles rather than two. This can be accomplished by dividing each of our intervals on the $X$ axis into halves (Fig. 9-4).

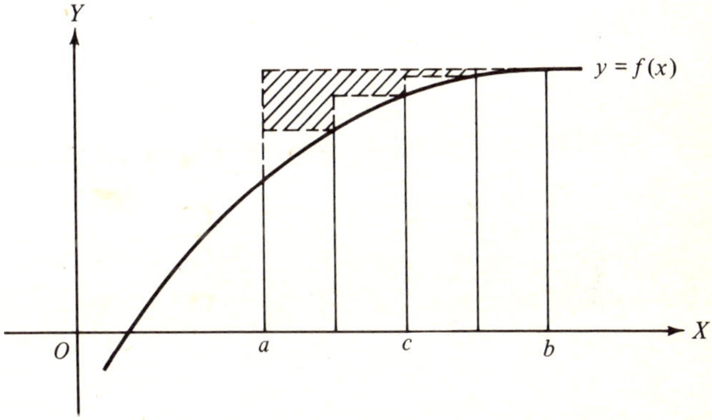


Fig. 9-4

We can continue improving our approximation by increasing the number of rectangles. In general, we can divide the interval from $x = a$ to $x = b$ into $n$ equal subintervals. We can denote the points of subdivision by (Fig. 9-5)

$$\begin{aligned} a &= x_0 \\ &x_1 \\ &x_2 \\ &x_3 \\ &\cdots \\ b &= x_n \end{aligned}$$

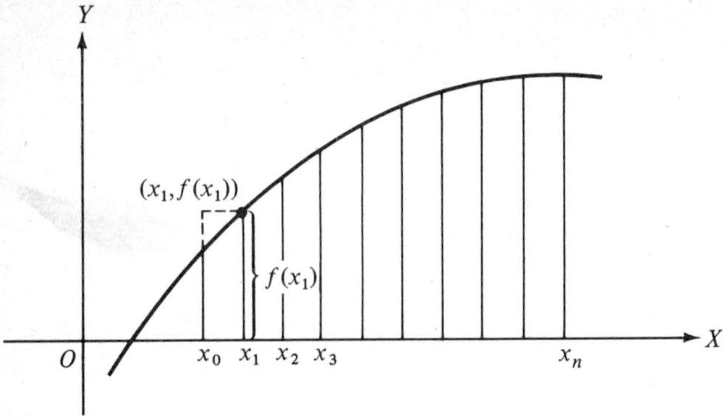


**Fig. 9-5**

Since all our subintervals are of equal length, we can denote the length of each subinterval by $\Delta x$. The area of the first rectangle is then $f(x_1)\,\Delta x$, the area of the second rectangle is $f(x_2)\,\Delta x$, etc. The sum $A_r$ of the areas of the rectangles is

$$\begin{aligned} A_r &= f(x_1)\,\Delta x + f(x_2)\,\Delta x + f(x_3)\,\Delta x + \cdots + f(x_n)\,\Delta x \\ &= \sum_{k=1}^{n} f(x_k)\,\Delta x \end{aligned}$$

The larger the value of $n$, the better our approximation of the actual area. In fact, we can now define the area $A$ as the limit of $A_r$ as $n$ approaches $\infty$.

$$A = \lim_{n \to \infty} A_r$$

$$\boxed{A = \lim_{n \to \infty} \sum_{k=1}^{n} f(x_k)\,\Delta x}$$

**Example 1** Find the area of the region bounded by the curves $y = 2x$, $y = 0$, $x = 0$, and $x = 4$.

*Solution* We shall divide the interval from $x = 0$ to $x = 4$ into $n$ subintervals (Fig. 9-6). To find the length of $\Delta x$ we must take the length of the entire interval and divide by $n$.

$$\Delta x = \frac{b - a}{n} = \frac{4 - 0}{n} = \frac{4}{n}$$

Our points of subdivision are then

$$\begin{array}{ll} x_0 = 0 & f(x) = 2x \\ x_1 = \dfrac{4}{n} & f(x_1) = 2 \cdot \dfrac{4}{n} \\ x_2 = 2 \cdot \dfrac{4}{n} & f(x_2) = 2 \cdot 2 \cdot \dfrac{4}{n} \\ x_3 = 3 \cdot \dfrac{4}{n} & f(x_3) = 2 \cdot 3 \cdot \dfrac{4}{n} \\ \cdots\cdots\cdots & \cdots\cdots\cdots \\ x_n = n \cdot \dfrac{4}{n} = 4 & f(x_n) = 2 \cdot n \cdot \dfrac{4}{n} \end{array}$$

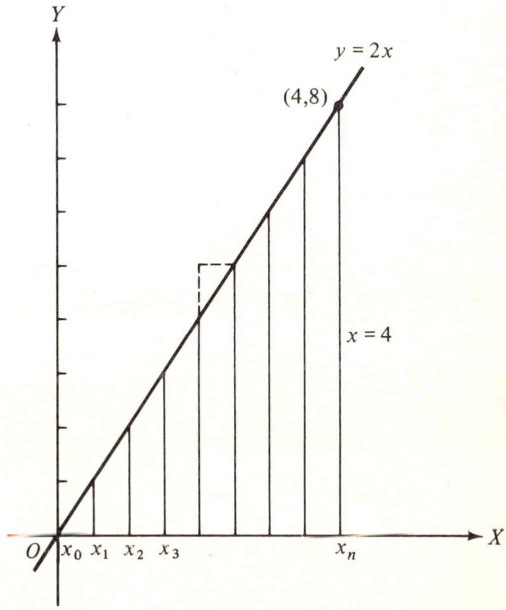


Fig. 9-6

$$\begin{aligned} A_r &= \sum_{k=1}^{n} f(x_k)\,\Delta x \\ &= f(x_1)\,\Delta x + f(x_2)\,\Delta x + f(x_3)\,\Delta x + \cdots + f(x_n)\,\Delta x \\ &= \left(2 \cdot \frac{4}{n}\right)\frac{4}{n} + \left(2 \cdot 2 \cdot \frac{4}{n}\right)\frac{4}{n} + \left(2 \cdot 3 \cdot \frac{4}{n}\right)\frac{4}{n} + \cdots + \left(2 \cdot n \cdot \frac{4}{n}\right)\frac{4}{n} \end{aligned}$$

We notice that each term has a factor of $2 \cdot \frac{4}{n} \cdot \frac{4}{n}$. If we factor out this product, we get

$$A_r = 2 \cdot \frac{4}{n} \cdot \frac{4}{n} (1 + 2 + 3 + 4 + \cdots + n)$$
$$= \frac{32}{n^2} (1 + 2 + 3 + 4 + \cdots + n)$$

Since $1 + 2 + 3 + 4 + \cdots + n = \dfrac{n(n+1)}{2}$,

$$A_r = \frac{32}{n^2} \frac{n(n+1)}{2}$$
$$= 16 \frac{n+1}{n}$$
$$= 16 \left(1 + \frac{1}{n}\right)$$

Thus, if we choose 10 rectangles, $n = 10$ and the area $A_r = 16(1 + \frac{1}{10})$. If $n = 100$, then $A_r = 16(1 + \frac{1}{100})$. To find the actual area $A$, we take the limit as $n$ approaches $\infty$.

$$A = \lim_{n \to \infty} \sum_{k=1}^{n} f(x_k) \Delta x$$
$$= \lim_{n \to \infty} \left[16 \left(1 + \frac{1}{n}\right)\right] = 16$$

As a check we note that the region under consideration is a triangle, and the area of a triangle is $\frac{1}{2}bh$.

$$b = 4$$
$$h = 8$$
$$A = \tfrac{1}{2}(4)(8) = 16$$

**Example 2** Find the area of the region bounded by the curves $y = x^2$, $y = 0$, $x = 0$, and $x = 2$ (Fig. 9-7).

*Solution* As before, we shall divide the interval from $x = 0$ to $x = 2$ into $n$ subintervals. Then,

$$\Delta x = \frac{b - a}{n} = \frac{2 - 0}{n} = \frac{2}{n}$$

Our points of subdivision are

$$x_0 = 0 \qquad f(x_0) = 0$$
$$x_1 = \frac{2}{n} \qquad f(x_1) = \left(\frac{2}{n}\right)^2$$
$$x_2 = 2 \cdot \frac{2}{n} \qquad f(x_2) = \left(2 \cdot \frac{2}{n}\right)^2$$

$$x_3 = 3 \cdot \frac{2}{n} \qquad f(x_3) = \left(3 \cdot \frac{2}{n}\right)^2$$

$$\cdots\cdots\cdots \qquad \cdots\cdots\cdots$$

$$x_n = n \cdot \frac{2}{n} = 2 \qquad f(x_n) = \left(n \cdot \frac{2}{n}\right)^2$$

$$\begin{aligned} A_r &= \sum_{k=1}^{n} f(x_k)\,\Delta x \\ &= f(x_1)\,\Delta x + f(x_2)\,\Delta x + f(x_3)\,\Delta x + \cdots + f(x_n)\,\Delta x \\ &= \left(\frac{2}{n}\right)^2 \cdot \frac{2}{n} + \left(2 \cdot \frac{2}{n}\right)^2 \cdot \frac{2}{n} + \left(3 \cdot \frac{2}{n}\right)^2 \cdot \frac{2}{n} + \cdots + \left(n \cdot \frac{2}{n}\right)^2 \cdot \frac{2}{n} \end{aligned}$$

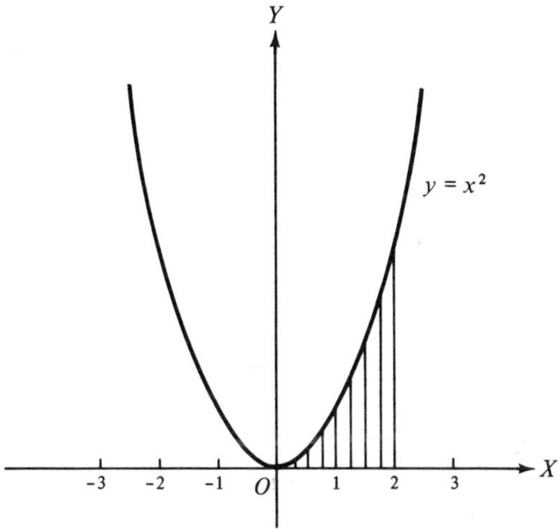


Fig. 9-7

Each term has a common factor of $\left(\frac{2}{n}\right)^2 \frac{2}{n}$. Therefore,

$$\begin{aligned} A_r &= \left(\frac{2}{n}\right)^2 \frac{2}{n} (1^2 + 2^2 + 3^2 + \cdots + n^2) \\ &= \frac{8}{n^3} (1^2 + 2^2 + 3^2 + \cdots + n^2) \end{aligned}$$

We can now use the formula

$$1^2 + 2^2 + 3^2 + \cdots + n^2 = \frac{n(n+1)(2n+1)}{6}$$

$$A_r = \frac{8}{n^3} \frac{n(n+1)(2n+1)}{6}$$

$$= \frac{8}{6}\frac{(n+1)(2n+1)}{n^2}$$
$$= \frac{4}{3}\frac{n+1}{n}\frac{2n+1}{n}$$
$$= \frac{4}{3}\left(1+\frac{1}{n}\right)\left(2+\frac{1}{n}\right)$$
$$A = \lim_{n\to\infty}\sum_{k=1}^{n} f(x_k)\,\Delta x$$
$$= \lim_{n\to\infty}\left[\frac{4}{3}\left(1+\frac{1}{n}\right)\left(2+\frac{1}{n}\right)\right]$$
$$= \tfrac{4}{3}(1)(2) = \tfrac{8}{3}$$

### EXERCISES

**1.** Use the limiting process to find the area of the region bounded by the curves $y = x$, $x = 0$, $x = 6$, and the $X$ axis. Use the formula for the area of a triangle to check your result.

**2.** Find the area of the region bounded by the curves $y = x^3$, $x = 0$, $x = 1$, and the $X$ axis.

**3.** Find the area of the region bounded by the curves $y = 4x^2$, $x = 0$, $x = 1$, and the $X$ axis.

## 9-3 THE DEFINITE INTEGRAL

We shall now consider one of the most fundamental concepts of the calculus, the *definite integral.* Let $y = f(x)$ be a function defined in the interval $a \le x \le b$. We shall take the interval from $x = a$ to $x = b$ and subdivide it into $n$ subintervals. This can be done by choosing $n - 1$ points in the interval between $x = a$ and $x = b$. To simplify the notation, we shall label the points of subdivision as $x_1 < x_2 < x_3 < \cdots < x_{n-1}$ and we will let $a = x_0$ and $b = x_n$ (Fig. 9-8).

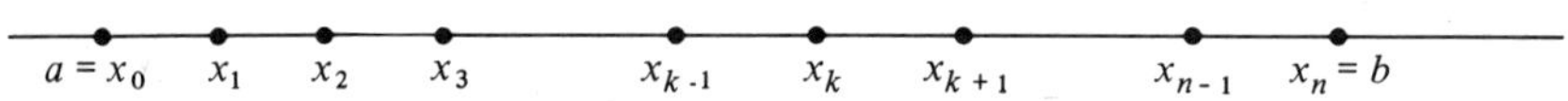


**Fig. 9-8**

We shall denote the length of the first subinterval as $\Delta x_1$, the length of the second subinterval as $\Delta x_2$, etc. Thus,

$$\Delta x_1 = x_1 - x_0$$
$$\Delta x_2 = x_2 - x_1$$
$$\cdots\cdots\cdots$$
$$\Delta x_n = x_n - x_{n-1}$$

We shall now choose a point in the first subinterval and denote its value as $\bar{x}_1$. This point can be the same as $x_0$ or $x_1$ or it can be any point between $x_0$ and $x_1$. We shall then choose a point in the second subinterval and denote it by $\bar{x}_2$.

We shall continue this process for all the $n$ subintervals until we have chosen $n$ points $\bar{x}_1, \bar{x}_2, \ldots, \bar{x}_n$. Since $y = f(x)$, corresponding to $\bar{x}_1, \bar{x}_2, \ldots, \bar{x}_n$, we will have $f(\bar{x}_1), f(\bar{x}_2), \ldots, f(\bar{x}_n)$. In each subinterval we shall form the product of the value of the function and the length of the interval. Thus, for the first interval we will have the product $(\bar{x}_1)\,\Delta x_1$, for the second interval $f(\bar{x}_2)\,\Delta x_2$, and in general for the $k$th interval $f(\bar{x}_k)\,\Delta x_k$.

Now we will form the sum of these products:

$$f(\bar{x}_1)\,\Delta x_1 + f(\bar{x}_2)\,\Delta x_2 + f(\bar{x}_3)\,\Delta x_3 + \cdots + f(\bar{x}_n)\,\Delta x_n = \sum_{k=1}^{n} f(\bar{x}_k)\,\Delta x_k$$

For a given function, this sum will depend on the number of subintervals, the choice of subdivision points, and the choice of points $\bar{x}_k$ within each subinterval.

Now, increase the number of subintervals in such a way so as to have the length of each $\Delta x_k$ approach 0 as $n \to \infty$. If

$$\lim_{n\to\infty} \sum_{k=1}^{n} f(\bar{x}_k)\,\Delta x_k$$

exists and is always the same no matter how the points of subdivision are chosen and no matter how the points $\bar{x}_k$ are chosen within the subintervals, then this limit is called the definite integral of $f(x)$ from $x = a$ to $x = b$ and is denoted by the symbol

$$\int_a^b f(x)\,dx$$

$$\boxed{\lim_{n\to\infty} \sum_{k=1}^{n} f(\bar{x}_k)\,\Delta x_k = \int_a^b f(x)\,dx}$$

In choosing the points of subdivision, it is often convenient to choose equal subintervals. If this is done, then we can use the symbol $\Delta x$ to represent the length of each subinterval and our sum then becomes

$$\sum_{k=1}^{n} f(\bar{x}_k)\,\Delta x$$

At this point the reader may be wondering whether the

$$\lim_{n\to\infty} \sum_{k=1}^{n} f(\bar{x}_k)\,\Delta x$$

ever does exist, and, if it does, under what circumstances will the limit be the same no matter what points of subdivision are chosen. The answer to this question is given by the following theorem, which will be stated here without proof.

**Theorem** *If $f(x)$ is a continuous function over the interval $a \leq x \leq b$, then $\int_a^b f(x)\,dx$ exists.*

Thus, in dealing with continuous functions, we are assured that the limit will exist and thus the definite integral will exist.

Another question which may be bothering the reader at this point is "When will one ever have an occasion to go through such a detailed process?" If one looks closely at our definition of the definite integral, it can be seen that this is exactly the process that was used in finding areas. In fact, the area $A$ of a region bounded by the curve $y = f(x) \geq 0$, the $X$ axis, and the lines $x = a$ and $x = b$ was given by the equation

$$A = \lim_{n \to \infty} \sum_{k=1}^{n} f(\bar{x}_k)\,\Delta x$$

Hence, we can say that

$$A = \int_a^b f(x)\,dx$$

We shall find that areas are not the only interpretations and applications of the definite integral. Whenever we are dealing with a continuous function $f(x)$ defined over the interval $a \leq x \leq b$ and we have occasion to form the sum $\sum_{k=1}^{n} f(\bar{x}_k)\,\Delta x_k$ and to find $\lim_{n \to \infty} \sum_{k=1}^{n} f(\bar{x}_k)\,\Delta x_k$, we are dealing with the definite integral $\int_a^b f(x)\,dx$. Our definition of the definite integral $\int_a^b f(x)\,dx$ applies only to the case where $a < b$. To take care of the other cases we make the following definitions.

$$\int_a^a f(x)\,dx = 0$$

$$\int_a^b f(x)\,dx = -\int_b^a f(x)\,dx \qquad \text{if } a > b$$

## 9-4 THE FUNDAMENTAL THEOREM OF INTEGRAL CALCULUS

No doubt the reader has noticed a strong similarity between the symbol used for the antiderivative [that is, $\int f(x)\,dx$] and the symbol used for the definite integral $\left[\text{that is, } \int_a^b f(x)\,dx\right]$. The justification for using the same symbol for two very different concepts is provided for us by a theorem called the *fundamental theorem of integral calculus.* This theorem provides us with a method for evaluating the definite integral and, at the same time, establishes a link between the concept of the derivative and the concept of the definite integral. We shall state the theorem here without proof. (A proof is given in Sec. 9-10.)

**Theorem** *If $f(x)$ is continuous in the interval $a \leq x \leq b$ and if $F(x)$ is an antiderivative of $f(x)$ [that is, $F'(x) = f(x)$], then*

$$\int_a^b f(x)\,dx = F(b) - F(a)$$

**Example 1** Evaluate $\int_1^3 x^2\,dx$.

*Solution* The antiderivative of $x^2$ is $\frac{x^3}{3} + C$. Thus

$$\begin{aligned}\int_1^3 x^2\,dx &= \frac{x^3}{3} + C\Big|_1^3 \\ &= (9 + C) - (\tfrac{1}{3} + C) \\ &= \tfrac{26}{3}\end{aligned}$$

We make the following observations in examining the solution:

1. The first step in the solution is to find an antiderivative. When the antiderivative is determined, a vertical line is written to the right of the antiderivative indicating the lower and upper limits.
2. In evaluating the antiderivative $F(x)$, we find that both $F(b)$ and $F(a)$ contain the same constant $C$ and since we are interested in $F(b) - F(a)$ the constant will always cancel out.

**Example 2** Evaluate $\int_2^4 (x^2 + 2x)\,dx$.

*Solution*

$$\begin{aligned}\int_2^4 (x^2 + 2x)\,dx &= \frac{x^3}{3} + x^2\Big|_2^4 \qquad \text{(\textit{Note:} we have omitted the constant } C) \\ &= (\tfrac{64}{3} + 16) - (\tfrac{8}{3} + 4) \\ &= \tfrac{92}{3}\end{aligned}$$

**Example 3** Evaluate $\int_0^2 \sqrt{4x + 1}\,dx$.

*Solution* We must first find the antiderivative. We shall apply the general power rule

$$\int u^n\,du = \frac{u^{n+1}}{n+1} + C$$

where $u = 4x + 1$
$n = \frac{1}{2}$
$du = 4dx$

$$\int_0^2 \sqrt{4x+1}\,dx = \frac{1}{4}\int_0^2 \sqrt{4x+1}\,4dx$$
$$= \frac{1}{4}\frac{(4x+1)^{\frac{3}{2}}}{\frac{3}{2}}\Bigg|_0^2$$
$$= \frac{1}{6}(4x+1)^{\frac{3}{2}}\Bigg|_0^2$$
$$= \frac{1}{6}(27-1) = \frac{26}{6} = \frac{13}{3}$$

**Example 4** Evaluate $\int_0^\pi \sin x\,dx$.

*Solution*

$$\int_0^\pi \sin x\,dx = -\cos x\Bigg|_0^\pi$$
$$= -(-1-1)$$
$$= 2$$

**EXERCISES**

Evaluate the following integrals.

1. $\int_1^4 4x^3\,dx$
2. $\int_1^3 (x^2 + 2x - 5)\,dx$
3. $\int_2^4 (x^2 + 2)\,dx$
4. $\int_1^3 (x^3 - 7)\,dx$
5. $\int_0^2 x\sqrt{2x^2+1}\,dx$
6. $\int_0^5 \frac{dx}{\sqrt{x+4}}$
7. $\int_2^7 \sqrt{x+2}\,dx$
8. $\int_1^3 \frac{dx}{x^2}$
9. $\int_2^3 (x-1)^2\,dx$
10. $\int_1^4 \frac{x+2}{\sqrt{x}}\,dx$
11. $\int_0^{\pi/2} \cos x\,dx$
12. $\int_{\pi/4}^{\pi} \sin x\,dx$
13. $\int_0^{\pi/3} \sin^2 x \cos x\,dx$
14. $\int_{\pi/6}^{\pi/3} \frac{6\cos x}{\sin^2 x}\,dx$
15. $\int_0^{\pi/2} \cos x \sin x\,dx$
16. $\int_{\pi/2}^{\pi} \sqrt{\sin x}\cos x\,dx$

## 9-5 FINDING AREAS USING THE DEFINITE INTEGRAL

It has already been pointed out that the area $A$ under the curve $y = f(x) \geq 0$, above the $X$ axis between $x = a$ and $x = b$, can be represented by $\int_a^b f(x)\,dx$. We now illustrate the use of integration to find areas.

**Example 1** Find the area of the region bounded by the curves $y = x^2$, $y = 0$, $x = 0$, and $x = 2$.

*Solution* This is the same problem that was evaluated in Sec. 2. This time we will find the area making use of the definite integral.

$$A = \int_0^2 x^2\,dx = \frac{x^3}{3}\bigg|_0^2 = \frac{8}{3} - 0 = \frac{8}{3}$$

This result checks with our previous result, but the labor involved using this method is considerably less.

**Example 2** Find the area under the curve $y = \cos x$, above the $X$ axis, between $x = 0$ and $x = \frac{\pi}{2}$ (Fig. 9-9).

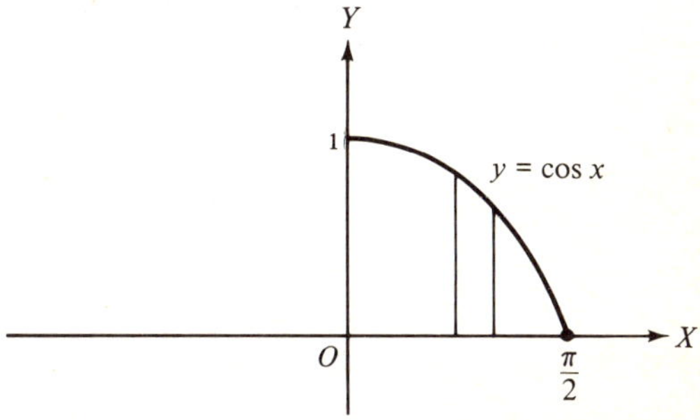


**Fig. 9-9**

*Solution*

$$A = \int_0^{\pi/2} \cos x\,dx = \sin x\bigg|_0^{\pi/2} = 1 - 0 = 1$$

**Example 3** Find the area between the curve $y = \sin x$ and the $X$ axis from $x = \frac{\pi}{2}$ to $x = \frac{3\pi}{2}$ (Fig. 9-10).

*Solution* We shall follow the procedure used in the previous examples. As before, we shall use the formula

$$\begin{aligned} A &= \int_a^b f(x)\,dx \\ &= \int_{\pi/2}^{3\pi/2} \sin x\,dx \\ &= -\cos x\bigg|_{\pi/2}^{3\pi/2} \\ &= -(0 - 0) \\ &= 0 \end{aligned}$$

Obviously something is wrong. A glance at Fig. 9-10 shows that there is, in fact, an area between $x = \frac{\pi}{2}$ and $x = \frac{3\pi}{2}$. A further glance at this figure shows

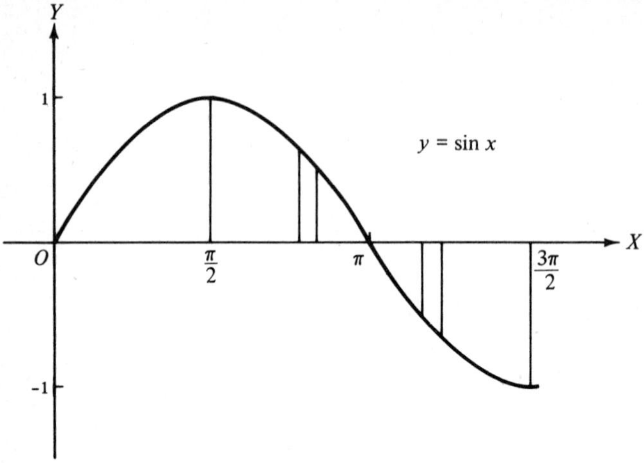


**Fig. 9-10**

us that part of the region is above the $X$ axis and part of the region is below the $X$ axis. It is this fact that is causing the difficulty. The definition of the definite integral tells us that

$$A = \int_a^b f(x)\,dx = \lim_{n\to\infty} \sum_{k=1}^{n} f(\bar{x}_k)\,\Delta x$$

A typical element of area is, then,

$$f(\bar{x}_k)\,\Delta x$$

which gives us the area of a typical rectangle. However, when the curve is below the $X$ axis, then $f(x)$ is negative and $f(\bar{x}_k)\,\Delta x$ will be negative. If the curve is completely below the $X$ axis for values of $x$ between $x = a$ and $x = b$. then the area of the region can be given as

$$A = -\int_a^b f(x)\,dx$$

In our example part of the region is above the $X$ axis and part is below the $X$ axis. Hence, we must consider the two regions separately.

$$\begin{aligned} A &= \int_{\pi/2}^{\pi} \sin x\,dx - \int_{\pi}^{3\pi/2} \sin x\,dx \\ &= -\cos x\Big|_{\pi/2}^{\pi} + \cos x\Big|_{\pi}^{3\pi/2} \\ &= -(-1+0) + [0-(-1)] \\ &= 1+1 \\ &= 2 \end{aligned}$$

**Example 4** Find the area between the curve $y = x^3$ and the $X$ axis between $x = -1$ and $x = 2$ (Fig. 9-11).

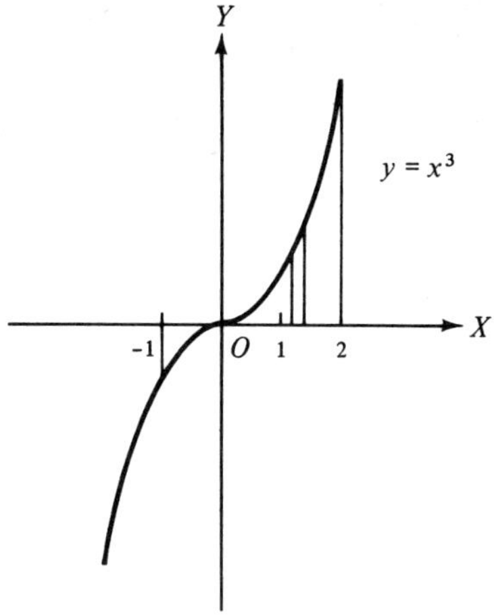


**Fig. 9-11**

*Solution* Figure 9-11 shows us that here again part of the region is below the $X$ axis and part is above. Hence,

$$\begin{aligned} A &= -\int_{-1}^{0} x^3\,dx + \int_0^2 x^3\,dx \\ &= -\frac{x^4}{4}\Big|_{-1}^{0} + \frac{x^4}{4}\Big|_0^2 \\ &= -(0 - \tfrac{1}{4}) + (4 - 0) \\ &= \tfrac{1}{4} + 4 \\ &= \tfrac{17}{4} \end{aligned}$$

**Example 5** Find the area between the curve $x = 4 - y^2$ and the $X$ axis (Fig. 9-12).

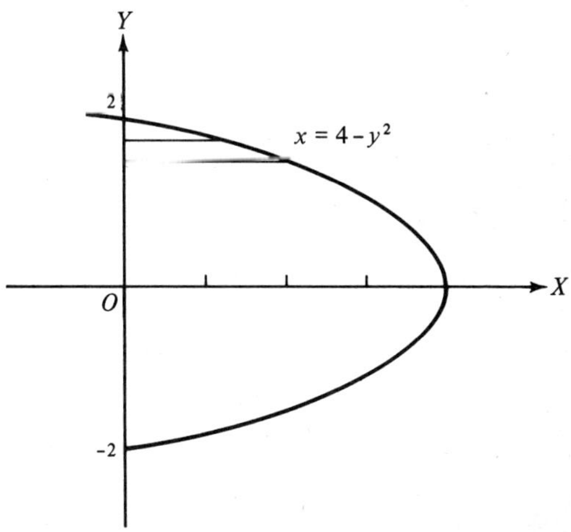


**Fig. 9-12**

*Solution* In this example our region is bounded by the $Y$ axis rather than the $X$ axis. A typical element will then be $f(\bar{y}_k)\,\Delta y$, and the area can be given as

$$A = \int_c^d f(y)\,dy$$

The $y$ values go from $y = -2$ to $y = 2$, and hence

$$A = \int_{-2}^{2} (4 - y^2)\,dy$$

$$\begin{aligned} A &= 4y - \frac{y^3}{3}\Big|_{-2}^{2} \\ &= (8 - \tfrac{8}{3}) - (-8 + \tfrac{8}{3}) \\ &= \tfrac{16}{3} + \tfrac{16}{3} \\ &= \tfrac{32}{3} \end{aligned}$$

An examination of Fig. 9-12 shows that the curve is symmetric with respect to the $X$ axis. Thus the area above the $X$ axis is identical to the area below the $X$ axis, and one can evaluate $A$ in the following way:

$$\begin{aligned} A &= 2\int_0^2 (4 - y^2)\,dy \\ &= 2\left(4y \frac{-y^3}{3}\right)\Big|_0^2 \\ &= 2[(8 - \tfrac{8}{3}) - 0] \\ &= 2(\tfrac{16}{3}) = \tfrac{32}{3} \end{aligned}$$

## EXERCISES

In each of the following exercises, sketch the curve and draw a typical rectangular element of the area to be found.

**1.** Find the area between the curve $y = 1 - x^2$ and the $X$ axis.

**2.** Find the area under the curve $y = x^2 - 4x + 6$, above the $X$ axis between $x = 0$ and $x = 3$.

**3. Find the area bounded by the curve $y = \cos x$, the $X$ axis, $x = 0$, and $x = \pi$.**

**4.** Find the area bounded by the curve $y = 4x - x^2$ and the $X$ axis.

In Exercises 5 to 8, find the area bounded by the given curve and the $Y$ axis.

**5.** $x = 3y - y^2$

**6.** $x = y^2 - y - 2$

**7.** $y^2 = x + 4$

**8.** $y^2 - 5y - x + 6 = 0$

**9.** Find the area between the curve $y = \sin x$ and the $X$ axis between $x = 0$ and $x = 2\pi$.

**10.** Find the area in the first quadrant bounded by the curve $y = \cos x$ and the coordinate axes.

**11.** Find the area of the region bounded by the curve $y = x^2 - 4x + 4$ and the coordinate axes.

**12.** Find the area in the first quadrant bounded by the curve $y = x^2$, the $Y$ axis, and the line $y = 2$.

**13.** Find the area bounded by the curve $x = \sqrt{y + 4}$ and the coordinate axes.

**14.** Find the area between the curve $y = x^2 - 4$ and the $X$ axis.

### 9-6 AREA BETWEEN TWO CURVES

We will next consider the problem of finding the area between two curves. Let $y = f(x) \geq 0$ and $y = g(x) \geq 0$ be two curves intersecting at $x = a$ and $x = b$. We wish to find the area between $y = f(x)$ and $y = g(x)$ between $x = a$ and $x = b$. We shall consider the case where the curve $y = f(x)$ is above the curve $y = g(x)$ (Fig. 9-13).

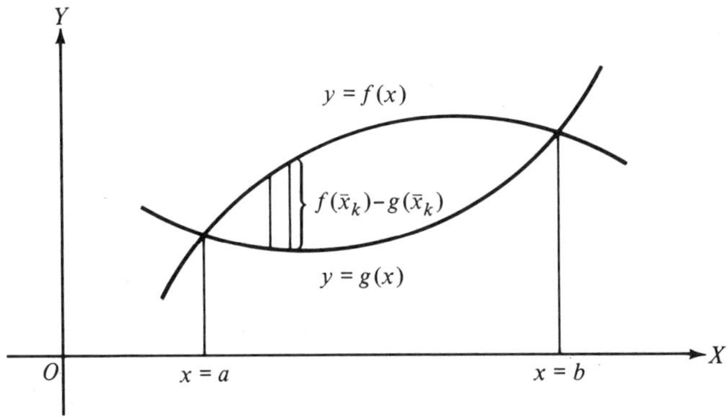


**Fig. 9-13**

The area below the curve $y = f(x)$, above the $X$ axis between $x = a$ and $x = b$, is given by $\int_a^b f(x)\,dx$.

Similarly, the area below the curve $y = g(x)$, above the $X$ axis between $x = a$ and $x = b$, is given by $\int_a^b g(x)\,dx$.

Hence, the area $A$ between the two curves is given by

$$\begin{aligned} A &= \int_a^b f(x)\,dx - \int_a^b g(x)\,dx \\ &= \int_a^b [f(x) - g(x)]\,dx \end{aligned}$$

In Fig. 9-13 we can see that a typical rectangular element has an area of $[f(\bar{x}_k) - g(\bar{x}_k)]\,\Delta x_k$, and that

$$A = \lim_{n\to\infty} \sum_{k=1}^{n} [f(\bar{x}_k) - g(\bar{x}_k)]\,\Delta x = \int_a^b [f(x) - g(x)]\,dx$$

It will be left to the reader to show that if the curve $y = f(x)$ is above the curve $y = g(x)$, then the area between the two curves is given by the formula $A = \int_a^b [f(x) - g(x)]\,dx$ even if we do not have the conditions $f(x) \geq 0$ and $g(x) \geq 0$.

**Example 1** Find the area between the parabola $y = x^2 - 4x + 4$ and the line $y = x$ (Fig. 9-14).

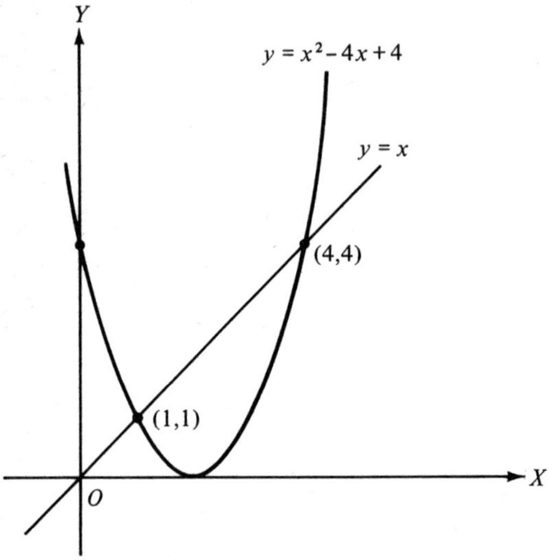


**Fig. 9-14**

*Solution* The first thing we must do is to find the points of intersection of the two curves $y = x$ and $y = x^2 - 4x + 4$.

$$\begin{aligned} x^2 - 4x + 4 &= x \\ x^2 - 5x + 4 &= 0 \\ (x - 1)(x - 4) &= 0 \\ x &= 1 \qquad x = 4 \end{aligned}$$

From Fig. 9-14 we can see that the line $y = x$ is above the parabola between $x = 1$ and $x = 4$. Thus, the area $A$ between the two curves is

$$\begin{aligned} A &= \int_1^4 [x - (x^2 - 4x + 4)]\, dx \\ &= \int_1^4 (-x^2 + 5x - 4)\, dx \\ &= -\frac{x^3}{3} + \frac{5x^2}{2} - 4x \Big|_1^4 \\ &= (-\tfrac{64}{3} + 40 - 16) - (-\tfrac{1}{3} + \tfrac{5}{2} - 4) \\ &= \tfrac{9}{2} \end{aligned}$$

**Example 2** Find the area of the region in the first quadrant bounded by the curves $y = \sin x$, $y = \cos x$, and the $Y$ axis (Fig. 9-15).

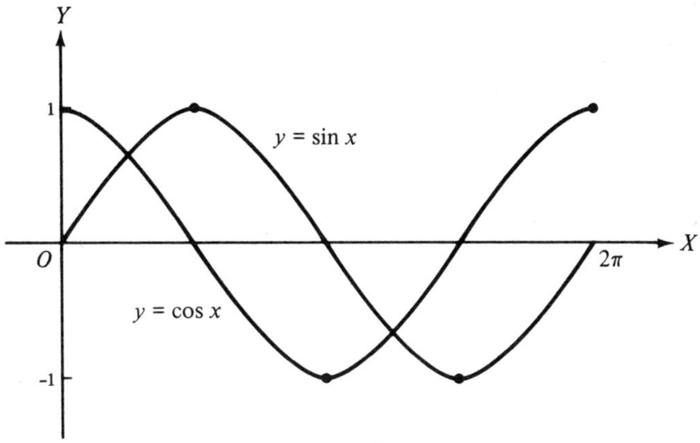


Fig. 9-15

*Solution* We must find the point or points of intersection of the curves $y = \sin x$ and $y = \cos x$.

$$\sin x = \cos x$$
$$\frac{\sin x}{\cos x} = 1$$
$$\tan x = 1$$
$$x = \frac{\pi}{4} \qquad x = \frac{5\pi}{4}$$

Since we are interested in the region bounded by the $Y$ axis, we shall only need the point of intersection for which $x = \frac{\pi}{4}$ (see Fig. 9-15). Hence,

$$\begin{aligned} A &= \int_0^{\pi/4} (\cos x - \sin x)\,dx \\ &= (\sin x + \cos x)\Big|_0^{\pi/4} \\ &= \left(\frac{\sqrt{2}}{2} + \frac{\sqrt{2}}{2}\right) - (0 + 1) \\ &= \sqrt{2} - 1 \end{aligned}$$

**Example 3** Find the area between the curves $y = x^3$ and $y = x$ (Fig. 9-16).

*Solution* To find the points of intersection, we set

$$\begin{aligned} x^3 &= x \\ x^3 - x &= 0 \\ x(x^2 - 1) &= 0 \\ x &= 0 \qquad x = \pm 1 \end{aligned}$$

We shall follow our standard procedure and use the formula

$$A = \int_a^b [f(x) - g(x)]\,dx$$

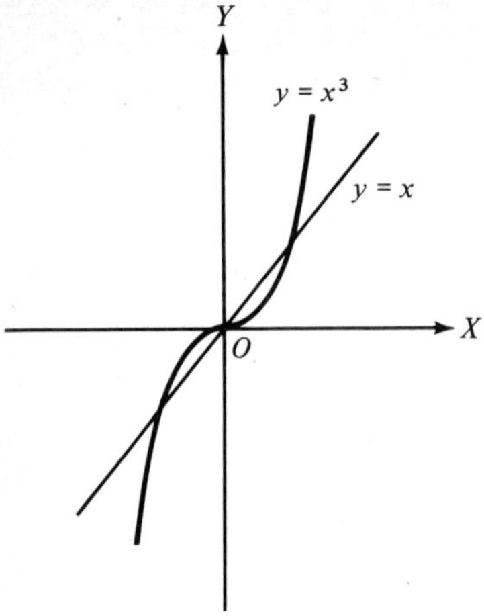


Fig. 9-16

The smallest $x$ value is $x = -1$ and the largest is $x = 1$. Using the formula,

$$\begin{aligned} A &= \int_{-1}^{1} (x^3 - x)\,dx \\ &= \frac{x^4}{4} - \frac{x^2}{2}\Big|_{-1}^{1} \\ &= (\tfrac{1}{4} - \tfrac{1}{2}) - (\tfrac{1}{4} - \tfrac{1}{2}) \\ &= -\tfrac{1}{4} + \tfrac{1}{4} \\ &= 0 \end{aligned}$$

We obviously have made some error along the way. If we had reversed the $f(x)$ and the $g(x)$, only the sign of the result would be reversed. An examination of Fig. 9-16 shows us that the two curves crisscross at $x = 0$. Between $x = -1$ and $x = 0$ the curve $y = x^3$ is above the curve $y = x$. Between $x = 0$ and $x = 1$ the curve $y = x$ is above the curve $y = x^3$. Thus, to find the area $A$ between the two curves we set

$$\begin{aligned} A &= \int_{-1}^{0} (x^3 - x)\,dx + \int_{0}^{1} (x - x^3)\,dx \\ &= \frac{x^4}{4} - \frac{x^2}{2}\Big|_{-1}^{0} + \frac{x^2}{2} - \frac{x^4}{4}\Big|_{0}^{1} \\ &= [0 - (\tfrac{1}{4} - \tfrac{1}{2})] + [(\tfrac{1}{2} - \tfrac{1}{4}) - 0] \\ &= \tfrac{1}{4} + \tfrac{1}{4} \\ &= \tfrac{1}{2} \end{aligned}$$

We can see from this example that when the two curves crosscross, it is necessary to separate the intervals where the curves cross each other.

**Example 4** Find the area between the parabola $y = x^2 - 6x + 5$ and the line $y = x - 5$ (Fig. 9-17).

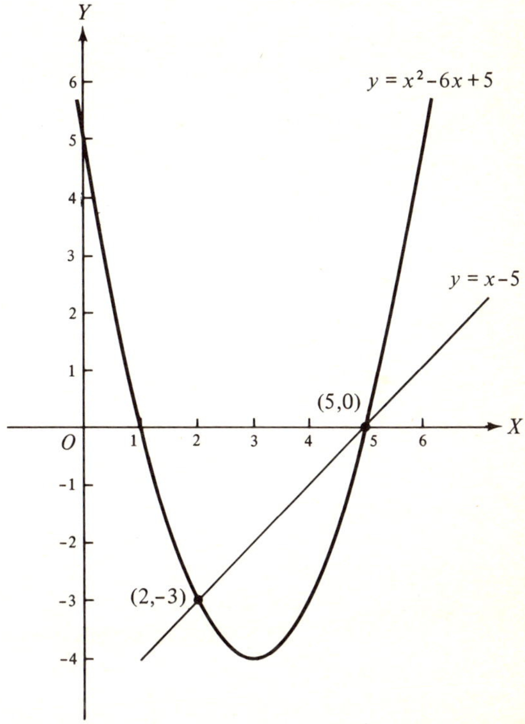


**Fig. 9-17**

*Solution* We first find the points of intersection

$$\begin{aligned} x^2 - 6x + 5 &= x - 5 \\ x^2 - 7x + 10 &= 0 \\ (x - 2)(x - 5) &= 0 \\ x &= 2 \qquad x = 5 \end{aligned}$$

Hence,

$$\begin{aligned} A &= \int_2^5 [(x - 5) - (x^2 - 6x + 5)]\,dx \\ &= \int_2^5 (-x^2 + 7x - 10)\,dx \\ &= -\frac{x^3}{3} + \frac{7x^2}{2} - 10x \Big|_2^5 \\ &= (-\tfrac{125}{3} + \tfrac{175}{2} - 50) - (-\tfrac{8}{3} + 14 - 20) \\ &= \tfrac{27}{6} \\ &= 4\tfrac{1}{2} \end{aligned}$$

We point out that in finding the area between the two curves, we have used the formula $A = \int_a^b [f(x) - g(x)]\,dx$, where $f(x)$ is the curve above and $g(x)$ is the curve below. The fact that our region (Fig. 9-17) lies entirely below the $X$ axis in no way alters the formula.

**EXERCISES**

Find the area bounded by each of the given pairs of curves.

**1.** $y = x^2$, $y = x$

**2.** $y = 4 - x^2$, $y = x + 2$

**3.** $y = \frac{1}{3}x^3$, $y = 3x$

**4.** $y = x^2 - 4$, $y = 4 - x^2$

**5.** $y = x^2$, $y = 4$

**6.** $2y - y^2 = x$, $y = x$

**7.** $x = y^2 - 4y$, $x = -3$

**8.** $y = x(x + 1)(x + 2)$, $y = 0$

**9.** $y = x^3$, $y = -x^2$

**10.** $y = x^2 - 4x$, $y = 2x - 8$

**11.** Find the area of the region bounded by the parabola $y = 2x^2$ and the lines $y = x$ and $y = 4x$.

**12.** Find the area of the region bounded by the curves $y = x - 2x^2$ and $y = -1$.

## ▶ 9-7 VOLUMES OF SOLIDS OF REVOLUTION

We have already considered the use of the definite integral in solving area problems. There are, however, a great many other problems for which the concept of the definite integral proves to be very useful.

We now consider the problem of finding volumes of solids of revolution. By definition, a solid of revolution is a solid which is generated by revolving a plane region about an axis in the plane.

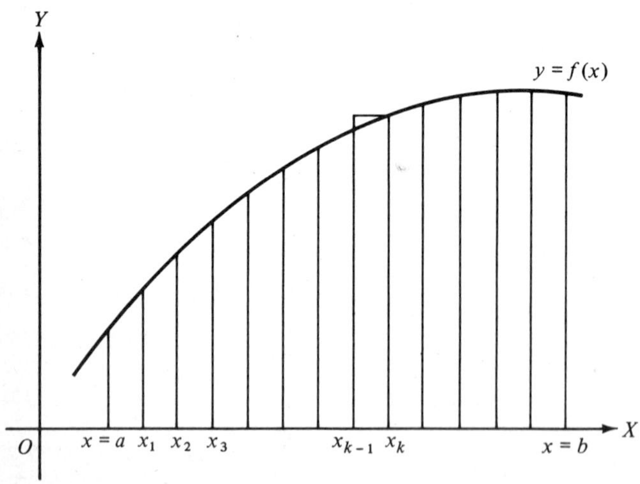


**Fig. 9-18**

Let us consider the region bounded by the curve $y = f(x)$, the $X$ axis, and the lines $x = a$ and $x = b$. We shall generate a solid of revolution by revolving this region about the $X$ axis, and we shall attempt to find the volume of this solid of revolution (Fig. 9-18).

We shall divide the interval from $x = a$ to $x = b$ into $n$ subintervals by choosing $n - 1$ intermediate points. We can denote the points of subdivision by

$$\begin{aligned} a &= x_0 \\ &\phantom{=} x_1 \\ &\phantom{=} x_2 \\ &\phantom{=} x_3 \\ &\cdots \\ b &= x_n \end{aligned}$$

The $k$th subinterval can be denoted by $\Delta x_k$. In the $k$th subinterval we shall choose an $x$ value $\bar{x}_k$ and form the rectangle with height $f(\bar{x}_k)$ and width $\Delta x_k$. If we revolve this rectangle about the $X$ axis, this will give us a cylindrical disk with thickness $\Delta x_k$ and radius $f(\bar{x}_k)$.

We shall repeat this procedure for each of the $n$ subintervals. This will give rise to $n$ cylindrical disks, and we can use the volume of the cylindrical disks as an approximation of the volume of the solid of revolution. This procedure will seem plausible if we think of the cylindrical disks as slices of the solid of revolution.

We recall that the volume of a cylinder is given by the formula $V = \pi r^2 h$. For the $k$th cylindrical disk, the height (thickness) is $\Delta x_k$, the radius is $f(\bar{x}_k)$, and the volume is $\pi[f(\bar{x}_k)]^2 \, \Delta x_k$ (Fig. 9-19). The vol-

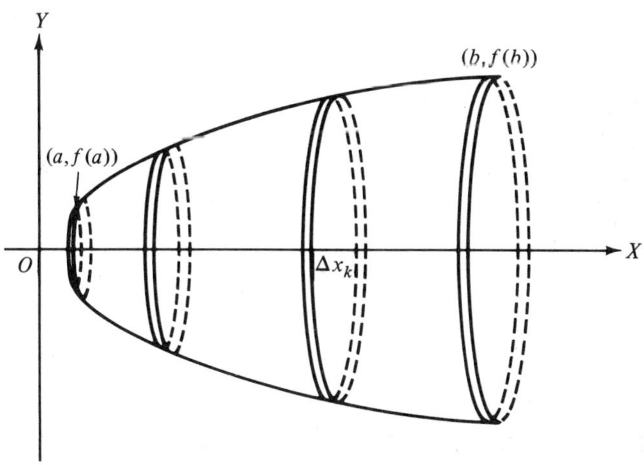


**Fig. 9-19**

ume $V_d$ of the $n$ cylindrical disks is

$$V_d = \sum_{k=1}^{n} \pi[f(\bar{x}_k)]^2 \, \Delta x_k$$

As in our discussion of areas we can improve our approximation by increasing the number of subintervals. To get the actual volume, we must consider the limit as $n \to \infty$ and all $\Delta x_k \to 0$. Thus,

$$V = \lim_{n\to\infty} \sum_{k=1}^{n} \pi[f(\bar{x}_k)]^2 \, \Delta x_k$$

But, $\pi[f(x)]^2$ is some function of $x$ and this limit is nothing more than the definite integral. Hence,

$$V = \lim_{n\to\infty} \sum_{k=1}^{n} \pi[f(x_k)]^2 \, \Delta x_k = \int_a^b \pi[f(x)]^2 \, dx$$

$$\boxed{V = \pi \int_a^b [f(x)]^2 \, dx}$$

**Example 1** Find the volume of the solid generated by revolving about the $X$ axis the region bounded by the curves $y = x^2$, $x = 2$, and the $X$ axis (Fig. 9-20).

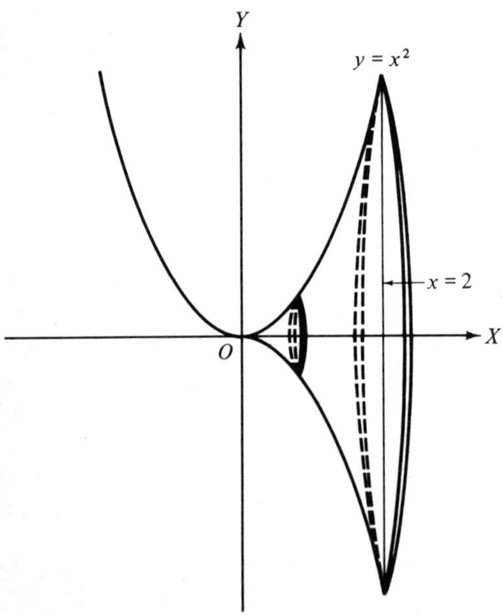


**Fig. 9-20**

*Solution*

$$V = \pi \int_a^b [f(x)]^2\, dx$$
$$= \pi \int_0^2 (x^2)^2\, dx$$
$$= \pi \int_0^2 x^4\, dx$$
$$= \pi \left.\frac{x^5}{5}\right|_0^2$$
$$= \frac{\pi}{5}(32 - 0)$$
$$= \frac{32\pi}{5}$$

**Example 2** Find the volume of the solid generated by revolving about the $Y$ axis the region bounded by the lines $y = 2x$, $y = 4$, and the $Y$ axis (Fig. 9-21).

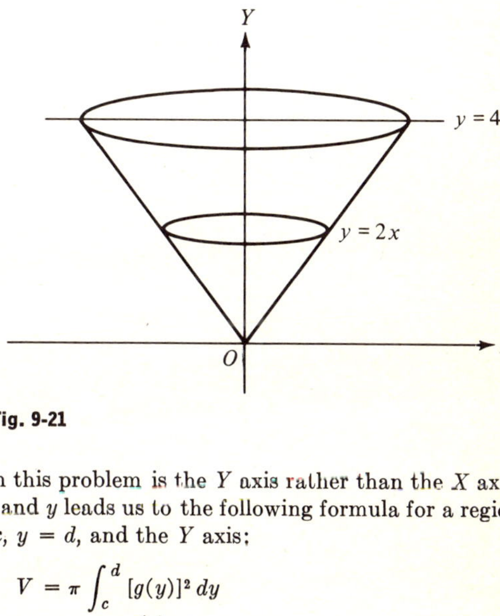


**Fig. 9-21**

*Solution* The axis of revolution in this problem is the $Y$ axis rather than the $X$ axis. Interchanging the roles of $x$ and $y$ leads us to the following formula for a region bounded by $x = g(y)$, $y = c$, $y = d$, and the $Y$ axis:

$$V = \pi \int_c^d [g(y)]^2\, dy$$
$$= \pi \int_0^4 \left(\frac{y}{2}\right)^2 dy$$
$$= \frac{\pi}{4} \int_0^4 y^2\, dy$$
$$= \frac{\pi}{4} \left.\left(\frac{y^3}{3}\right)\right|_0^4$$
$$= \frac{\pi}{12}(64 - 0)$$
$$= \frac{16\pi}{3}$$

As a check we note that our region is a right triangle and our solid is generated by revolving the right triangle about one of its legs. The solid generated in this way will be a right circular cone for which $V = \frac{1}{3}\pi r^2 h$. For the given cone, $h = 4$, $r = 2$, and

$$V = \tfrac{1}{3}\pi(4)(4) = \frac{16\pi}{3}$$

**Example 3** Find the volume of the solid generated by revolving about the line $y = 1$ the region bounded by the parabola $y = 5 - x^2$ and the line $y = 1$ (Fig. 9-22).

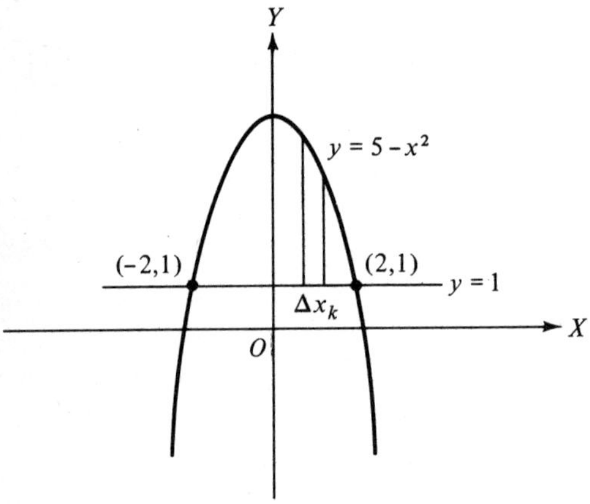


**Fig. 9-22**

*Solution* A typical rectangular element in our region will have a width of $\Delta x_k$ and a height of $f(\bar{x}_k) - 1$. Thus, a typical cylindrical disk will have a volume of $\pi[f(\bar{x}_k) - 1]^2\,\Delta x_k$. Hence,

$$\begin{aligned}
V &= \pi \int_{-2}^{2} [(5 - x^2) - 1]^2\,dx \\
&= \pi \int_{-2}^{2} (4 - x^2)^2\,dx \\
&= \pi \int_{-2}^{2} (16 - 8x^2 + x^4)\,dx \\
&= \pi \left(16x - \frac{8x^3}{3} + \frac{x^5}{5}\right)\Bigg|_{-2}^{2} \\
&= \pi[(32 - \tfrac{64}{3} + \tfrac{32}{5}) - (-32 + \tfrac{64}{3} - \tfrac{32}{5})] \\
&= \pi(\tfrac{256}{15} + \tfrac{256}{15}) \\
&= \frac{512\pi}{15}
\end{aligned}$$

**Example 4** Find the volume of the solid generated by revolving about the $X$ axis the region bounded by the parabola $y = 5 - x^2$ and the line $y = 1$.

*Solution* We note that the given region is exactly the same as the region in the previous example (Fig. 9-22). In this example, however, our axis is the $X$ axis rather than the line $y = 1$. If we examine a typical element of volume, we find that we have a cylindrical washer, with a hollow section introduced by the region between the $X$ axis and the line $y = 1$ (Fig. 9-23). The volume of a typical washer element for our region can be given as

$$\pi\{[f(x)]^2 - 1\}\,\Delta x_k$$

Hence,

$$\begin{aligned} V &= \pi \int_{-2}^{2} [(5 - x^2)^2 - 1]\,dx \\ &= \pi \int_{-2}^{2} (24 - 10x^2 + x^4)\,dx \\ &= \pi \left( 24x - \frac{10x^3}{3} + \frac{x^5}{5} \right) \Bigg|_{-2}^{2} \\ &= \pi\left(\tfrac{416}{15} + \tfrac{416}{15}\right) \\ &= \frac{832\pi}{15} \end{aligned}$$

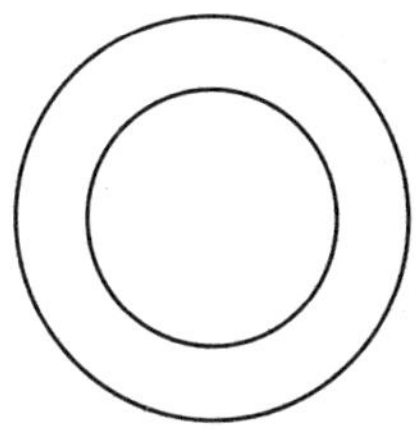

Fig. 9-23

The results in this example can be generalized to give us a formula for the volume of a solid generated by revolving about the $X$ axis a region bounded by the curves $y = f(x)$ and $y = g(x)$ (Fig. 9-24).

$$\boxed{V = \pi \int_a^b \{[f(x)]^2 - [g(x)]^2\}\,dx}$$

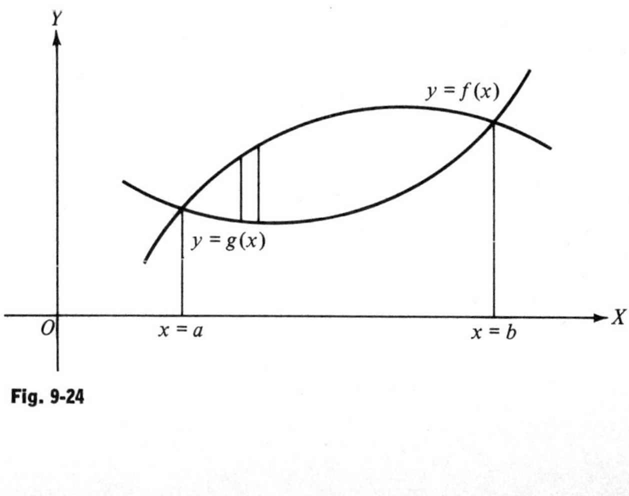


Fig. 9-24

**EXERCISES**

In Exercises 1 to 8 find the volume of the solid generated by revolving about the $X$ axis the region bounded by the given curves.

**1.** $y = x,\ y = 0,\ x = 4$

**2.** $y = 4 - x^2,\ y = 0$

**3.** $y = x^2,\ y = 2 - x^2$

**4.** $y = 4x^2,\ y = 4x$

**5.** $x = 4 - y^2,\ x = 0,\ y = 0$

**6.** $y = x^2 - 4x,\ y = 0$

**7.** $y = \sqrt{a^2 - x^2}$

**8.** $y = x^2,\ x = y^2$

In Exercises 9 to 12 find the volume of the solid generated by revolving about the $Y$ axis the region bounded by the given curves.

**9.** $y = 1 - x^2,\ y = 0,\ x = 0$

**10.** $y = \dfrac{h}{r}x,\ y = h,\ x = 0$

**11.** $y = x,\ y = x^2$

**12.** $x = 2y - y^2,\ x = 0$

**13.** Find the volume of the solid generated by revolving about the line $y = 4$ the region bounded by the parabola $y = x^2$ and the line $y = 4$.

**14.** Find the volume of the solid generated by revolving about the line $x = 1$ the region bounded by the parabola $y = x^2$ and the lines $y = 0$ and $x = 1$.

## ▶ 9-8 THE METHOD OF CYLINDRICAL SHELLS

In some problems involving the volume of a solid of revolution, it is more convenient to use the method of cylindrical shells rather than the method of cylindrical disks. We shall now examine the method of cylindrical shells.

Consider a region bounded by the curve $y = f(x)$, the $X$ axis, and the lines $x = a$ and $x = b$. We shall generate a solid by revolving this region about the $Y$ axis (Fig. 9-25).

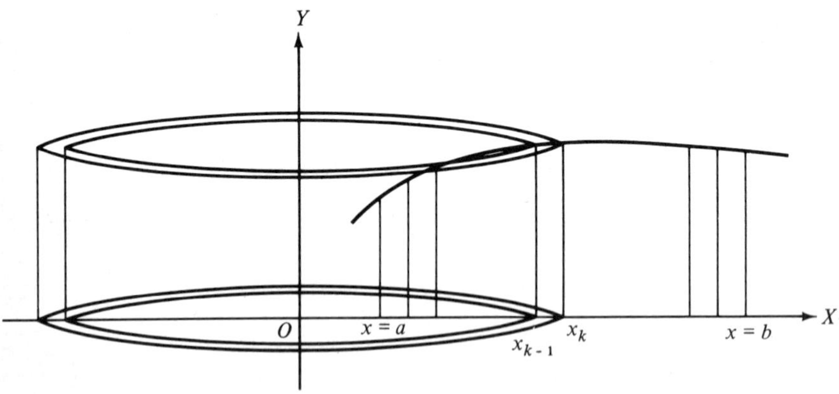


**Fig. 9-25**

To find the volume of the solid, we shall divide the interval from $x = a$ to $x = b$ into $n$ subintervals in the same way that we have done previously. We wish to form the corresponding rectangles, revolve them about the $Y$ axis, and use the volumes of the rectangles as an approximation for the volume of the solid. We note that when a rectangular element is revolved about the $Y$ axis, it will generate a cylindrical shell, that is, a thin hollow circular cylinder. The volume of a cylindrical shell with height $h$, inner radius $r_1$, and outer radius $r_2$ can be found as follows:

$$\begin{aligned} V &= \pi r_2^2 h - \pi r_1^2 h \\ &= \pi h(r_2^2 - r_1^2) \\ &= \pi h(r_2 + r_1)(r_2 - r_1) \\ &= 2\pi h \frac{r_2 + r_1}{2} (r_2 - r_1) \end{aligned}$$

$\dfrac{r_2 + r_1}{2}$ is the average radius (i.e., the value exactly midway between $r_1$ and $r_2$), and we shall let

$$\bar{r} = \frac{r_1 + r_2}{2}$$

We shall also let

$$\Delta r = r_2 - r_1$$

Thus,

$$V = 2\pi \bar{r} h \, \Delta r$$

In the $k$th subinterval $\Delta x_k$, we can choose $\bar{x}_k$ as the midpoint of that interval, and we can use $f(\bar{x}_k)$ as the height of our rectangle. If this rectangle is now revolved about the $Y$ axis, the volume of the corresponding cylindrical shell will be

$$2\pi \bar{x}_k f(\bar{x}_k) \, \Delta x_k$$

If this procedure is repeated for all $n$ subintervals, the sum of the volumes corresponding to the cylindrical shells will then be

$$V_s = \sum_{k=1}^{n} 2\pi \bar{x}_k f(\bar{x}_k) \, \Delta x_k$$

Here again, $V_s$ is only an approximation of the volume of the solid of revolution. To get the actual volume, we must increase the number of subintervals and consider the limit as $n \to \infty$ and all $\Delta x_k \to 0$. Thus,

$$\begin{aligned} V &= \lim_{n \to \infty} \sum_{k=1}^{n} 2\pi \bar{x}_k f(\bar{x}_k) \, \Delta x_k \\ &= \int_a^b 2\pi x f(x) \, dx \end{aligned}$$

$$\boxed{V = 2\pi \int_a^b x f(x) \, dx}$$

**Example 1** Find the volume of the solid generated by revolving about the $Y$ axis the region bounded by the parabola $y = x^2$, the $Y$ axis, and the line $x = 2$.

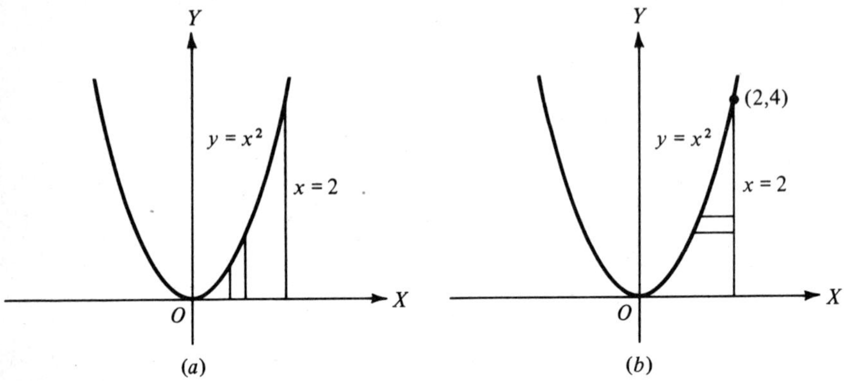


**Fig. 9-26**

*Solution: Method* 1 Using the method of cylindrical shells (Fig. 9-26*a*),

$$
\begin{aligned}
V &= 2\pi \int_0^2 x \cdot x^2\,dx \\
&= 2\pi \int_0^2 x^3\,dx \\
&= 2\pi \cdot \frac{x^4}{4}\Big|_0^2 \\
&= 2\pi(4 - 0) \\
&= 8\pi
\end{aligned}
$$

*Solution: Method* 2 Using the method of cylindrical disks (Fig. 9-26*b*),

$$
\begin{aligned}
V &= \pi \int_c^d \{[g(y)^2 - f(y)^2]\}\,dy \\
&= \pi \int_0^4 (4 - y)\,dy \\
&= \pi\left(4y - \frac{y^2}{2}\right)\Big|_0^4 \\
&= \pi(16 - 8) \\
&= 8\pi
\end{aligned}
$$

**Example 2** Find the volume of the solid generated by revolving about the $Y$ axis the region bounded by the parabola $y = -x^2 + 6x - 8$ and the $X$ axis.

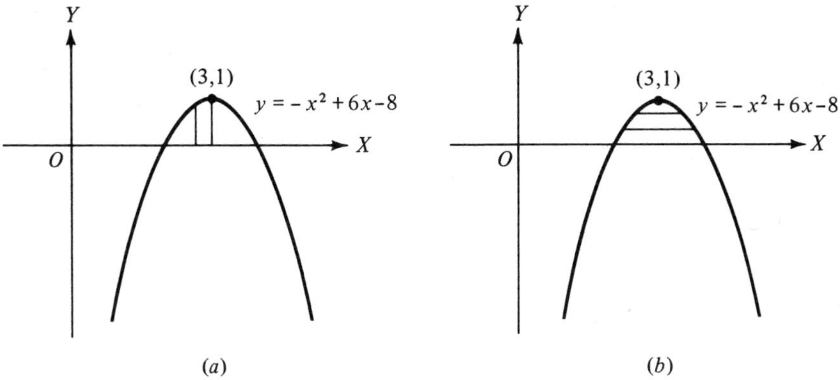


**Fig. 9-27**

*Solution: Method* 1 We shall use the method of cylindrical shells (Fig. 9-27*a*). The curve $y = -x^2 + 6x - 8$ cuts the $X$ axis at $x = 2$ and $x = 4$.

$$
\begin{aligned}
V &= 2\pi \int_2^4 x(-x^2 + 6x - 8)\,dx \\
&= 2\pi \int_2^4 (-x^3 + 6x^2 - 8x)\,dx \\
&= 2\pi \left(-\frac{x^4}{4} + 2x^3 - 4x^2\right)\Bigg|_2^4 \\
&= 2\pi[(-64 + 128 - 64) - (-4 + 16 - 16)] \\
&= 8\pi
\end{aligned}
$$

*Solution: Method* 2 In using the method of cylindrical disks, we find that a typical element of volume will be a cylindrical washer (Fig. 9-27*b*). We must use the formula

$$V = \pi \int_c^d \{[f(y)]^2 - [g(y)]^2\}\,dy$$

To find $f(y)$ and $g(y)$, we must solve for $x$.

$$
\begin{aligned}
y &= -x^2 + 6x - 8 \\
x^2 - 6x + y + 8 &= 0 \\
x &= \frac{6 \pm \sqrt{36 - 4(y + 8)}}{2} \\
&= \frac{6 \pm \sqrt{4 - 4y}}{2} \\
&= 3 \pm \sqrt{1 - y}
\end{aligned}
$$

Thus, $\qquad f(y) = 3 + \sqrt{1 - y} \qquad g(y) = 3 - \sqrt{1 - y}$

The vertex of the parabola is the point (3,1), and thus the $y$ values go from 0 to 1.

$$\begin{aligned} V &= \pi \int_0^1 [(3 + \sqrt{1-y})^2 - (3 - \sqrt{1-y})^2]\, dy \\ &= \pi \int_0^1 12\sqrt{1-y}\, dy \\ &= 12\pi \int_0^1 \sqrt{1-y}\, dy \\ &= -12\pi \frac{(1-y)^{\frac{3}{2}}}{\frac{3}{2}} \Big|_0^1 \\ &= -8\pi(0-1) \\ &= 8\pi \end{aligned}$$

The reader can readily see the advantage of using the method of cylindrical shells in this example.

**EXERCISES**

In Exercises 1 to 4 find the volume of the solid generated by revolving about the $Y$ axis the region bounded by the given curves.

**1.** $y = 4x^2$, $y = 0$, $x = 4$

**2.** $y = 2x - x^2$, $y = 0$

**3.** $y = 4 - x^2$, $y = 0$, $x = 0$

**4.** $y = 2x^2$, $x = 1$, $x = 3$, $y = 0$

In Exercises 5 to 8 find the volume of the solid generated by revolving about the $X$ axis the region bounded by the given curves.

**5.** $x = 4y - y^2$, $x = 0$

**6.** $x = y^2$, $y = 2$, $x = 0$

**7.** $x = \sqrt{1-y}$, $y = 0$

**8.** The smaller region bounded by $x = 3y - y^2$, $x = 0$, $y = 1$

**9.** A sphere can be generated by revolving a semicircle about its diameter. Using the circle $x^2 + y^2 = 4$, use the above procedure to generate a sphere and find the corresponding volume.

**10.** Find the volume of the solid generated by revolving about the $X$ axis the region bounded by the parabola $y = x^2$ and the line $y = x$.

## ▶ 9-9 THE TRAPEZOIDAL RULE

At this point, our methods of integration are quite limited, and we cannot evaluate an integral such as $\int \frac{dx}{x}$. (Why can't we use the power rule?) Though a method for the previous integral will be developed later, there are integrals such as $\int \frac{\sin x}{x}\, dx$ for which standard integration techniques are not available. In such cases, it is often convenient to use methods of approximate integration. In this section we shall consider one such method known as the *trapezoidal rule.*

The basis for the trapezoidal rule is the fact that we may represent a definite integral by an area under a curve. With this in mind, an approximation of the area will serve as an approximation for the definite integral.

To evaluate $\int_a^b f(x)\,dx$ we shall consider the area under the curve $y = f(x)$, above the $X$ axis between $x = a$ and $x = b$. We shall divide the interval from $x = a$ to $x = b$ into $n$ subintervals by choosing $n - 1$ equally spaced points $x_1, \ldots, x_{n-1}$ between $x = a$ and $x = b$. Thus,

$$\begin{aligned} a &= x_0 \\ x_1 &= x_0 + \Delta x \\ x_2 &= x_1 + \Delta x \\ &\cdots\cdots\cdots \\ b = x_n &= x_{n-1} + \Delta x \qquad \text{where } \Delta x = \frac{b-a}{n} \end{aligned}$$

We now connect the points $[x_0,f(x_0)]$ and $[x_1,f(x_1)]$ with a straight-line segment to form a trapezoid, and we repeat this procedure for all $n$ subintervals (Fig. 9-28). We shall use these $n$ trapezoids as an approximation for the area under the curve. In doing this, we are using the chord instead of the arc of $y = f(x)$ as an upper boundary of the region.

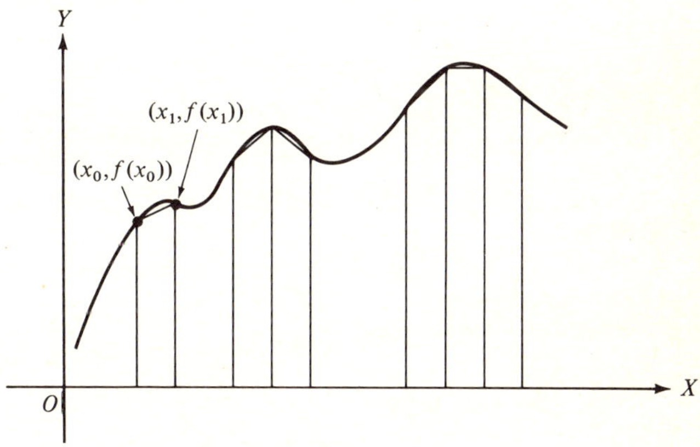


Fig. 9-28

We recall that the formula for the area of a trapezoid with height $h$ and parallel bases $b_1$ and $b_2$ is

$$A = \tfrac{1}{2}h(b_1 + b_2)$$

Applying this formula to our $n$ trapezoids, we note that for each rectangle $h = \Delta x = \dfrac{b-a}{n}$,

$$\begin{aligned} A_1 &= \tfrac{1}{2}[f(x_0) + f(x_1)]\,\Delta x \\ A_2 &= \tfrac{1}{2}[f(x_1) + f(x_2)]\,\Delta x \\ &\cdots\cdots\cdots\cdots\cdots \\ A_n &= \tfrac{1}{2}[f(x_{n-1}) + f(x_n)]\,\Delta x \end{aligned}$$

$$\sum_{k=1}^{n} A_k = \tfrac{1}{2}[f(x_0) + 2f(x_1) + 2f(x_2) + \cdots + 2f(x_{n-1}) + f(x_n)]\,\Delta x$$

$$\boxed{\int_a^b f(x)\,dx \approx \tfrac{1}{2}[f(x_0) + 2f(x_1) + 2f(x_2) + \cdots + 2f(x_{n-1}) + f(x_n)]\,\Delta x}$$

**Example 1** Use the trapezoidal rule to approximate the value of $\displaystyle\int_1^3 \frac{dx}{x}$.

*Solution* We can choose any number for $n$, the number of subintervals between $x = a$ and $x = b$. The larger the value of $n$, the greater the accuracy of our approximation. We shall choose $n = 4$. Hence,

$$\Delta x = \frac{b-a}{n} = \frac{3-1}{4} = \frac{2}{4} = \frac{1}{2}$$

$$\begin{aligned} x_0 &= 1 & f(x_0) &= 1 \\ x_1 &= \tfrac{3}{2} & f(x_1) &= \tfrac{2}{3} \\ x_2 &= 2 & f(x_2) &= \tfrac{1}{2} \\ x_3 &= \tfrac{5}{2} & f(x_3) &= \tfrac{2}{5} \\ x_4 &= 3 & f(x_4) &= \tfrac{1}{3} \end{aligned}$$

$$\sum_{k=1}^{4} A_k = \tfrac{1}{2}[1 + 2(\tfrac{2}{3}) + 2(\tfrac{1}{2}) + 2(\tfrac{2}{5}) + \tfrac{1}{3}](\tfrac{1}{2})$$

$$\int_1^3 \frac{dx}{x} \approx 1.1167$$

The value of $\displaystyle\int_1^3 \frac{dx}{x}$ correct to four decimal places is actually 1.0986, and our error is 0.0181.

**Example 2** Use the trapezoidal rule to approximate the value of $\displaystyle\int_0^2 x^2\,dx$.

*Solution* We shall choose $n = 4$. Thus,

$$\Delta x = \frac{2-0}{4} = \frac{2}{4} = \frac{1}{2}$$

$$\begin{aligned} x_0 &= 0 & f(x_0) &= 0 \\ x_1 &= \tfrac{1}{2} & f(x_1) &= \tfrac{1}{4} \\ x_2 &= 1 & f(x_2) &= 1 \\ x_3 &= \tfrac{3}{2} & f(x_3) &= \tfrac{9}{4} \\ x_4 &= 2 & f(x_4) &= 4 \end{aligned}$$

$$\sum_{k=1}^{4} A_k = \tfrac{1}{2}[0 + 2(\tfrac{1}{4}) + 2(1) + 2(\tfrac{9}{4}) + 4](\tfrac{1}{2})$$

$$\int_0^2 x^2\,dx \approx 2.75$$

The exact value of $\int_0^2 x^2\,dx$ can be found using the power rule:

$$\int_0^2 x^2\,dx = \frac{x^3}{3}\Big|_0^2 = \frac{8}{3} = 2.67$$

**EXERCISES**

Using the trapezoidal rule, find the values of the given definite integrals using the number of intervals given. Compare your result with the exact value of the integral wherever possible.

**1.** $\int_1^3 x^3\,dx,\ n = 4$

**2.** $\int_0^2 x^2\,dx,\ n = 8$. Compare with the result for $n = 4$.

**3.** $\int_2^8 \frac{dx}{1+x},\ n = 6$

**4.** $\int_0^1 \frac{dx}{1+x^2},\ n = 4$

**5.** $\int_0^{\pi/2} \cos x\,dx,\ n = 3$

**6.** $\int_2^8 \frac{dx}{1+x},\ n = 12$. Compare with the result in Exercise 3.

**7.** $\int_1^{10} \frac{dx}{x},\ n = 9$

**8.** $\int_1^2 \frac{dx}{x^3},\ n = 3$

## ▶ 9-10 PROOF OF THE FUNDAMENTAL THEOREM

Before proving the fundamental theorem, we shall review the definition of the definite integral and the mean value theorem for derivatives. The mean value theorem states that if $f(x)$ is a function which is continuous in the closed interval $a \le x \le b$ and differentiable in the open interval $a < x < b$, then there exists a number $c$ between $a$ and $b$ such that

$$f(b) - f(a) = f'(c)(b - a)$$

The definite integral was defined as a limit of a sum. We considered a function $f(x)$ defined over an interval $a \le x \le b$. We divided the interval into $n$ subintervals, we chose points $\bar{x}_1, \ldots, \bar{x}_n$ in the $n$ subintervals, and we formed the sum

$$\sum_{k=1}^{n} f(\bar{x}_k)\,\Delta x_k$$

We then considered the limit as $n \to \infty$ and all $\Delta x_k \to 0$. If

$$\lim_{n\to\infty} \sum_{k=1}^{n} f(\bar{x}_k)\,\Delta x_k$$

exists and is always the same no matter how the points $\bar{x}_k$ are chosen within the subintervals, then this limit is called the definite integral of $f(x)$ from $x = a$ to $x = b$ and is denoted by the symbol

$$\int_a^b f(x)\,dx$$

We have also assumed the theorem which states that $\lim_{n\to\infty} \sum_{k=1}^{n} f(\bar{x}_k)\,\Delta x_k$ will always exist and be independent of the points of subdivision so long as $f(x)$ is continuous over the interval $a \le x \le b$. We now proceed to the fundamental theorem.

**The fundamental theorem of integral calculus** *If $f(x)$ is continuous in the interval $a \le x \le b$ and if $F(x)$ is an antiderivative of $f(x)$ [that is, $F'(x) = f(x)$], then*

$$\int_a^b f(x)\,dx = F(b) - F(a)$$

*Proof* Let $F(x)$ be an antiderivative of $f(x)$. Let $x_1, \ldots, x_{n-1}$ be any $n - 1$ subdivision points between $x = a$ and $x = b$, and let $\Delta x_1, \ldots, \Delta x_n$ be the corresponding subintervals. Since

$$F'(x) = f(x)$$

$F(x)$ is a function which is differentiable and hence continuous in the interval $a \le x \le b$. The above properties will thus apply to each of the $n$ subintervals $\Delta x_1, \ldots, \Delta x_n$. We can then apply the mean value theorem for derivatives to the function $F(x)$ for each of the $n$ subintervals. In the first subinterval the mean value theorem tells us that there exists a number $\bar{x}_1$, $x_0 < \bar{x}_1 < x_1$, such that $F(x_1) - F(x_0) = F'(\bar{x}_1)\,\Delta x_1$. Repeating this procedure in all $n$ subintervals, we get

$$\begin{aligned}
F(x_1) - F(a) &= F'(\bar{x}_1)\,\Delta x_1 = f(\bar{x}_1)\,\Delta x_1 \\
&\searrow \\
F(x_2) - F(x_1) &= F'(\bar{x}_2)\,\Delta x_2 = f(\bar{x}_2)\,\Delta x_2 \\
&\searrow \\
F(x_3) - F(x_2) &= F'(\bar{x}_3)\,\Delta x_3 = f(\bar{x}_3)\,\Delta x_3 \\
&\cdots\cdots\cdots\cdots\cdots\cdots \\
F(x_{n-1}) - F(x_{n-2}) &= F'(\bar{x}_{n-1})\,\Delta x_{n-1} = f(\bar{x}_{n-1})\,\Delta x_{n-1} \\
&\searrow \\
F(b) - F(x_{n-1}) &= F'(\bar{x}_n)\,\Delta x_n = f(\bar{x}_n)\,\Delta x_n
\end{aligned}$$

where $\bar{x}_1, \ldots, \bar{x}_n$ are points such that $x_{k-1} < \bar{x}_k < x_k$. We shall now add the left sides of the equations to each other and the right sides of the equations to each other. We note that the terms on the left side cancel out in pairs, except for $F(b)$ in the last equation and $-F(a)$ in the first equation. Thus,

$$F(b) - F(a) = \sum_{k=1}^{n} f(\bar{x}_k)\,\Delta x_k$$

We now consider the limit as $n \to \infty$ in a way such that all $\Delta x_k$ approach 0. The expression on the left is independent of $n$ and will remain unchanged. Thus,

$$F(b) - F(a) = \lim_{n\to\infty} \sum_{k=1}^{n} f(\bar{x}_k)\,\Delta x_k$$

Since $f(x)$ is continuous, the limit exists and is independent of the points of subdivision. Hence,

$$F(b) - F(a) = \lim_{n\to\infty} \sum_{k=1}^{n} f(\bar{x}_k)\,\Delta x_k = \int_a^b f(x)\,dx$$

$$\boxed{\int_a^b f(x)\,dx = F(b) - F(a)} \quad \blacksquare$$

# 10
# Transcendental Functions

## 10-1 REVIEW OF TRIGONOMETRIC FUNCTIONS

The trigonometric functions were defined in Chap. 8. We recall that if $\theta$ is an angle in standard position, $P(x,y)$ is a point on the terminal side of angle $\theta$, and $r$ is the distance from the vertex of the angle to point $P$ (Fig. 10-1), then

$$\sin\theta = \frac{y}{r} \qquad \csc\theta = \frac{r}{y}$$

$$\cos\theta = \frac{x}{r} \qquad \sec\theta = \frac{r}{x}$$

$$\tan\theta = \frac{y}{x} \qquad \cot\theta = \frac{x}{y}$$

We also have derived the following formulas for differentiation and integration.

$$
\begin{aligned}
D_x \sin x &= \cos x & D_x \cot x &= -\csc^2 x \\
D_x \cos x &= -\sin x & D_x \sec x &= \sec x \tan x \\
D_x \tan x &= \sec^2 x & D_x \csc x &= -\csc x \cot x
\end{aligned}
$$

$$
\begin{aligned}
\int \sin x\, dx &= -\cos x + C \\
\int \cos x\, dx &= \sin x + C \\
\int \sec^2 x\, dx &= \tan x + C \\
\int \csc^2 x\, dx &= -\cot x + C \\
\int \sec x \tan x\, dx &= \sec x + C \\
\int \csc x \cot x\, dx &= -\csc x + C
\end{aligned}
$$

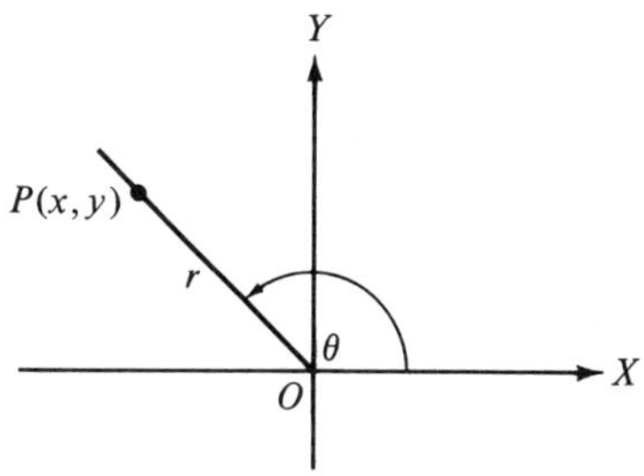


Fig. 10-1

**EXERCISES**

Find $\dfrac{dy}{dx}$ in Exercises 1 to 8.

**1.** $y = \sin 2x$

**2.** $y = \cos^2 x$

**3.** $y = \sin 3x \cos 3x$

**4.** $y = 1 + \tan^2 4x$

**5.** $y = \sin^2 2x - \cos^2 2x$

**6.** $y = \csc^3 4x$

**7.** $y = \sec x \tan x$

**8.** $y = \dfrac{1 + \sin x}{\cos x}$

Evaluate the integrals in Exercises 9 to 16.

**9.** $\int \sin 3x\, dx$

**10.** $\int \cos^2 x \sin x\, dx$

**11.** $\displaystyle\int \frac{\sin x}{\cos^2 x}\, dx$

**12.** $\int (1 + \sin t)^{\frac{3}{2}} \cos t\, dt$

**13.** $\int \sec 2x \tan 2x\, dx$

**14.** $\int \sqrt{1 - \sin x} \cos x\, dx$

**15.** $\displaystyle\int \left(\sec^2 \frac{1}{x}\right) \frac{dx}{x^2}$

**16.** $\displaystyle\int \frac{\sin x}{\sqrt{2 - \cos x}}\, dx$

## 10-2 INVERSE sine AND INVERSE cosine FUNCTIONS

If we are given the equation $y = 2x - 4$, we can think of this as a rule of correspondence for a function where $y$ is a function of $x$. We can, in fact, make the following table for this function:

| $x$ | $y$ |
|---|---|
| 0 | $-4$ |
| 1 | $-2$ |
| 2 | 0 |
| 3 | 2 |
| 4 | 4 |

If we solve for $x$ in terms of $y$, we get

$$2x = y + 4$$

$$x = \frac{y+4}{2}$$

We would now like to consider the variable $x$ as a function of the variable $y$. We notice that when $y = -4$, $x = 0$; when $y = -2$, $x = 1$; when $y = 0$, $x = 2$. This is precisely the same set of values that we have calculated in our table of values for $y$ as a function of $x$. In fact, any set of values for $x$ and $y$ that will satisfy the first equation will also satisfy the second equation.

In general, when we are given $y$ as a function of $x$ and we are able to solve for $x$ in terms of $y$ so that $x$ is a function of $y$, the two functions are referred to as *inverse functions.*

We now seek to find the inverse function for the function

$$x = \sin y$$

Unlike the algebraic functions, we cannot readily solve for $y$ in terms of $x$. The best we can do is to describe $y$ in terms of $x$ by saying that $y$ is the angle whose sine is equal to $x$. We shall denote this expression with the symbol

$$y = \arcsin x$$

This can be thought of as "$y$ is the angle whose sine is equal to the number $x$."

**Example 1** Find $\theta$, if $\theta = \arcsin \frac{1}{2}$.

*Solution* We wish to find the angle whose sine is equal to $\frac{1}{2}$. Certainly,

$$\sin \frac{\pi}{6} = \frac{1}{2}$$

and thus

$$\arcsin \frac{1}{2} = \frac{\pi}{6}$$

However,
$$\sin\frac{5\pi}{6} = \frac{1}{2}$$
$$\sin\frac{13\pi}{6} = \frac{1}{2}$$
$$\sin\frac{17\pi}{6} = \frac{1}{2}$$
$$\cdots\cdots$$

Thus, we cannot find a unique angle whose sine is equal to $\frac{1}{2}$, since there are infinitely many such angles (Fig. 10-2).

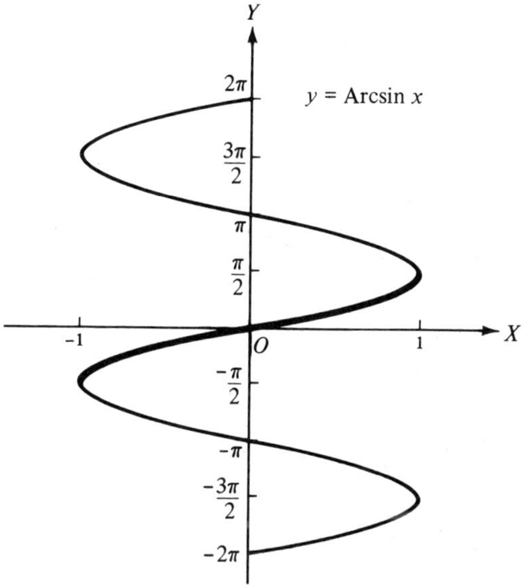


**Fig. 10-2**

If $\theta$ is an angle in the first quadrant, then we can say that

$$\theta = \frac{\pi}{6} + 2n\pi$$

If $\theta$ is an angle in the second quadrant, then

$$\theta = \frac{5\pi}{6} + 2n\pi$$

The definition of a function requires that if $y = \arcsin x$ is to be a function, then for every value of $x$ we must have exactly one value of $y$. To attain this end, we shall define the principal values of arcsin $x$ to be the values between $\frac{-\pi}{2}$ and $\frac{\pi}{2}$.

$$\boxed{\frac{-\pi}{2} \leq \text{Arcsin } x \leq \frac{\pi}{2}}$$

Thus, whenever $x$ is a positive number, our angle is in the first quadrant and is between 0 and $\frac{\pi}{2}$. If $x$ is a negative number, then our angle is in the fourth quadrant and it is between $\frac{-\pi}{2}$ and 0. To indicate that principal values are being used, a capital letter will be used for the letter A in the word Arcsin.

**Example 2**

$$\text{Arcsin } \frac{1}{2} = \frac{\pi}{6}$$

$$\text{Arcsin } \frac{\sqrt{3}}{2} = \frac{\pi}{3}$$

$$\text{Arcsin } \left(\frac{-\sqrt{2}}{2}\right) = \frac{-\pi}{4} \qquad \left(\textit{Note: } \frac{7\pi}{4} \text{ is not an acceptable answer}\right)$$

The inverse function for the function $x = \cos y$ is

$$y = \text{Arccos } x$$

where the principal values of the Arccos $x$ are between 0 and $\pi$ (Fig. 10-3).

$$\boxed{0 \leq \text{Arccos } x \leq \pi}$$

If $x$ is positive, our angle is in the first quadrant and is between 0 and $\frac{\pi}{2}$.

If $x$ is negative, then our angle is in the second quadrant and is between $\frac{\pi}{2}$ and $\pi$.

**Example 3**

$$\text{Arccos } 0 = \frac{\pi}{2}$$

$$\text{Arccos } \left(-\frac{1}{2}\right) = \frac{2\pi}{3}$$

$$\text{Arccos } 1 = 0$$

$$\text{Arccos } \left(-\frac{\sqrt{2}}{2}\right) = \frac{3\pi}{4}$$

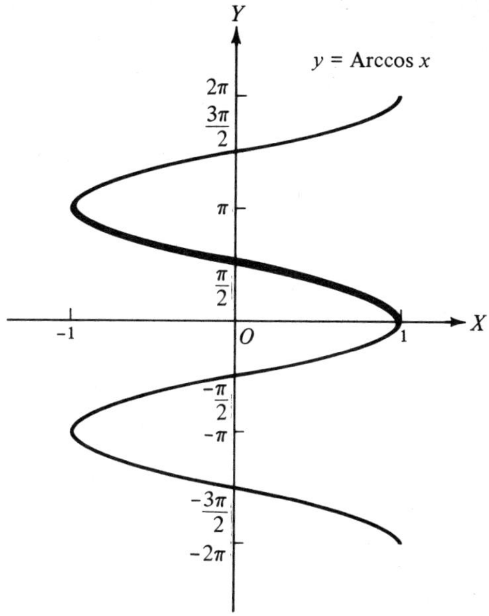


Fig. 10-3

**EXERCISES**

Find the values of the following expressions.

**1.** $\text{Arcsin}\,(-\frac{1}{2})$

**2.** $\text{Arccos}\,\frac{\sqrt{3}}{2}$

**3.** $\text{Arccos}\,\frac{\sqrt{2}}{2}$

**4.** $\text{Arcsin}\left(-\frac{\sqrt{3}}{2}\right)$

**5.** $\text{Arcsin}\,0$

**6.** $\text{Arccos}\left(\tan\frac{\pi}{4}\right)$

Find the inverse function for each of the following functions.

**7.** $x = \sin 2y$

**8.** $x = \frac{1}{2}\cos y$

**9.** $x = \text{Arcsin}\,2y$

**10.** $x = 2\,\text{Arccos}\,2y$

## 10-3 THE DERIVATIVE OF THE Arcsin AND Arccos FUNCTIONS

To find the derivative of $y = \text{Arcsin}\,x$, we shall use the inverse function

$$x = \sin y$$

Differentiating implicitly with respect to $x$, we get

$$1 = (\cos y)y'$$

$$y' = \frac{1}{\cos y}$$

We wish to find $y'$ in terms of $x$. We know that $\sin y = x = \frac{x}{1}$. If we construct a right triangle with hypotenuse 1 and side $x$ opposite angle $y$, we find that the remaining side will have a length of $\sqrt{1 - x^2}$ (Fig. 10-4). For this triangle, $\cos y = \frac{\sqrt{1 - x^2}}{1} = \sqrt{1 - x^2}$. Hence,

$$\begin{aligned} y' &= \frac{1}{\cos y} \\ &= \frac{1}{\sqrt{1 - x^2}} \end{aligned}$$

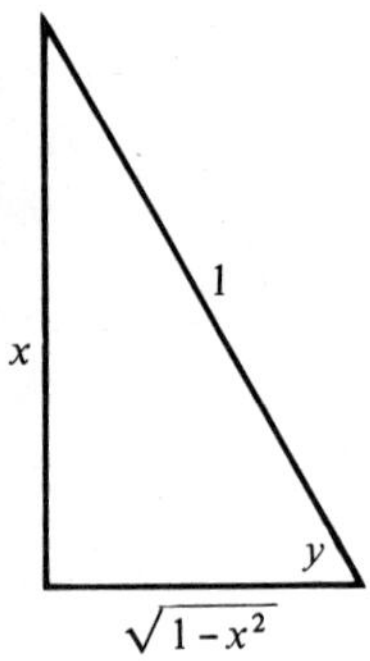


Fig. 10-4

If $y = \text{Arcsin } u$, where $u$ is a function of $x$, then, using the chain rule,

$$\boxed{y' = \frac{1}{\sqrt{1 - u^2}} D_x u}$$

The formula for the derivative of $y = \text{Arccos } x$ can be found in a similar way.

$$\begin{aligned} x &= \cos y \\ 1 &= -(\sin y)y' \\ y' &= -\frac{1}{\sin y} \end{aligned}$$

Constructing our triangle (Fig. 10-5), we find that

$$\sin y = \sqrt{1 - x^2}$$

Hence,
$$y' = -\frac{1}{\sqrt{1 - x^2}}$$

As before,

$$D_x \text{ Arccos } u = -\frac{1}{\sqrt{1-u^2}} \cdot D_x u$$

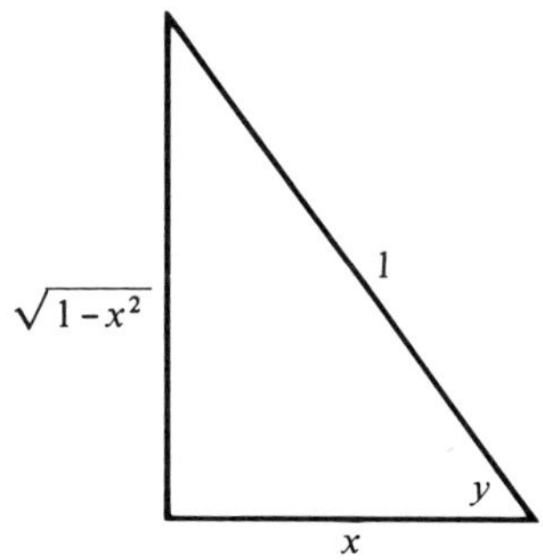


**Fig. 10-5**

**Example 1**

$$(a)\ D_x \text{ Arcsin } 2x = \frac{1}{\sqrt{1-4x^2}}\,2 = \frac{2}{\sqrt{1-4x^2}}$$

$$(b)\ D_x \text{ Arccos } x^2 = -\frac{1}{\sqrt{1-x^4}}\,2x = \frac{-2x}{\sqrt{1-x^4}}$$

$$(c)\ D_x \text{ Arcsin } \frac{x}{2} = \frac{1}{\sqrt{1-\dfrac{x^2}{4}}}\,\frac{1}{2} = \frac{1}{\sqrt{4-x^2}}$$

**Example 2** A boy flies a kite which is 100 ft above the ground. If the string is being let out at a rate of 10 ft/sec and the wind carries the kite horizontally directly away from the boy, find the rate of change of the angle of elevation of the kite when the angle is 30°.

*Solution* We have already solved this problem in Sec. 8-3. We shall now solve the problem making use of the inverse sine function. Let $\alpha$ be the angle of elevation and let $s$ be the amount of string. We wish to find $\dfrac{d\alpha}{dt}$ given that $\dfrac{ds}{dt} = 10$ ft/sec (Fig. 10-6).

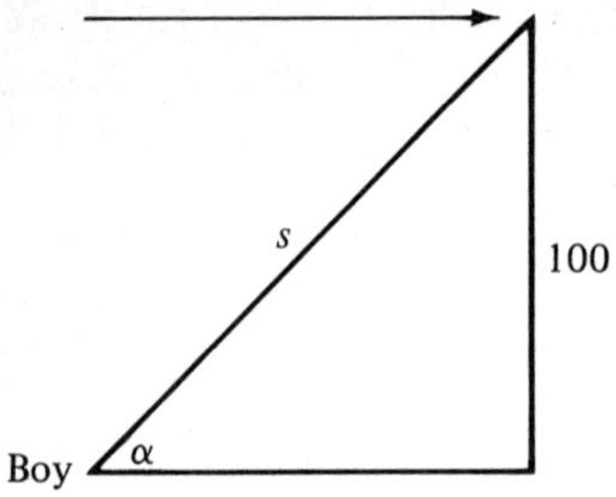


**Fig. 10-6**

When $\alpha = 30°$, $s = 200$. (Why?) We now express $\alpha$ as a function of $s$:

$$\alpha = \text{Arcsin}\, \frac{100}{s}$$

$$\frac{d\alpha}{dt} = \frac{1}{\sqrt{1 - \left(\frac{100}{s}\right)^2}} \left(-\frac{100}{s^2}\right) \frac{ds}{dt}$$

$$= \frac{1}{\sqrt{1 - \left(\frac{100}{200}\right)^2}} \left(-\frac{100}{40{,}000}\right) (10)$$

$$= \frac{1}{\sqrt{1 - \frac{1}{4}}} \frac{-1}{40}$$

$$= \frac{-1}{20\sqrt{3}}$$

$$= -\frac{\sqrt{3}}{60}$$

This checks with our previous result in Sec. 8-3.

**EXERCISES**

Differentiate the following functions.

**1.** $y = \text{Arcsin}\,(x + \frac{1}{2})$

**2.** $y = \text{Arccos}\, 2x$

**3.** $y = \text{Arccos}\, \sqrt{x}$

**4.** $y = \text{Arcsin}\, \frac{1}{x}$

**5.** $y = \text{Arcsin}\, x^2$

**6.** $y = \text{Arccos}\, \frac{1}{2}x^2$

**7.** $y = x\, \text{Arcsin}\, x$

**8.** $y = \text{Arcsin}\, \frac{1}{3}x - \sqrt{9 - x^2}$

**9.** $y = \text{Arcsin}\, x - x\sqrt{1 - x^2}$

**10.** $y = \text{Arccos}\, \frac{1}{2}x - \sqrt{4 - x^2}$

## 10-4 THE EXPONENTIAL FUNCTION

A function in the form $y = b^x$ where $b$ is a constant is referred to as an *exponential function.* It is important for the reader to distinguish between the equation $y = x^2$ and the equation $y = 2^x$. The latter function is an exponential function in the form $y = b^x$, with base $b$ equal to 2 and with a variable exponent $x$. The equation $y = x^2$ has a variable base $x$ and a constant exponent 2.

The function $y = b^x$ (Fig. 10-7) is defined for all real values of $x$, and the student should be familiar with the following rules for exponents.

1. $$b^m b^n = b^{(m+n)}$$

2. $$\frac{b^m}{b^n} = b^{(m-n)}$$

3. $$(b^m)^n = b^{mn}$$

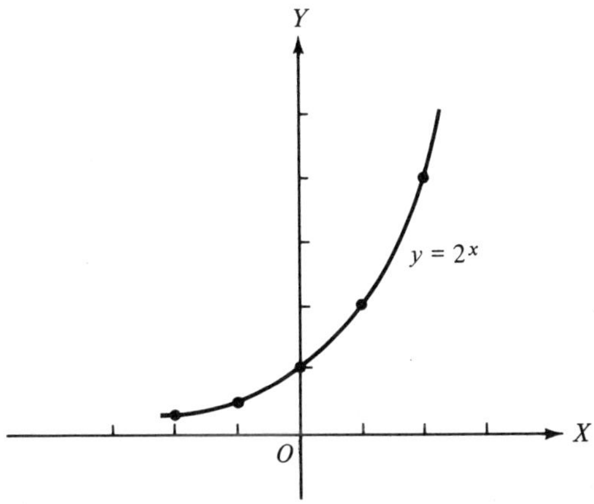


Fig. 10-7

## 10-5 THE LOGARITHMIC FUNCTION

As in the case of the algebraic functions and the trigonometric functions we shall now try to find the inverse function of the exponential function $x = b^y$.

If we wish to describe the variable $y$ in terms of the variable $x$, we can say that $y$ is the power (exponent) to which the base $b$ must be raised in order to obtain the number $x$. We shall denote this by

$$y = \log_b x$$

and this will be read as "$y$ equals log to the base $b$ of $x$."

**Example 1** Find $x$, if $x = \log_2 16$.

*Solution*

$$x = \log_2 16$$

Changing to exponential form,

$$2^x = 16$$

To what power must we raise the number 2 to get 16?

$$x = 4$$

**Example 2** Find $x$ if $3 = \log_{10} x$.

*Solution*

$$3 = \log_{10} x$$
$$x = 10^3$$
$$x = 1{,}000$$

There are two numbers which are often used for the base $b$. These are the rational number 10 and the irrational number $e$. In more advanced works the number $e$ is defined as a limit:

$$e = \lim_{h \to 0} (1 + h)^{1/h} = 2.718 \ldots$$

A proof of the existence of this limit is beyond the scope of this text.

When the number 10 is used as a base for logarithms, the logarithms are referred to as *common logarithms*. When the number $e$ is used as a base, the logarithms are referred to as *natural logarithms*. The natural logarithms $\log_e x$ will be denoted by $\ln x$, and for most of our work with logarithms we shall be using base $e$.

Corresponding to the rules for exponents given in Sec. 4 we have the following rules for logarithms.

1. $$\log_b AB = \log_b A + \log_b B$$
2. $$\log_b \frac{A}{B} = \log_b A - \log_b B$$
3. $$\log_b A^r = r \log_b A$$

To prove the above rules, we shall let

$$m = \log_b A$$
$$n = \log_b B$$

Changing these statements to exponential statements, we get

$$A = b^m$$
$$B = b^n$$
$$AB = b^m b^n$$
$$AB = b^{m+n}$$

Changing back to logarithms

$$\log_b AB = m + n = \log_b A + \log_b B$$

Similarly,

$$\frac{A}{B} = \frac{b^m}{b^n}$$

$$\frac{A}{B} = b^{m-n}$$

$$\log_b \frac{A}{B} = m - n = \log_b A - \log_b B$$

and

$$A^r = (b^m)^r$$
$$A^r = b^{rm}$$
$$\log_b A^r = rm = r \log_b A$$

**Example 3**

(a) $\ln xe^x = \ln x + \ln e^x$
$= \ln x + x$

(b) $\ln x^2 = 2 \ln x$

(c) $\ln \frac{e^x}{x} = \ln e^x - \ln x$
$= x - \ln x$

**EXERCISES**

In Exercises 1 to 6 find the value of $x$.

**1.** $x = \log_{10} 100$

**2.** $x = \log_2 32$

**3.** $x = \log_3 \frac{1}{9}$

**4.** $\log_2 x = -3$

**5.** $\log_5 x = \frac{3}{2}$

**6.** $x = \ln e^2$

In Exercises 7 to 10 rewrite the given expressions applying the rules for logarithms.

**7.** $y = \ln (x^2e^{2x})$

**8.** $y = \ln \sqrt{\frac{x}{x+1}}$

**9.** $y = 2 \ln \sqrt{x}$

**10.** $y = \ln \frac{(1 + x)^2 \sqrt{1 - x}}{x^3}$

## 10-6 THE DERIVATIVE OF THE LOGARITHMIC FUNCTION

We now proceed to find the derivative of $y = \log_b x$:

$$y = \log_b x$$
$$\Delta y = \log_b (x + \Delta x) - \log_b x$$
$$= \log_b \frac{x + \Delta x}{x}$$

$$= \log_b \left(1 + \frac{\Delta x}{x}\right)$$

$$\frac{\Delta y}{\Delta x} = \frac{1}{\Delta x} \log_b \left(1 + \frac{\Delta x}{x}\right)$$

If we now consider the limit as $\Delta x$ approaches 0, we get the indeterminate form $\frac{0}{0}$. We recall that

$$e = \lim_{h \to 0} (1 + h)^{1/h}$$

If we wish, we can think of $\dfrac{\Delta x}{x}$ as the $h$ term; however, we still lack the $\dfrac{1}{h}$ power. To remedy this we shall rewrite $\dfrac{\Delta y}{\Delta x}$ in the following way:

$$\frac{\Delta y}{\Delta x} = \frac{x}{x} \frac{1}{\Delta x} \log_b \left(1 + \frac{\Delta x}{x}\right)$$

$$= \frac{1}{x} \frac{x}{\Delta x} \log_b \left(1 + \frac{\Delta x}{x}\right)$$

$$= \frac{1}{x} \log_b \left(1 + \frac{\Delta x}{x}\right)^{x/\Delta x}$$

Thus, if we think of $\dfrac{\Delta x}{x}$ as the $h$ term, then $\dfrac{x}{\Delta x}$ would be the $\dfrac{1}{h}$ term. Hence,

$$\lim_{\Delta x \to 0} \frac{\Delta y}{\Delta x} = \frac{1}{x} \log_b e$$

If we specify the base $b$ as the number $e$, then $\log_e e = 1$ and

$$\boxed{D_x \ln x = \frac{1}{x}}$$

If we have the function $y = \ln u$ where $u$ is a function of $x$, we can apply the chain rule to get

$$\boxed{D_x \ln u = \frac{1}{u} \cdot D_x u}$$

**Example 1**

(a) $D_x \ln x^2 = \dfrac{1}{x^2} \cdot 2x$

$$= \frac{2}{x}$$

$$(b)\quad D_x \ln \frac{1}{x} = \frac{1}{\frac{1}{x}}\left(-\frac{1}{x^2}\right)$$
$$= -\frac{1}{x}$$

$$(c)\quad D_x \ln \cos x = \frac{1}{\cos x}(-\sin x)$$
$$= -\tan x$$

We recall that our power formula for antiderivatives does not apply when $n = -1$.

$$\int u^n\,du = \frac{u^{n+1}}{n+1} + C \qquad n \neq -1$$

However, since $D_x \ln u = \frac{1}{u} D_x u$, we can reverse the procedure and say

$$\int \frac{du}{u} = \ln u + C$$

If $u$ is negative, then

$$\int \frac{du}{u} = \int \frac{-du}{-u} = \ln(-u) + C$$

Hence, whether $u$ is positive or negative, we can conclude that

$$\boxed{\int \frac{du}{u} = \ln|u| + C}$$

This will take care of the one case not provided for by the power formula.

**Example 2**

$$(a)\quad \int \frac{dx}{x+1} = \ln|x+1| + C$$

$$(b)\quad \int \frac{dx}{x-5} = \ln|x-5| + C$$

**Example 3** Evaluate $\int \frac{dx}{5x+3}$.

*Solution* If we let $u = 5x + 3$, then $du = 5dx$. Hence,

$$\int \frac{dx}{5x+3} = \frac{1}{5}\int \frac{5dx}{5x+3} = \frac{1}{5}\ln|5x+3| + C$$

**EXERCISES**

Differentiate the following functions.

**1.** $y = \ln (x + 1)$

**2.** $y = \ln \sqrt{x}$

**3.** $y = \ln (x^2 - 1)^2$

**4.** $y = \ln \dfrac{x}{x + 1}$

**5.** $y = \ln \sin x$

**6.** $y = \ln (\ln x)$

**7.** $y = x \ln x$

**8.** $y = \dfrac{\ln x}{x}$

**9.** $y = \sqrt{\ln x}$

**10.** $y = \ln \dfrac{1 + x}{1 - x}$

**11.** $y = \ln \dfrac{5}{x}$

**12.** $y = \ln \dfrac{x^2}{x^2 - 1}$

**13.** $y = \ln \tan x$

**14.** $y = \ln \cos^2 x$

**15.** $y = x^2 \ln \dfrac{1}{x^2}$

**16.** $y = x \ln \sqrt{2 - x}$

Evaluate the following integrals.

**17.** $\displaystyle\int \frac{dx}{x + 7}$

**18.** $\displaystyle\int \frac{dx}{x - 2}$

**19.** $\displaystyle\int \frac{dx}{4x + 3}$

**20.** $\displaystyle\int \frac{dx}{5x + 10}$

**21.** $\displaystyle\int \frac{4dx}{3x - 4}$

**22.** $\displaystyle\int \frac{\cos x \, dx}{\sin x}$

**23.** $\int \tan x \, dx \quad \left(\textit{Hint: } \tan x = \dfrac{\sin x}{\cos x}\right)$

**24.** $\displaystyle\int \frac{dx}{x \ln \frac{1}{x}}$

## 10-7 THE DERIVATIVE OF THE EXPONENTIAL FUNCTION

To find the derivative of the exponential function $y = e^x$, we shall consider the inverse function

$$x = \ln y$$

Differentiating with respect to $x$,

$$1 = \frac{1}{y} y'$$

$$y' = y = e^x$$

Thus,

$$\boxed{D_x e^x = e^x}$$

For the more general exponential function $y = e^u$, where $u$ is a function of $x$, we apply the chain rule.

$$D_x e^u = e^u \, D_x u$$

**Example 1**

$$(a)\ D_x e^{x^2} = e^{x^2} \cdot 2x = 2xe^{x^2}$$

$$(b)\ D_x e^{-1/x} = e^{-1/x} \cdot \frac{1}{x^2} = \frac{e^{-1/x}}{x^2}$$

$$(c)\ D_x x^2 e^{-2x} = x^2 \cdot e^{-2x}(-2) + e^{-2x} \cdot 2x = e^{-2x}(2x - 2x^2)$$

The rule for the derivative of $e^u$ immediately leads us to the following antiderivative result.

$$\int e^u \, du = e^u + C$$

**Example 2**

$$(a)\ \int e^{2x}\,dx = \tfrac{1}{2}\int e^{2x} \cdot 2dx = \tfrac{1}{2}e^{2x} + C$$

$$(b)\ \int e^{-\frac{1}{2}x}\,dx = -2\int e^{-\frac{1}{2}x}(-\tfrac{1}{2})\,dx = -2e^{-\frac{1}{2}x} + C$$

$$(c)\ \int xe^{x^2}\,dx = \tfrac{1}{2}\int e^{x^2} \cdot 2x\,dx = \tfrac{1}{2}e^{x^2} + C$$

**Example 3** Find the maxima and minima and inflection points for the curve $y = e^{-2x^2}$ (Fig. 10-8).

*Solution*

$$\begin{aligned} y &= e^{-2x^2} \\ y' &= e^{-2x^2}(-4x) \\ &= -4xe^{-2x^2} \end{aligned}$$

Our only critical value is then $x = 0$.

$$\begin{aligned} y'' &= -4x \cdot e^{-2x^2}(-4x) - 4e^{-2x^2} \\ &= 4e^{-2x^2}(4x^2 - 1) \\ y''(0) &= -4 \end{aligned}$$

Thus, there is a relative maximum at $x = 0$. To find the inflection points, we set $y'' = 0$:

$$\begin{aligned} y'' &= 4e^{-2x^2}(4x^2 - 1) = 0 \\ 4x^2 &= 1 \\ x^2 &= \tfrac{1}{4} \\ x &= \pm\tfrac{1}{2} \end{aligned}$$

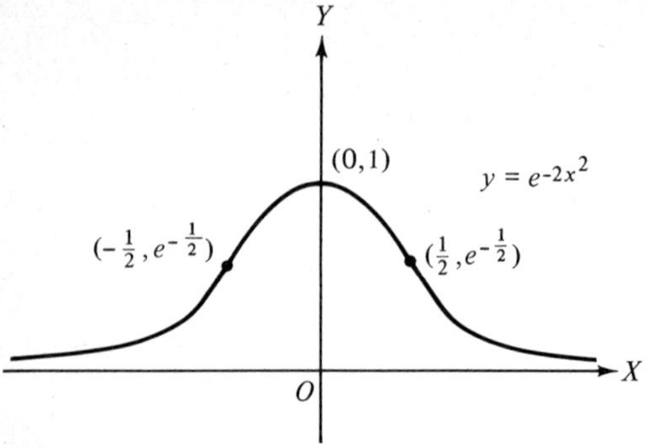


**Fig. 10-8**

**EXERCISES**

Differentiate the following functions.

**1.** $y = e^{2x}$

**2.** $y = e^{-3x^2}$

**3.** $y = e^{-1/x^2}$

**4.** $y = e^{x^2+1}$

**5.** $y = x^2e^{1/x}$

**6.** $y = e^{\sqrt{x+1}}$

**7.** $y = \dfrac{e^x}{x}$

**8.** $y = e^{x^2} \ln x^2$

**9.** $y = \dfrac{\ln x}{e^x}$

**10.** $y = \ln e^x$

**11.** $y = e^{\sin x}$

**12.** $y = \sin xe^{\cos x}$

Evaluate the following integrals.

**13.** $\int e^{3x}\, dx$

**14.** $\int xe^{-x^2}\, dx$

**15.** $\displaystyle\int \frac{e^{-1/x}}{x^2}\, dx$

**16.** $\int e^{\sin x} \cos x\, dx$

**17.** $\int e^{\cos x} \sin x\, dx$

**18.** $\displaystyle\int \frac{e^{\ln x}}{x}\, dx$

**19.** $\displaystyle\int_4^9 \frac{e^{\sqrt{x}}}{\sqrt{x}}\, dx$

**20.** $\displaystyle\int_0^1 \frac{e^x}{e^x + 1}\, dx$

**21.** Find the maxima and minima and points of inflection for the following curves.

(a) $y = e^{-x}$ (b) $y = xe^x$

(c) $y = xe^{-x}$ (d) $y = xe^{-x^2}$

## ▶ 10-8 LOGARITHMIC DIFFERENTIATION

It is often convenient to use logarithms as an aid in differentiating complicated expressions. We shall now illustrate the method of logarithmic differentiation which employs the rules for logarithms discussed in Sec. 10-5.

**Example 1** Find the derivative of $y = \sqrt[3]{\dfrac{x+2}{x-2}}$.

$$\begin{aligned}
\ln y &= \ln \sqrt[3]{\frac{x+2}{x-2}} \\
&= \tfrac{1}{3} \ln (x+2) - \tfrac{1}{3} \ln (x-2) \\
\frac{1}{y} y' &= \frac{1}{3} \frac{1}{x+2} - \frac{1}{3} \frac{1}{x-2} \\
&= \frac{1}{3(x+2)} - \frac{1}{3(x-2)} \\
&= \frac{-4}{3(x+2)(x-2)} \\
y' &= \frac{-4}{3(x+2)(x-2)} y \\
&= \frac{-4}{3(x+2)(x-2)} \sqrt[3]{\frac{x+2}{x-2}} \\
&= \frac{-4}{3(x+2)^{\frac{2}{3}}(x-2)^{\frac{4}{3}}}
\end{aligned}$$

**Example 2** Find the derivative of $y = x^{\ln x}$.

*Solution*

$$\begin{aligned}
y &= x^{\ln x} \\
\ln y &= \ln x^{\ln x} \\
&= (\ln x)(\ln x) \\
&= (\ln x)^2 \\
\frac{1}{y} y' &= 2(\ln x) \frac{1}{x} \\
y' &= \frac{2 \ln x}{x} y \\
&= \frac{(2 \ln x)x^{\ln x}}{x}
\end{aligned}$$

**EXERCISES**

Use logarithmic differentiation to find $\dfrac{dy}{dx}$.

**1.** $y = (x+1)^2(2x-1)$

**2.** $y = x\sqrt{x+3}$

**3.** $y = \sqrt{\dfrac{x+1}{x-2}}$

**4.** $y = \dfrac{\sqrt{x+2}\,(x+4)}{(x-3)^2}$

**5.** $y = \dfrac{x\sqrt[3]{x+3}}{2x-1}$

**6.** $y = \dfrac{\sqrt{x+1}}{\sqrt[3]{x-1}}$

**7.** $y = x^{\sin x}$

**8.** $y = (\sin x)^{\cos x}$

## ▶ 10-9 HYPERBOLIC FUNCTIONS

In applied mathematics it is found that certain combinations of exponential functions occur quite frequently, and consequently separate names and definitions have been given to these functions. The functions are called *hyperbolic functions* and they are defined as follows.

$$\text{Hyperbolic sine of } x = \sinh x = \frac{e^x - e^{-x}}{2}$$

$$\text{Hyperbolic cosine of } x = \cosh x = \frac{e^x + e^{-x}}{2}$$

$$\text{Hyperbolic tangent of } x = \tanh x = \frac{e^x - e^{-x}}{e^x + e^{-x}}$$

$$\text{Hyperbolic cotangent of } x = \coth x = \frac{e^x + e^{-x}}{e^x - e^{-x}}$$

$$\text{Hyperbolic secant of } x = \operatorname{sech} x = \frac{2}{e^x + e^{-x}}$$

$$\text{Hyperbolic cosecant of } x = \operatorname{csch} x = \frac{2}{e^x - e^{-x}}$$

The trigonometric functions that we have studied are sometimes referred to as *circular functions* because of the relationship between the definitions of the trigonometric functions and a circle with center at the origin. There is a similar relationship between the hyperbolic functions and the conic section called a rectangular hyperbola, and hence the name hyperbolic functions. It will also be found that the hyperbolic functions have many properties analogous to the trigonometric functions. We shall now illustrate several of these properties.

$$\frac{\sinh x}{\cosh x} = \frac{\dfrac{e^x - e^{-x}}{2}}{\dfrac{e^x + e^{-x}}{2}} = \frac{e^x - e^{-x}}{e^x + e^{-x}} = \tanh x$$

$$\boxed{\tanh x = \frac{\sinh x}{\cosh x}}$$

$$\frac{\cosh x}{\sinh x} = \frac{\dfrac{e^x + e^{-x}}{2}}{\dfrac{e^x - e^{-x}}{2}} = \frac{e^x + e^{-x}}{e^x - e^{-x}} = \coth x$$

$$\boxed{\coth x = \frac{\cosh x}{\sinh x}}$$

$$\operatorname{sech} x = \frac{2}{e^x + e^{-x}} = \frac{1}{\cosh x} \qquad \operatorname{csch} x = \frac{2}{e^x - e^{-x}} = \frac{1}{\sinh x}$$

$$\boxed{\operatorname{sech} x = \frac{1}{\cosh x} \qquad \operatorname{csch} x = \frac{1}{\sinh x}}$$

$$\begin{aligned} \cosh^2 x - \sinh^2 x &= \left(\frac{e^x + e^{-x}}{2}\right)^2 - \left(\frac{e^x - e^{-x}}{2}\right)^2 \\ &= \tfrac{1}{4}(e^{2x} + 2 + e^{-2x}) - \tfrac{1}{4}(e^{2x} - 2 + e^{-2x}) \\ &= 1 \end{aligned}$$

$$\boxed{\cosh^2 x - \sinh^2 x = 1}$$

$$\begin{aligned} \tanh^2 x + \operatorname{sech}^2 x &= \frac{\sinh^2 x}{\cosh^2 x} + \frac{1}{\cosh^2 x} \\ &= \frac{\sinh^2 x + 1}{\cosh^2 x} \\ &= \frac{\cosh^2 x}{\cosh^2 x} = 1 \end{aligned}$$

$$\boxed{\tanh^2 x + \operatorname{sech}^2 x = 1}$$

$$\begin{aligned} \sinh(-x) &= \frac{e^{(-x)} - e^{-(-x)}}{2} \\ &= \frac{e^{-x} - e^{x}}{2} \\ &= -\frac{(e^x - e^{-x})}{2} = -\sinh x \end{aligned}$$

$$\boxed{\sinh(-x) = -\sinh x}$$

Similar identities will be left to the student in the exercises.

We shall now consider differentiation and integration of the hyperbolic functions.

$$\sinh x = \frac{e^x - e^{-x}}{2} = \frac{e^x}{2} - \frac{e^{-x}}{2}$$

$$D_x \sinh x = \frac{e^x}{2} + \frac{e^{-x}}{2} = \frac{e^x + e^{-x}}{2}$$

$$\boxed{D_x \sinh x = \cosh x}$$

$$\cosh x = \frac{e^x + e^{-x}}{2} = \frac{e^x}{2} + \frac{e^{-x}}{2}$$

$$D_x \cosh x = \frac{e^x}{2} - \frac{e^{-x}}{2} = \frac{e^x - e^{-x}}{2}$$

$$\boxed{D_x \cosh x = \sinh x}$$

$$\tanh x = \frac{\sinh x}{\cosh x}$$

$$\begin{aligned} D_x \tanh x &= \frac{\cosh x\, D_x \sinh x - \sinh x\, D_x \cosh x}{\cosh^2 x} \\ &= \frac{\cosh^2 x - \sinh^2 x}{\cosh^2 x} \\ &= \frac{1}{\cosh^2 x} \\ &= \operatorname{sech}^2 x \end{aligned}$$

$$\boxed{D_x \tanh x = \operatorname{sech}^2 x}$$

It will be left as an exercise to show that

$$\boxed{\begin{aligned} D_x \coth x &= -\operatorname{csch}^2 x \\ D_x \operatorname{sech} x &= -\operatorname{sech} x \tanh x \\ D_x \operatorname{csch} x &= -\operatorname{csch} x \coth x \end{aligned}}$$

**Example 1**

(a) $D_x \sinh x^2 = (\cosh x^2)2x$
$\qquad = 2x \cosh x^2$

(b) $D_x \tanh (1 - 2x) = \operatorname{sech}^2 (1 - 2x)(-2)$
$\qquad = -2 \operatorname{sech}^2 (1 - 2x)$

The following integration formulas follow immediately from our derivative formulas.

$$\begin{aligned}
\int \sinh u\, du &= \cosh u + C\\
\int \cosh u\, du &= \sinh u + C\\
\int \operatorname{sech}^2 u\, du &= \tanh u + C\\
\int \operatorname{csch}^2 u\, du &= -\coth u + C\\
\int \operatorname{sech} u \tanh u\, du &= -\operatorname{sech} u + C\\
\int \operatorname{csch} u \coth u\, du &= -\operatorname{csch} u + C
\end{aligned}$$

**Example 2**

$$\begin{aligned}
(a)\ \int \cosh 2x\, dx &= \tfrac{1}{2}\int (\cosh 2x)\, 2dx\\
&= \tfrac{1}{2}\sinh 2x + C\\
(b)\ \int x \sinh x^2\, dx &= \tfrac{1}{2}\int (\sinh x^2) 2x\, dx\\
&= \tfrac{1}{2}\cosh x^2 + C
\end{aligned}$$

**EXERCISES**

Verify the following identities.

**1.** $\cosh(-x) = \cosh x$

**2.** $\coth^2 x - \operatorname{csch}^2 x = 1$

**3.** $\sinh(x + y) = \sinh x \cosh y + \cosh x \sinh y$

**4.** $\cos(x + y) = \cosh x \cosh y + \sinh x \sinh y$

**5.** $\sinh 2x = 2 \sinh x \cosh x$

**6.** $\cosh 2x = \cosh^2 x + \sinh^2 x$

Verify the following rules for differentiation.

**7.** $D_x \coth x = -\operatorname{csch}^2 x$

**8.** $D_x \operatorname{sech} x = -\operatorname{sech} x \tanh x$

**9.** $D_x \operatorname{csch} x = -\operatorname{csch} x \coth x$

Differentiate the following functions.

**10.** $y = \sinh 3x$

**11.** $y = \cosh x^2$

**12.** $y = \tanh(x - 1)$

**13.** $y = \sinh(2x + 1)$

**14.** $y = \operatorname{sech}(1 - x^2)$

**15.** $y = \cosh 2x - \sinh 2x$

**16.** $y = \coth^2 x$

**17.** $y = x \operatorname{csch} x$

Evaluate the following integrals.

**18.** $\int \sinh \frac{1}{2}x\, dx$

**19.** $\int \cosh(1 - x)\, dx$

**20.** $\int \operatorname{sech}^2 2x\, dx$

**21.** $\int 3x \sinh x^2\, dx$

**22.** $\int x^2 \operatorname{csch}^2 x^3\, dx$

**23.** $\int \operatorname{sech} 3x \tanh 3x\, dx$

**24.** $\int \sinh^2 2x \cosh 2x\, dx$

**25.** $\int \tanh x\, dx$

# 11
# Methods of Integration

## 11-1 REVIEW OF THE POWER FORMULA AND INTEGRATION OF TRANSCENDENTAL FUNCTIONS

In our previous work we have developed the following formulas:

$$\int u^n \, du = \frac{u^{n+1}}{n+1} + C \qquad n \neq -1$$

$$\int \sin u \, du = -\cos u + C$$

$$\int \cos u \, du = \sin u + C$$

$$\int \sec^2 u \, du = \tan u + C$$

$$\int \csc^2 u \, du = -\cot u + C$$

$$\int \tan u \sec u \, du = \sec u + C$$

$$\int \cot u \csc u \, du = -\csc u + C$$

$$\int e^u \, du = e^u + C$$

$$\int \frac{du}{u} = \ln |u| + C$$

**Example**

(a) $\int \sin^3 x \cos x\, dx = \dfrac{\sin^4 x}{4} + C$

(b) $\displaystyle\int \frac{dx}{2x+4} = \frac{1}{2}\int \frac{2\,dx}{2x+4} = \tfrac{1}{2}\ln|2x+4| + C$

(c) $$\int \frac{e^x\,dx}{\sqrt{1+e^x}} = \int (1+e^x)^{-\frac{1}{2}} e^x\,dx = \frac{(1+e^x)^{\frac{1}{2}}}{\frac{1}{2}} + C = 2\sqrt{1+e^x} + C$$

(d) $$\int \frac{\sec^2 x\,dx}{\sqrt{1+\tan x}} = \int (1+\tan x)^{-\frac{1}{2}} \sec^2 x\,dx = \frac{(1+\tan x)^{\frac{1}{2}}}{\frac{1}{2}} + C = 2\sqrt{1+\tan x} + C$$

**EXERCISES**

Evaluate the following integrals.

**1.** $\int (1-x^3)^2 x^2\,dx$

**2.** $\displaystyle\int \frac{1+x}{\sqrt{x}}\,dx$

**3.** $\displaystyle\int \frac{x\,dx}{1+x^2}$

**4.** $\int (1-x^2)^{\frac{3}{2}} x\,dx$

**5.** $\int \sin^2 2x \cos 2x\,dx$

**6.** $\displaystyle\int \frac{dx}{e^{2x}}$

**7.** $\int \sec^2 3x\,dx$

**8.** $\displaystyle\int \frac{1}{\ln x}\frac{dx}{x}$

**9.** $\int \cot x\,dx$

**10.** $\int (1-\sin^2 x)\cos x\,dx$

**11.** $\int \tan 3x \sec 3x\,dx$

**12.** $\int \sec^3 x(\tan x \sec x)\,dx$

**13.** $\int (1+e^{x^2}) x e^{x^2}\,dx$

**14.** $\displaystyle\int \frac{\cos x\,dx}{1+\sin x}$

**15.** $\int e^{\sin x}\cos x\,dx$

**16.** $\int \csc^2 2x\,dx$

**17.** $\int x e^{(1-x^2)}\,dx$

**18.** $\int \tan 2x\,dx$

**19.** $\int \sin 3x \cos 3x\,dx$

**20.** $\int \sqrt{2x-x^2}\,(1-x)\,dx$

**21.** $\displaystyle\int \frac{1+\sin x}{x-\cos x}\,dx$

**22.** $\int \cot\,(1-x)\,dx$

**23.** $\displaystyle\int \frac{e^{2/x}}{x^2}\,dx$

**24.** $\int (1+\sec^2 2x)\,dx$

**25.** $\displaystyle\int \frac{x^2\,dx}{1+x^3}$

**26.** $\int (2-\sin^2 x)\sin x \cos x\,dx$

**27.** $\int (1-e^x)^3 e^x\,dx$

**28.** $\displaystyle\int \frac{\sin x}{\cos^2 x}\,dx$

**29.** $\displaystyle\int \frac{x^2\,dx}{e^{x^3}}$

**30.** $\displaystyle\int \frac{\cos x\,dx}{1-\sin x}$

## 11-2 INTEGRATION BY PARTS

If $u$ and $v$ are functions of $x$, our product rule for differentials tells us that

$$\begin{aligned} d(uv) &= u\,dv + v\,du \\ u\,dv &= d(uv) - v\,du \end{aligned}$$

Integrating both sides,

$$\int u\,dv = uv - \int v\,du$$

We thus have a method of reversing the product rule. The above method is called *integration by parts.*

**Example 1** Evaluate $\int x \sin x\,dx$.

*Solution* Let

$$\begin{aligned} u &= x \qquad \text{and} \qquad dv = \sin x\,dx \\ \text{then} \qquad du &= dx \qquad\qquad\qquad v = -\cos x + C' \\ \int x \sin x\,dx &= x(-\cos x + C') - \int(-\cos x + C')\,dx \\ &= (-x\cos x + C'x) + (\sin x - C'x + C) \\ &= -x\cos x + \sin x + C \end{aligned}$$

We note that the constant $C'$ found in the antiderivative of $dv$ appears in both the terms $uv$ and $v\,du$. In one term we have $C'u$ and in the other, $-\int C'\,du = -C'u$. Hence, the constant $C'$ will always cancel out, and it will not be necessary to include it in finding the solution.

**Example 2** Evaluate $\int xe^{-x}\,dx$.

*Solution* Let

$$\begin{aligned} u &= x \qquad \text{and} \qquad dv = e^{-x}\,dx \\ \text{then} \qquad du &= dx \qquad\qquad\qquad v = -e^{-x} \\ \int xe^{-x}\,dx &= -xe^{-x} - \int(-e^{-x})\,dx \\ &= -xe^{-x} - e^{-x} + C \end{aligned}$$

where $C$ is our constant of integration for the given integral.

**Example 3** Evaluate $\int x^2e^x\,dx$.

*Solution* Let

$$\begin{aligned} u &= x^2 \qquad \text{and} \qquad dv = e^x\,dx \\ \text{then} \qquad du &= 2x\,dx \qquad\qquad\qquad v = e^x \\ \int x^2e^x\,dx &= x^2e^x - 2\int xe^x\,dx \end{aligned}$$

We must repeat the method of integration by parts to evaluate

$$\int xe^x\,dx$$

Let

$$\begin{aligned} u &= x \qquad \text{and} \qquad dv = e^x\,dx \\ \text{then} \qquad du &= dx \qquad\qquad\qquad v = e^x \\ \int xe^x &= xe^x - \int e^x\,dx \\ &= xe^x - e^x + C \\ \text{Hence,} \qquad \int x^2e^x\,dx &= x^2e^x - 2(xe^x - e^x) + C' \end{aligned}$$

**EXERCISES**

Evaluate the following integrals.

**1.** $\int xe^{2x}\,dx$

**2.** $\int x \cos x\,dx$

**3.** $\int x \ln x\,dx$

**4.** $\int \ln x\,dx$

**5.** $\int x \csc^2 x\,dx$

**6.** $\displaystyle\int \frac{x\,dx}{e^x}$

**7.** $\int x^2 e^{-x}\,dx$

**8.** $\int x^2 \sin x\,dx$

**9.** $\int \ln (x + 2)\,dx$

**10.** $\int x^2 \cos 2x\,dx$

**11.** $\int x^2 \ln x\,dx$

**12.** $\int x \sec^2 x\,dx$

## 11-3 $\int$ sec $x\,dx$

To evaluate $\int \sec x\,dx$, we point out that

$$\begin{aligned}\int \sec x\,dx &= \int \sec x \frac{\sec x + \tan x}{\sec x + \tan x}\,dx \\ &= \int \frac{\sec^2 x + \sec x \tan x}{\sec x + \tan x}\,dx \\ &= \ln|\sec x + \tan x| + C\end{aligned}$$

$$\boxed{\int \sec x\,dx = \ln|\sec x + \tan x| + C}$$

## 11-4 $\int$ sec$^3$ $x\,dx$

We shall use the method of integration by parts to evaluate $\int \sec^3 x\,dx$.

$$\int \sec^3 x\,dx = \int \sec x \sec^2 x\,dx$$

Now, let $u = \sec x$ and $dv = \sec^2 x\,dx$. Then

$$du = \sec x \tan x\,dx \qquad v = \tan x$$

Hence, $$\int \sec^3 x\,dx = \sec x \tan x - \int \sec x \tan^2 x\,dx$$

We must still evaluate $\int \sec x \tan^2 x\,dx$, which does not appear to be a simpler task than evaluating the original integral. Since

$$\begin{aligned}\tan^2 x &= \sec^2 x - 1 \\ \sec x \tan^2 x &= \sec x(\sec^2 x - 1) = \sec^3 x - \sec x \\ \int \sec^3 x\,dx &= \sec x \tan x - \int(\sec^3 x - \sec x)\,dx \\ &= \sec x \tan x - \int \sec^3 x\,dx + \int \sec x\,dx\end{aligned}$$

Transposing $\int \sec^3 x\,dx$ from the right side to the left side gives us

$$\begin{aligned}2\int \sec^3 x\,dx &= \sec x \tan x + \int \sec x\,dx \\ &= \sec x \tan x + \ln|\sec x + \tan x| + C'\end{aligned}$$

$$\boxed{\int \sec^3 x\,dx = \tfrac{1}{2}(\sec x \tan x + \ln|\sec x + \tan x|) + C}$$

**EXERCISES**

Apply the same procedure used in finding $\int \sec^3 x\, dx$ to evaluate

**1.** $\int e^x \sin x\, dx$

**2.** $\int e^x \cos x\, dx$

## 11-5 THE METHOD OF TRIGONOMETRIC SUBSTITUTION

Trigonometric substitutions are sometimes used to evaluate integrals. In particular, we shall be using the following trigonometric identities:

$$\begin{aligned} 1 + \tan^2 x &= \sec^2 x \\ \sec^2 x - 1 &= \tan^2 x \\ 1 - \sin^2 x &= \cos^2 x \end{aligned}$$

These identities will be used to evaluate integrals involving $\sqrt{a^2 + u^2}$, $\sqrt{u^2 - a^2}$, and $\sqrt{a^2 - u^2}$.

1. (Fig. 11-1*a*.) For $\sqrt{a^2 + u^2}$, let

$$\begin{aligned} u &= a \tan \theta \\ \text{then} \qquad du &= a \sec^2 \theta\, d\theta \\ \sqrt{a^2 + u^2} &= \sqrt{a^2 + a^2 \tan^2 \theta} = a\sqrt{1 + \tan^2 \theta} = a \sec \theta \end{aligned}$$

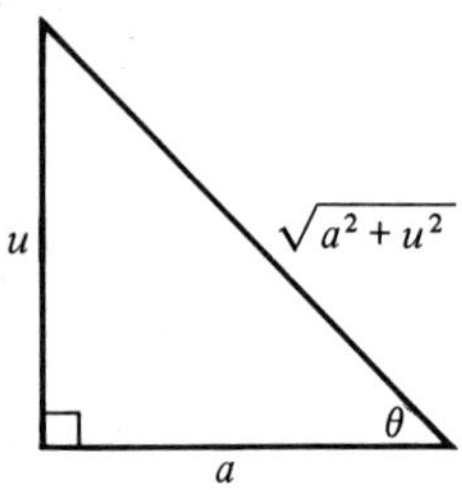


**Fig. 11-1a**

2. (Fig. 11-1*b*.) For $\sqrt{u^2 - a^2}$, let

$$\begin{aligned} u &= a \sec \theta \\ \text{then} \qquad du &= a \sec \theta \tan \theta\, d\theta \\ \sqrt{u^2 - a^2} &= \sqrt{a^2 \sec^2 \theta - a^2} = a\sqrt{\sec^2 \theta - 1} = a \tan \theta \end{aligned}$$

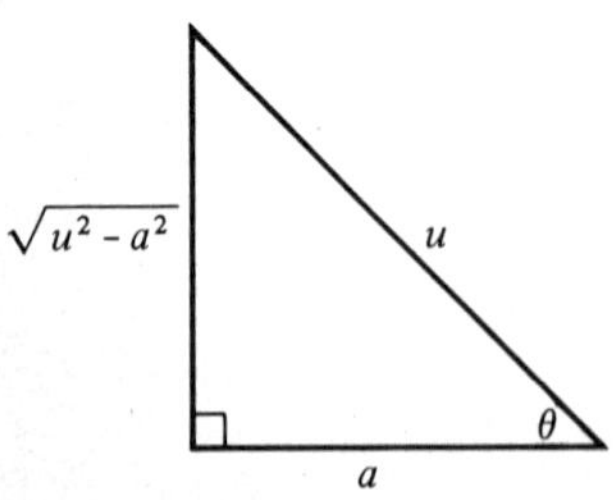


**Fig. 11-1b**

3. (Fig. 11-1*c*.) For $\sqrt{a^2 - u^2}$, let

$$u = a \sin \theta$$

then

$$du = a \cos \theta \, d\theta$$

$$\sqrt{a^2 - u^2} = \sqrt{a^2 - a^2 \sin^2 \theta} = a\sqrt{1 - \sin^2 \theta} = a \cos \theta$$

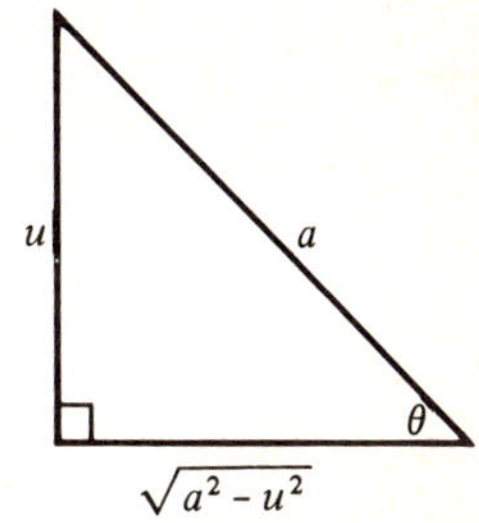


Fig. 11-1c

For $\sqrt{a^2 + u^2}$ substitute $u = a \tan \theta$.
For $\sqrt{u^2 - a^2}$ substitute $u = a \sec \theta$.
For $\sqrt{a^2 - u^2}$ substitute $u = a \sin \theta$.

**Example 1** Evaluate $\int \sqrt{4 + x^2}\, dx$.

*Solution* Let $x = 2 \tan \theta$. Then

$$dx = 2 \sec^2 \theta \, d\theta$$

$$\sqrt{4 + x^2} = \sqrt{4 + 4 \tan^2 \theta} = 2\sqrt{1 + \tan^2 \theta} = 2 \sec \theta$$

$$\int \sqrt{4 + x^2}\, dx = \int 2 \sec \theta \cdot 2 \sec^2 \theta \, d\theta$$

$$= 4 \int \sec^3 \theta \, d\theta$$

$$= 4(\tfrac{1}{2})(\sec \theta \tan \theta + \ln |\sec \theta + \tan \theta|) + C'$$

We wish to have our result in terms of $x$. Since $x = 2 \tan \theta$,

$$\tan \theta = \frac{x}{2}$$

We can construct a right triangle with angle $\theta$, side opposite $\theta$ equal to $x$, and adjacent side equal to 2 (Fig. 11-2). Using the pythagorean theorem, we find

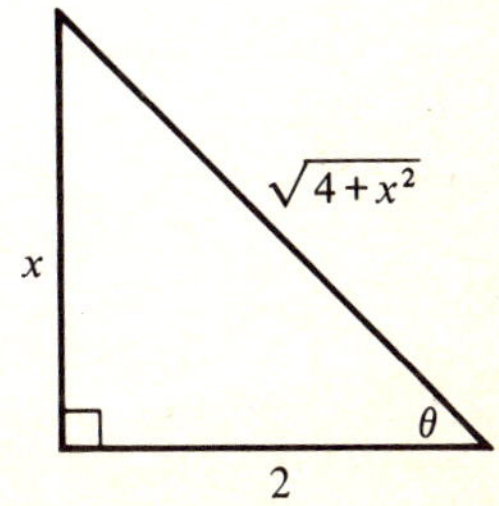


Fig. 11-2

that the hypotenuse equals $\sqrt{4 + x^2}$. Hence,

$$\tan\theta = \frac{x}{2}$$

$$\sec\theta = \frac{\sqrt{4 + x^2}}{2}$$

$$\begin{aligned}\int \sqrt{4 + x^2}\,dx &= 2\left(\frac{\sqrt{4 + x^2}}{2}\cdot\frac{x}{2} + \ln\left|\frac{\sqrt{4 + x^2} + x}{2}\right|\right) + C' \\ &= \tfrac{1}{2}x\sqrt{4 + x^2} + 2\ln|\sqrt{4 + x^2} + x| - 2\ln 2 + C' \\ &= \tfrac{1}{2}x\sqrt{4 + x^2} + 2\ln|\sqrt{4 + x^2} + x| + C\end{aligned}$$

**Example 2** Evaluate $\int x\sqrt{1 - x^2}\,dx$ (Fig. 11-3).

*Solution* Using the method of trigonometric substitution, let

$$x = \sin\theta$$
$$dx = \cos\theta\,d\theta$$

then

$$\sqrt{1 - x^2} = \sqrt{1 - \sin^2\theta} = \cos\theta$$

$$\begin{aligned}\int x\sqrt{1 - x^2}\,dx &= \int \sin\theta\cdot\cos\theta\cdot\cos\theta\,d\theta \\ &= \int \cos^2\theta\sin\theta\,d\theta \\ &= -\frac{\cos^3\theta}{3} + C\end{aligned}$$

Since $x = \sin\theta$,

$$\cos\theta = \sqrt{1 - x^2}$$

$$\int x\sqrt{1 - x^2}\,dx = -\frac{(1 - x^2)^{\frac{3}{2}}}{3} + C$$

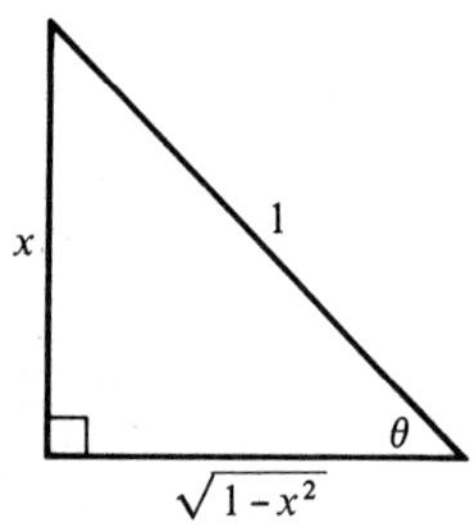


Fig. 11-3

As a check, we note that our integral can also be evaluated by using the power rule:

$$\begin{aligned}\int x\sqrt{1 - x^2}\,dx &= -\tfrac{1}{2}\int \sqrt{1 - x^2}\,(-2x)\,dx \\ &= -\frac{1}{2}\,\frac{(1 - x^2)^{\frac{3}{2}}}{\frac{3}{2}} + C \\ &= -\frac{(1 - x^2)^{\frac{3}{2}}}{3} + C\end{aligned}$$

**Example 3** Evaluate $\int \sqrt{x^2 - 9}\, dx$ (Fig. 11-4).

*Solution* Let $x = 3 \sec \theta$. Then

$$\begin{aligned}
dx &= 3 \sec \theta \tan \theta \, d\theta \\
\sqrt{x^2 - 9} &= \sqrt{9 \sec^2 \theta - 9} = 3\sqrt{\sec^2 \theta - 1} = 3 \tan \theta \\
\int \sqrt{x^2 - 9}\, dx &= \int 3 \tan \theta \cdot 3 \sec \theta \tan \theta \, d\theta \\
&= 9 \int \tan^2 \theta \sec \theta \, d\theta \\
&= 9 \int (\sec^2 \theta - 1) \sec \theta \, d\theta \\
&= 9 \int (\sec^3 \theta - \sec \theta) \, d\theta \\
&= 9 \int \sec^3 d\theta - 9 \int \sec \theta \, d\theta \\
&= 9(\tfrac{1}{2})(\sec \theta \tan \theta + \ln |\sec \theta + \tan \theta|) - 9 \ln |\sec \theta + \tan \theta| + C' \\
&= \tfrac{9}{2} \sec \theta \tan \theta - \tfrac{9}{2} \ln |\sec \theta + \tan \theta| + C' \\
&= \frac{9}{2} \cdot \frac{x}{3} \cdot \frac{\sqrt{x^2 - 9}}{3} - \frac{9}{2} \ln \left| \frac{x}{3} + \frac{\sqrt{x^2 - 9}}{3} \right| + C' \\
&= \tfrac{1}{2} x \sqrt{x^2 - 9} - \tfrac{9}{2} \ln |x + \sqrt{x^2 - 9}| + \tfrac{9}{2} \ln 3 + C' \\
&= \tfrac{1}{2} x \sqrt{x^2 - 9} - \tfrac{9}{2} \ln |x + \sqrt{x^2 - 9}| + C
\end{aligned}$$

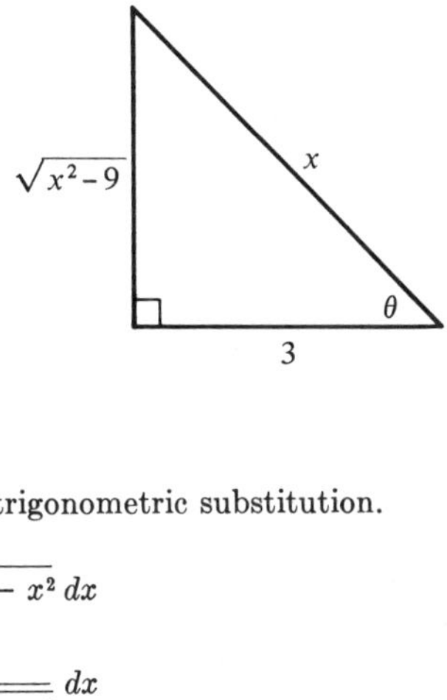


**Fig. 11-4**

**EXERCISES**

Evaluate the following integrals using the method of trigonometric substitution.

**1.** $\displaystyle\int \frac{dx}{x^2 + 1}$

**2.** $\int x \sqrt{4 - x^2}\, dx$

**3.** $\int x(1 - x^2)^{\frac{3}{2}}\, dx$

**4.** $\displaystyle\int \frac{x}{\sqrt{4 - x^2}}\, dx$

**5.** $\int \sqrt{9 + x^2}\, dx$

**6.** $\int \sqrt{x^2 - a^2}\, dx$

**7.** $\int (16 - x^2)^{-\frac{3}{2}}\, dx$

**8.** $\displaystyle\int \frac{dx}{x^2 \sqrt{x^2 + 1}}$

## 11-6 THE METHOD OF PARTIAL FRACTIONS

In our study of algebra we have often had occasion to combine several fractions into one fraction with a common denominator. We shall now consider the reverse process of breaking up a fraction into sums of fractions whose denominators have either linear or irreducible quadratic factors. This process is known as the *method of partial fractions*. In algebra it can be shown that any polynomial in $x$ with real coefficients can be expressed as a product of real linear and quadratic factors.

**Example 1**

$$\frac{5}{x+1} + \frac{7}{x+2} = \frac{5(x+2) + 7(x+1)}{(x+1)(x+2)} = \frac{12x+17}{(x+1)(x+2)}$$

In employing the method of partial fractions, we are given an expression such as

$$\frac{12x+17}{(x+1)(x+2)}$$

and we must determine constants $A$ and $B$ such that

$$\frac{12x+17}{(x+1)(x+2)} = \frac{A}{x+1} + \frac{B}{x+2}$$

Assuming that we did not know the values of $A$ and $B$, we shall proceed to solve for $A$ and $B$.

$$\begin{aligned} \frac{12x+17}{(x+1)(x+2)} &= \frac{A(x+2) + B(x+1)}{(x+1)(x+2)} \\ 12x + 17 &= Ax + 2A + Bx + B \\ &= (A+B)x + 2A + B \end{aligned}$$

Since the two sides of our equation must be the same for all values of $x$, the coefficients of like powers of $x$ must be equal. Thus,

$$\begin{aligned} A + B &= 12 \\ 2A + B &= 17 \\ A = 5 \qquad &\text{and} \qquad B = 7 \end{aligned}$$

**Example 2** Express $\dfrac{5x+4}{x^2+x-2}$ as a sum of partial fractions.

*Solution*

$$\frac{5x+4}{x^2+x-2} = \frac{5x+4}{(x-1)(x+2)}$$

Let
$$\frac{5x+4}{(x-1)(x+2)} = \frac{A}{x-1} + \frac{B}{x+2}$$

$$\begin{aligned} 5x + 4 &= A(x+2) + B(x-1) \\ 5x + 4 &= (A+B)x + 2A - B \\ A + B &= 5 \\ 2A - B &= 4 \end{aligned}$$

Hence,
$$\begin{aligned} 3A &= 9 \\ A = 3 \qquad & B = 2 \end{aligned}$$

An alternate way of solving for $A$ and $B$ in the equation

$$5x + 4 = A(x+2) + B(x-1)$$

is to note that this equation must hold for all values of $x$. In particular:

When $x = 1$, we get

$$\begin{aligned} 5(1) + 4 &= A(1 + 2) + B(0) \\ 9 &= 3A \\ A &= 3 \end{aligned}$$

When $x = -2$, we get

$$\begin{aligned} -10 + 4 &= A(0) + B(-3) \\ -6 &= -3B \\ B &= 2 \end{aligned}$$

In employing the method of partial fractions to a fraction in the form $\dfrac{f(x)}{g(x)}$, it is important that the fraction be a proper fraction, i.e., a fraction where the degree of the numerator is less than the degree of the denominator. If we are given an improper fraction, then we can perform long division to obtain a mixed form consisting of a polynomial and a proper fraction.

**Example 3**

$$\frac{x^3 + 2x - 4}{x^2 + 1} = x + \frac{x - 4}{x^2 + 1}$$

where $\dfrac{x - 4}{x^2 + 1}$ is a proper fraction.

We shall first consider those fractions whose denominators can be factored into linear factors. For every factor in the form $x - a$, where $(x - a)^m$ is the highest power of $x - a$ in the denominator, we shall assign the following sum of $m$ partial fractions:

$$\frac{A_1}{x - a} + \frac{A_2}{(x - a)^2} + \frac{A_3}{(x - a)^3} + \cdots + \frac{A_m}{(x - a)^m}$$

**Example 4** Express $\dfrac{x + 2}{x^2(x + 1)}$ as a sum of partial fractions.

*Solution* We have two linear factors $x$ and $x + 1$. The highest power of $x$ is 2 and the highest power of $x + 1$ is 1. Hence,

$$\begin{aligned} \frac{x + 2}{x^2(x + 1)} &= \frac{A}{x} + \frac{B}{x^2} + \frac{C}{x + 1} \\ x + 2 &= Ax(x + 1) + B(x + 1) + Cx^2 \\ &= Ax^2 + Ax + Bx + B + Cx^2 \\ &= (A + C)x^2 + (A + B)x + B \\ A + C &= 0 \\ A + B &= 1 \\ B &= 2 \end{aligned}$$

Therefore $\qquad A = -1 \qquad B = 2 \qquad C = 1$

$$\frac{x+2}{x^2(x+1)} = \frac{-1}{x} + \frac{2}{x^2} + \frac{1}{x+1}$$

**Example 5** Evaluate $\int \frac{x+2}{x^2(x+1)}\,dx$.

*Solution* Using the result of the previous example,

$$\frac{x+2}{x^2(x+1)} = \frac{-1}{x} + \frac{2}{x^2} + \frac{1}{x+1}$$

$$\begin{aligned}
\int \frac{x+2}{x^2(x+1)}\,dx &= -\int \frac{dx}{x} + 2\int \frac{dx}{x^2} + \int \frac{dx}{x+1} \\
&= -\ln|x| + 2\,\frac{x^{-1}}{-1} + \ln|x+1| + C \\
&= -\ln|x| - \frac{2}{x} + \ln|x+1| + C \\
&= \ln\left|\frac{x+1}{x}\right| - \frac{2}{x} + C
\end{aligned}$$

**Example 6** Evaluate $\int \frac{x\,dx}{x^2+2x+1}$.

*Solution*

$$\begin{aligned}
\int \frac{x\,dx}{x^2+2x+1} &= \int \frac{x\,dx}{(x+1)^2} \\
\frac{x}{(x+1)^2} &= \frac{A}{x+1} + \frac{B}{(x+1)^2} \\
x &= A(x+1) + B \\
&= Ax + A + B \\
A &= 1 \\
A + B &= 0 \\
A &= 1 \qquad B = -1 \\
\int \frac{x\,dx}{(x+1)^2} &= \int \frac{dx}{x+1} - \int \frac{dx}{(x+1)^2} \\
&= \int \frac{dx}{x+1} - \int (x+1)^{-2}\,dx \\
&= \ln|x+1| - \frac{(x+1)^{-1}}{-1} + C \\
&= \ln|x+1| + \frac{1}{x+1} + C
\end{aligned}$$

**Example 7** Evaluate $\int \frac{x^2 + 2x + 5}{x(x+1)^2}\,dx$.

*Solution*

$$\begin{aligned}
\frac{x^2 + 2x + 5}{x(x+1)^2} &= \frac{A}{x} + \frac{B}{x+1} + \frac{C}{(x+1)^2} \\
x^2 + 2x + 5 &= A(x+1)^2 + Bx(x+1) + Cx \\
&= A(x^2 + 2x + 1) + B(x^2 + x) + Cx \\
&= (A + B)x^2 + (2A + B + C)x + A \\
A + B &= 1 \\
2A + B + C &= 2 \\
A &= 5 \\
A = 5 \qquad B &= -4 \qquad C = -4 \\
\int \frac{x^2 + 2x + 5}{x(x+1)^2} &= 5\int \frac{dx}{x} - 4\int \frac{dx}{x+1} - 4\int \frac{dx}{(x+1)^2} \\
&= 5 \ln |x| - 4 \ln |x + 1| + \frac{4}{x+1} + C
\end{aligned}$$

**Example 8** Evaluate $\int \frac{x^2 + 2x}{x^2 - 1}\,dx$.

*Solution* This is an improper fraction. Using long division,

$$\begin{aligned}
\frac{x^2 + 2x}{x^2 - 1} &= 1 + \frac{2x + 1}{x^2 - 1} \\
\frac{2x + 1}{x^2 - 1} &= \frac{A}{x - 1} + \frac{B}{x + 1} \\
2x + 1 &= A(x + 1) + B(x - 1) \\
&= (A + B)x + (A - B) \\
A + B &= 2 \\
A - B &= 1 \\
A &= \tfrac{3}{2} \qquad B = \tfrac{1}{2} \\
\int \frac{x^2 + 2x}{x^2 - 1}\,dx &= \int dx + \frac{3}{2}\int \frac{dx}{x - 1} + \frac{1}{2}\int \frac{dx}{x + 1} \\
&= x + \tfrac{3}{2} \ln |x - 1| + \tfrac{1}{2} \ln |x + 1| + C
\end{aligned}$$

Thus far we have only considered a rule for repeated linear factors. If

$$(x^2 + bx + c)^n$$

is the highest power of the irreducible quadratic factor $x^2 + bx + c$, then to this factor we shall assign the following sum of $n$ partial fractions:

$$\frac{B_1x + C_1}{x^2 + bx + c} + \frac{B_2x + C_2}{(x^2 + bx + c)^2} + \cdots + \frac{B_nx + C_n}{(x^2 + bx + c)^n}$$

**Example 9** Evaluate $\int \frac{x^3\,dx}{(x^2+1)^2}$.

*Solution*

$$\begin{aligned}
\frac{x^3}{(x^2+1)^2} &= \frac{Ax+B}{x^2+1} + \frac{Cx+D}{(x^2+1)^2} \\
x^3 &= (Ax+B)(x^2+1) + Cx + D \\
&= Ax^3 + Bx^2 + Ax + B + Cx + D \\
&= Ax^3 + Bx^2 + (A+C)x + B + D \\
A &= 1 \\
B &= 0 \\
A + C &= 0 \\
B + D &= 0 \\
A = 1 \quad B &= 0 \qquad C = -1 \qquad D = 0 \\
\int \frac{x^3\,dx}{(x^2+1)^2} &= \int \frac{x\,dx}{x^2+1} - \int \frac{x\,dx}{(x^2+1)^2} \\
&= \frac{1}{2}\int \frac{2x\,dx}{x^2+1} - \frac{1}{2}\int (x^2+1)^{-2}\,2x\,dx \\
&= \tfrac{1}{2}\ln(x^2+1) - \frac{1}{2}\,\frac{(x^2+1)^{-1}}{-1} + C \\
&= \tfrac{1}{2}\ln(x^2+1) + \frac{1}{2(x^2+1)} + C
\end{aligned}$$

**Example 10** Evaluate $\int \frac{2x^2}{(x-2)(x^2+4)}\,dx$.

*Solution* In our denominator we have one linear and one quadratic factor. Thus,

$$\begin{aligned}
\frac{2x^2}{(x-2)(x^2+4)} &= \frac{A}{x-2} + \frac{Bx+C}{x^2+4} = \frac{Ax^2 + 4A + Bx^2 - 2Bx + Cx - 2C}{(x-2)(x^2+4)} \\
2x^2 &= (A+B)x^2 + (-2B+C)x + 4A - 2C \\
A + B &= 2 \\
-2B + C &= 0 \\
4A - 2C &= 0 \\
A = 1 \qquad B &= 1 \qquad C = 2 \\
\int \frac{2x^2}{(x-2)(x^2+4)}\,dx &= \int \frac{dx}{x-2} + \int \frac{x+2}{x^2+4}\,dx \\
&= \int \frac{dx}{x-2} + \int \frac{x\,dx}{x^2+4} + 2\int \frac{dx}{x^2+4} \\
&= \ln|x-2| + \tfrac{1}{2}\ln(x^2+4) + 2\int \frac{dx}{x^2+4}
\end{aligned}$$

Using the trigonometric substitution $x = 2\tan\theta$, we found

$$\int \frac{dx}{x^2+4} = \frac{1}{2}\operatorname{Arctan}\frac{x}{2} + C$$

Hence, $\int \frac{2x^2}{(x-2)(x^2+4)}\,dx = \ln|x-2| + \tfrac{1}{2}\ln(x^2+4) + \operatorname{Arctan}\frac{x}{2} + C$

**EXERCISES**

Evaluate the following integrals.

1. $\displaystyle\int \frac{x+3}{(x+1)(x+2)}\,dx$

2. $\displaystyle\int \frac{x-1}{(x-2)(x-3)}\,dx$

3. $\displaystyle\int \frac{x\,dx}{(x-1)^2}$

4. $\displaystyle\int \frac{(x+2)\,dx}{x^2(x-1)}$

5. $\displaystyle\int \frac{(3x+4)\,dx}{x(x+1)(x+2)}$

6. $\displaystyle\int \frac{x^2-x+3}{x^3-2x^2+x}\,dx$

7. $\displaystyle\int \frac{x^3\,dx}{(x^2+2)^2}$

8. $\displaystyle\int \frac{x^3\,dx}{x^2+4}$

9. $\displaystyle\int \frac{x^2-3x-4}{x^2-4}\,dx$

10. $\displaystyle\int \frac{x^2-4x+7}{(x+1)(x^2+1)}\,dx$

# 12
# Conic Sections

## 12-1 INTERSECTION OF A CONE BY A PLANE

When a circular cone is intersected by a plane, the resulting curves are referred to as *conic sections*. These conic sections consist of the *circle*, the *ellipse*, the *parabola*, and the *hyperbola*.

1. If the intersecting plane is perpendicular to the axis of the circular cone, the resulting curve is a circle (Fig. 12-1*a*).
2. If the intersecting plane is not perpendicular to the axis of the cone but cuts entirely across one nappe of the cone forming a closed curve, the resulting curve is an ellipse (Fig. 12-1*b*).
3. If the intersecting plane is parallel to an element of the cone, the resulting curve is a parabola (Fig. 12-1*c*).
4. If the intersecting plane cuts both nappes of the cone, the resulting curve is a hyperbola (Fig. 12-1*d*).
5. If the intersecting plane passes through the vertex of the cone, the resulting curve may be either a single point, a single straight line, or a pair of intersecting straight lines.

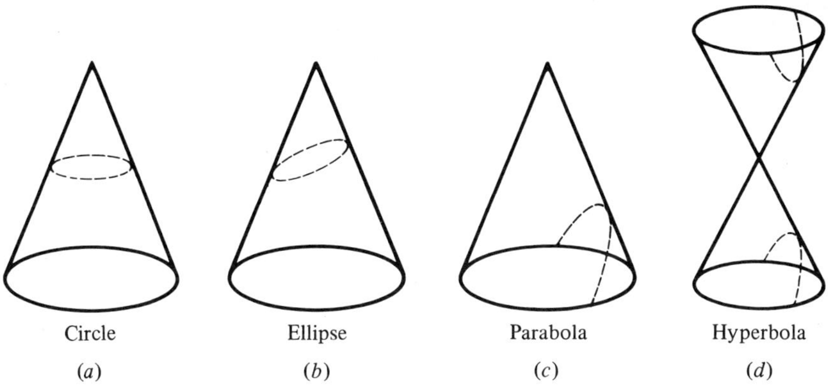


**Fig. 12-1**

## 12-2 THE CIRCLE

A circle is defined as a locus of points in a plane equidistant from a fixed point in the plane. The fixed point is called the *center* of the circle and the equal distance is called the *radius* of the circle. We shall now proceed to develop a standard form for the equation of a circle.

Let the center of the circle $C$ be a point in a plane with coordinates $(h,k)$, let $r$ be the equal distance, and let $P(x,y)$ be any point satisfying the definition (Fig. 12-2). Then,

$$\overline{PC} = r$$

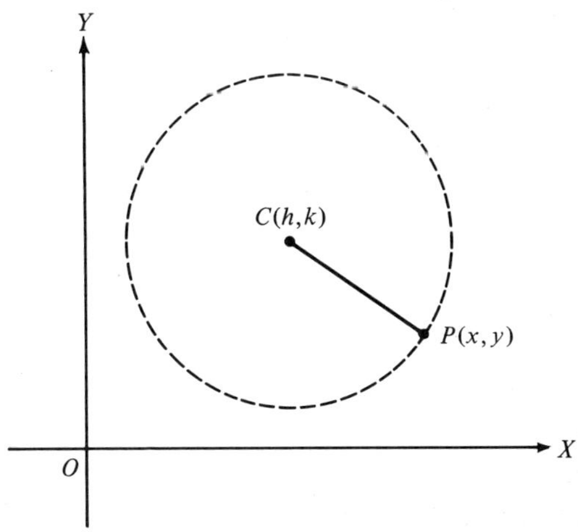


**Fig. 12-2**

Using the distance formula,

$$\sqrt{(x-h)^2+(y-k)^2} = r$$
$$(x-h)^2+(y-k)^2 = r^2$$

$$\boxed{(x-h)^2+(y-k)^2 = r^2}$$

When we choose the origin (0,0) as the center of our circle, our equation reduces to

$$x^2+y^2 = r^2$$

If the form $(x-h)^2+(y-k)^2 = r^2$ is multiplied out, we get

$$x^2-2hx+h^2+y^2-2ky+k^2 = r^2$$
$$x^2+y^2-2hx-2ky+h^2+k^2-r^2 = 0$$
$$x^2+y^2+Cx+Dy+E = 0$$

which is called the general form of the equation of a circle.

**Example 1** Determine an equation of a circle with center (2,4) and radius 5.

*Solution*

$$(x-h)^2+(y-k)^2 = r^2$$
$$h = 2 \qquad k = 4 \qquad r = 5$$

Hence, $$(x-2)^2+(y-4)^2 = 25$$

**Example 2** Find the center and the radius of the circle whose equation is

$$(x+3)^2+(y-2)^2 = 16$$

*Solution* Since the equation is in the standard form $(x-h)^2+(y-k)^2 = r^2$, we can conclude that $h = -3$, $k = 2$, $r = 4$. Thus the center is the point $(-3,2)$ and the radius equals 4.

**Example 3** Find the center and the radius of the circle whose equation is

$$x^2+y^2+6x-4y-3 = 0$$

*Solution* We shall try to rewrite our equation in the standard form $(x-h)^2+(y-k)^2 = r^2$. To accomplish this we shall use the method of completing the square. We recall that when we have a perfect square $(x-a)^2$, this can be expanded so that

$$(x-a)^2 = x^2-\underline{2a}x+\underline{a^2}$$

We note that if we square half the coefficient of $x$ we get the constant term $a^2$. Thus, to make the expression

$$x^2+6x$$

into a perfect square, we must take half of 6, square it, and add to $x^2 + 6x$:

$$x^2 + 6x + 9 = (x + 3)^2$$

Similarly, to make $y^2 - 4y$ into a perfect square we must square half the coefficient of $y$.

$$y^2 - 4y + 4 = (y - 2)^2$$

Returning to our equation,

$$\begin{aligned} x^2 + y^2 + 6x - 4y - 3 &= 0 \\ x^2 + 6x + y^2 - 4y &= 3 \\ x^2 + 6x + 9 + y^2 - 4y + 4 &= 3 + 9 + 4 \\ (x + 3)^2 + (y - 2)^2 &= 16 \end{aligned}$$

The equation is now in standard form, and we have found that

$$h = -3 \qquad k = 2 \qquad r = 4$$

**EXERCISES**

Find the center and radius of the following circles.

**1.** $(x + 2)^2 + (y - 3)^2 = 9$

**2.** $(x - 3)^2 + (y + 10)^2 = 25$

**3.** $(x + 10)^2 + (y - 10)^2 = 100$

**4.** $x^2 + y^2 + 4x + 2y - 4 = 0$

**5.** $x^2 + y^2 + 10x - 8y - 8 = 0$

**6.** $2x^2 + 2y^2 + 8x - 4y - 15 = 0$

Find the equations of the following circles.

**7.** Center at $(2,-5)$ and radius 3

**8.** Center at the origin and radius $\sqrt{7}$

**9.** Center at the origin and passing through the point $(-4,3)$

**10.** Center at $(3,4)$ and tangent to the $Y$ axis

## 12-3 THE PARABOLA

A parabola is defined as a locus of points in the plane that are equidistant from a fixed point in the plane and a fixed line in the plane. The fixed point is called the *focus* of the parabola and the fixed line is called the *directrix*. The point midway between the focus and the directrix is called the *vertex*.

We shall first consider the case where the vertex is at the origin and the directrix is perpendicular to the $X$ axis. We shall denote the distance between the focus and the directrix by $P$. The focus $F$ will then have coordinates $\left(\frac{P}{2}, 0\right)$, and the directrix will have the equation $x = -\frac{P}{2}$ (Fig. 12-3).

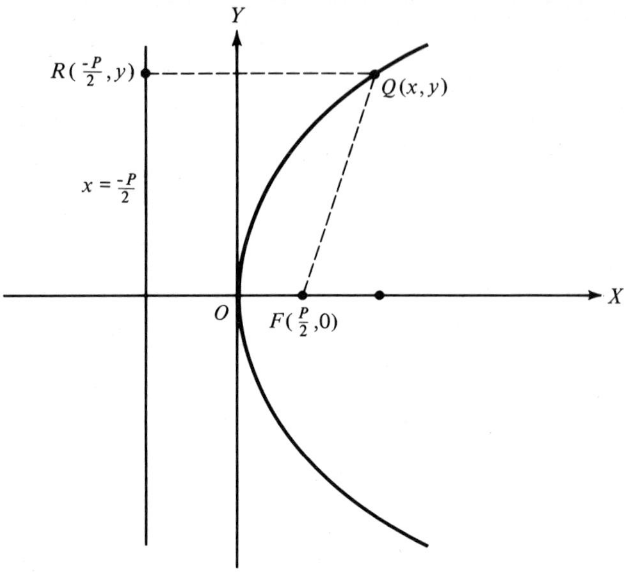


Fig. 12-3

Let $Q(x,y)$ be any point on the parabola. From the definition, the distance from $Q$ to $F$ must be the same as the distance from the point $Q$ to the directrix $x = -\frac{P}{2}$. Hence,

$$\overline{FQ} = \overline{QR}$$

$$\sqrt{\left(x - \frac{P}{2}\right)^2 + y^2} = x + \frac{P}{2}$$

Squaring both sides,

$$\left(x - \frac{P}{2}\right)^2 + y^2 = \left(x + \frac{P}{2}\right)^2$$

$$x^2 - Px + \frac{P^2}{4} + y^2 = x^2 + Px + \frac{P^2}{4}$$

$$y^2 = 2Px$$

When $P > 0$, then $x$ must be $\geq 0$. The only intercept is at the origin and the curve is symmetric with respect to the $X$ axis. Thus, when $P > 0$, the parabola opens to the right. When $P < 0$, then $x \leq 0$ and the parabola opens to the left (Fig. 12-4).

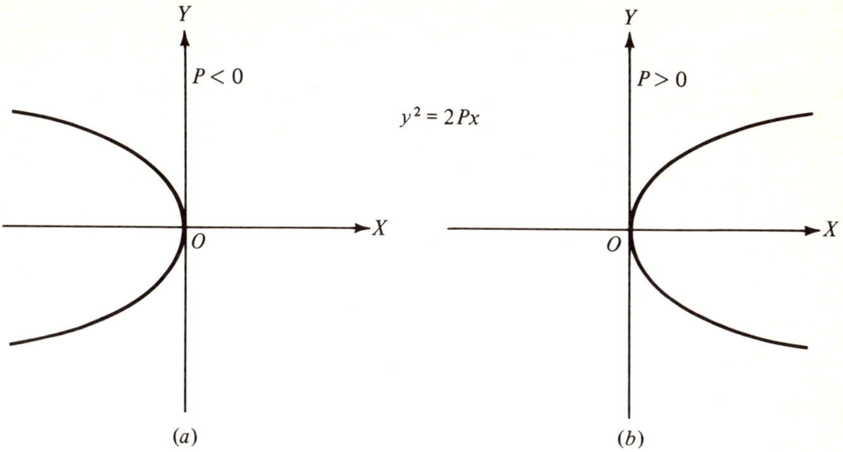


**Fig. 12-4**

If we choose the vertex at the origin and the directrix perpendicular to the $Y$ axis, we merely reverse the roles of $x$ and $y$ (Fig. 12-5), and the resulting equation will then be

$$x^2 = 2Py$$

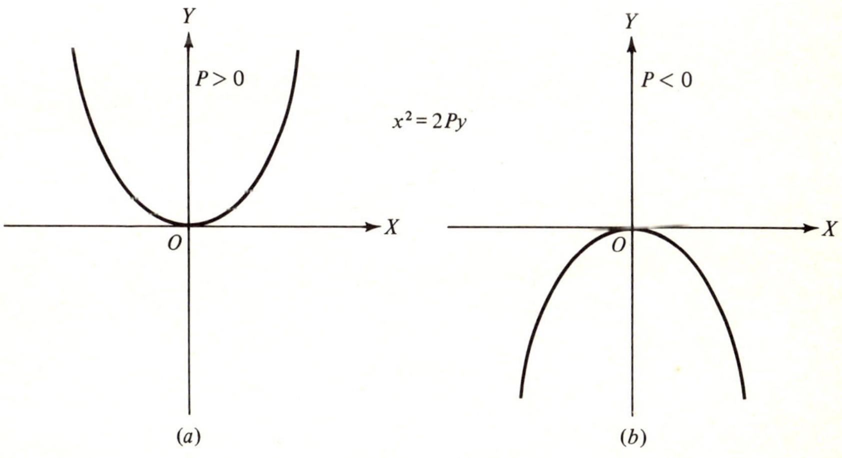


**Fig. 12-5**

**Example 1** Determine the coordinates of the focus and the equation of the directrix for the parabola $y^2 = 8x$ (Fig. 12-6).

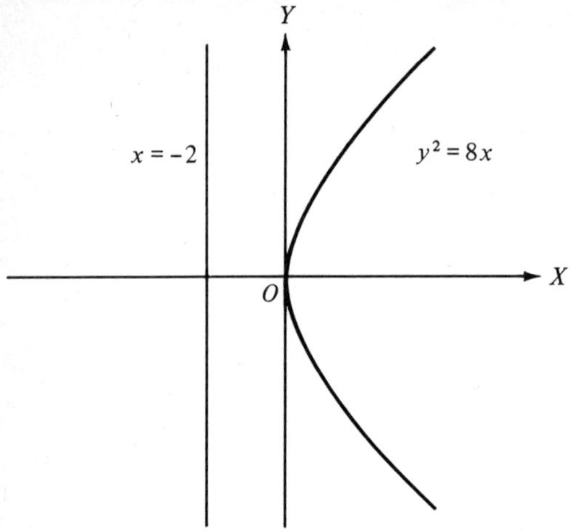


Fig. 12-6

*Solution* The equation is in the form $y^2 = 2Px$.

$$2P = 8$$
$$P = 4$$
$$\frac{P}{2} = 2$$

Thus, the parabola opens to the right with:

Focus: (2,0)
Directrix: $x = -2$

**Example 2** Determine the coordinates of the focus and the equation of the directrix for the parabola $x^2 = -12y$ (Fig. 12-7).

*Solution* We have the form $x^2 = 2Py$ with $P = -6$. Since $P$ is negative, the curve opens downward with focus at $(0, -3)$ and directrix $y = 3$.

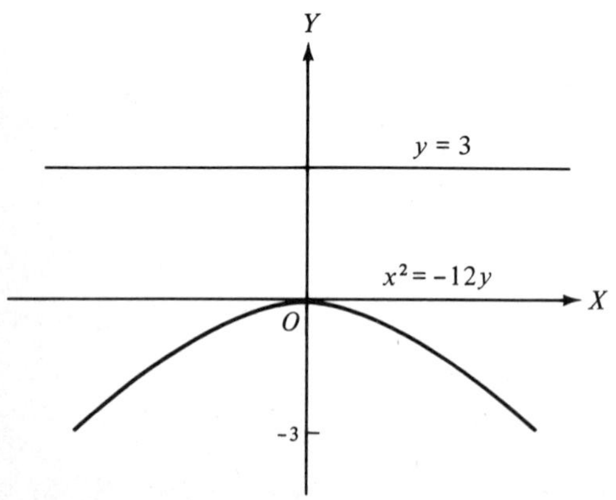


Fig. 12-7

**Example 3** Find an equation for the parabola with focus at (0,4) and vertex at the origin (Fig. 12-8).

*Solution* Since the vertex is at the origin and the focus is at (0,4), our equation will be in the form

$$x^2 = 2Py \qquad \text{with } P > 0$$

The length of the segment from the vertex to the focus is 4. Thus,

$$\frac{P}{2} = 4$$

$$P = 8$$

$$2P = 16$$

Hence,

$$x^2 = 16y$$

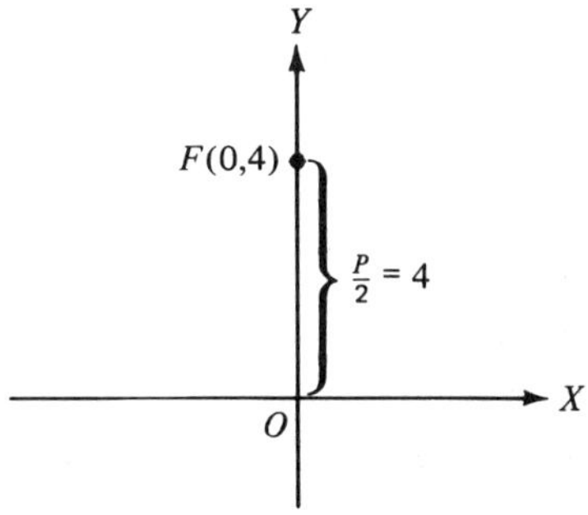


Fig. 12-8

The line dividing the parabola into two symmetric halves is called the *axis* of the parabola. A line segment passing through the focus and connecting two points on the parabola is called a *focal chord*. The focal chord perpendicular to the axis of the parabola is called the *latus rectum*. To find the length of the latus rectum, we shall consider the parabola $y^2 = 2Px$ (Fig. 12-9).

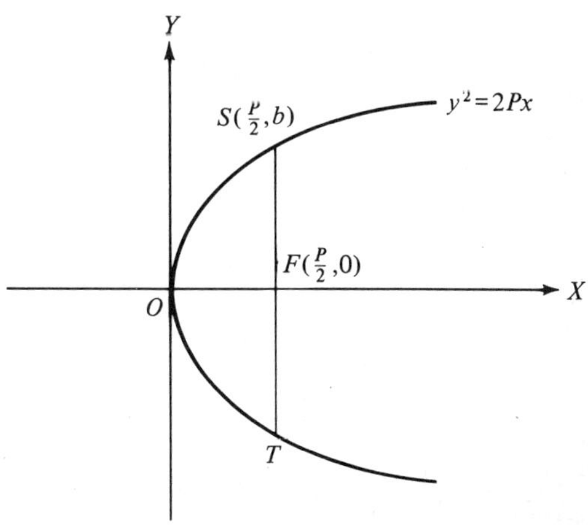


Fig. 12-9

Let $\overline{ST}$ be the latus rectum of the parabola $y^2 = 2Px$. The $x$ coordinate for the point $S$ is $\frac{P}{2}$, and we shall denote the $y$ coordinate with the letter $b$. Since $S$ is a point on the curve, its coordinates satisfy the equation $y^2 = 2Px$. Thus,

$$\begin{aligned} b^2 &= 2P\left(\frac{P}{2}\right) \\ &= P^2 \\ b &= \pm P \end{aligned}$$

Hence, the length of the latus rectum $\overline{ST}$ is $2P$. We can show that the result is the same for the parabola in the form $x^2 = 2Py$.

**Example 4** Determine the length of the latus rectum for the parabola $y^2 = 32x$.

*Solution* We have the form $y^2 = 2Px$ with $2P = 32$. Thus, the latus rectum has a length of 32.

Thus far we have only considered parabolas with vertex at the origin. We shall next consider the equation of a parabola with vertex not at the origin but with axis of symmetry parallel to one of the coordinate axes.

Consider a parabola with vertex at the point $(h,k)$ and axis parallel to the $X$ axis (Fig. 12-10). If the point $(h,k)$ were at the origin, we would have no difficulty in writing an equation for the parabola. We shall temporarily construct a new set of axes $X'$ and $Y'$, so that the point

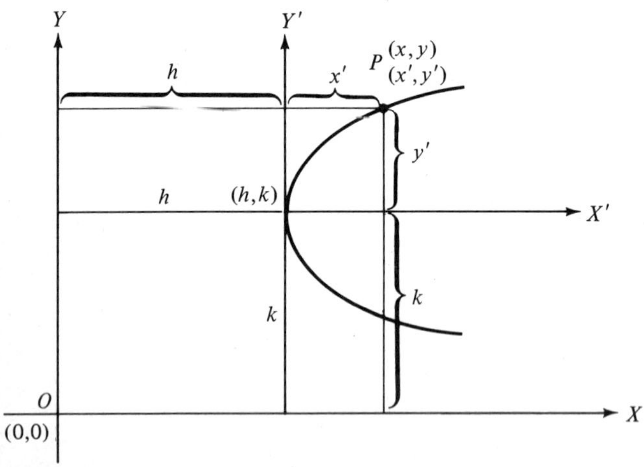


**Fig. 12-10**

$(h,k)$ is at the origin of the new coordinate system. In the new coordinate system our equation becomes

$$(y')^2 = 2P(x')$$

We must now find the relationship between the new coordinates and our original set of coordinates. Let $P$ be a point on the parabola with coordinates $(x,y)$ in the original coordinate system and $(x',y')$ in the new coordinate system. From Fig. 12-10 we find that

$$x' + h = x$$
$$y' + k = y$$

or

$$x' = x - h$$
$$y' = y - k$$

Hence, since $(y')^2 = 2P(x')$, by substitution we get

$$\boxed{(y - k)^2 = 2P(x - h)}$$

Similarly we can show that if the vertex of the parabola is at $(h,k)$ and the axis of symmetry is parallel to the $Y$ axis, the equation of the parabola has the form

$$\boxed{(x - h)^2 = 2P(y - k)}$$

**Example 5** Write an equation for the parabola with vertex at the point (2,4) and focus at the point (6,4) (Fig. 12-11).

*Solution* The vertex and the focus both have a $y$ coordinate of 4, and thus the line $y = 4$ is our axis of symmetry. The equation of the parabola will then have the form

$$(y - k)^2 = 2P(x - h)$$

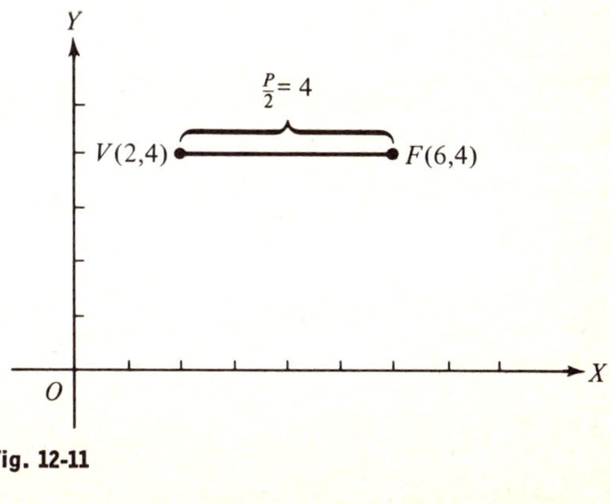


**Fig. 12-11**

Since the vertex is at (2,4), we can fill in the values of $h$ and $k$:

$$(y - 4)^2 = 2P(x - 2)$$

We must now determine the value of $P$. The distance from the vertex (2,4) to the focus (6,4) is 4. Thus,

$$\frac{P}{2} = 4$$
$$P = 8$$
$$2P = 16$$

Our equation is then

$$(y - 4)^2 = 16(x - 2)$$

**Example 6** Determine an equation for the parabola with vertex at the point $(3,-1)$ and directrix $y = 2$ (Fig. 12-12).

*Solution* The directrix is above the vertex, and thus our equation will have the form

$$(x - h)^2 = -2P(y - k)$$

The vertex is at $(3,-1)$ and thus the equation becomes

$$(x - 3)^2 = -2P(y + 1)$$

The distance from the vertex to the directrix is 3. Thus

$$\frac{P}{2} = 3$$
$$P = 6$$
$$2P = 12$$

Our equation is then

$$(x - 3)^2 = -12(y + 1)$$

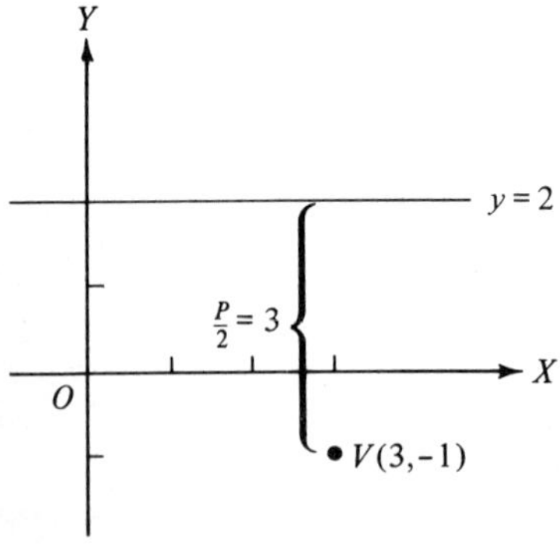


**Fig. 12-12**

**Example 7** Determine the coordinates of the focus and the vertex for the parabola $x^2 - 2x - 8y - 15 = 0$ (Fig. 12-13).

*Solution* We will try to change the equation into the form $(x - h)^2 = 2P(y - k)$. To accomplish this we will have to use the method of completing the square.

$$x^2 - 2x - 8y - 15 = 0$$
$$x^2 - 2x = 8y + 15$$
$$x^2 - 2x + 1 = 8y + 15 + 1$$
$$(x - 1)^2 = 8(y + 2)$$

Thus, $h = 1 \qquad k = -2$

Our vertex is at $(1, -2)$. We must now determine the coordinates of the focus.

$$2P = 8$$
$$\frac{P}{2} = 2$$

Our axis of symmetry is $x = 1$. (Why not $y = -2$?) Therefore the coordinates of the focus will be (1,0).

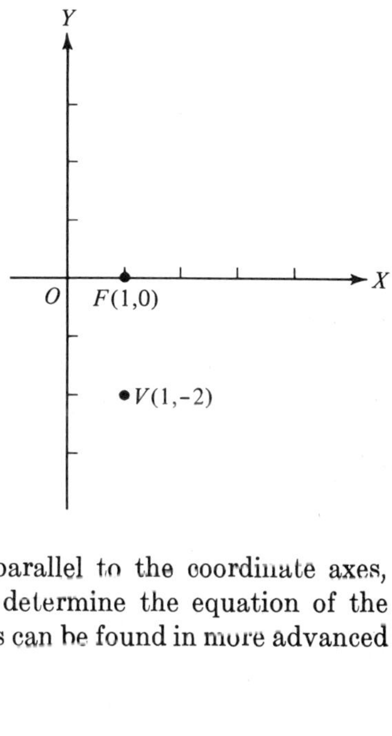


Fig. 12-13

If the axis of the parabola is not parallel to the coordinate axes, a rotation of axes can be performed to determine the equation of the parabola. A discussion of rotation of axes can be found in more advanced texts on analytic geometry and calculus.

**EXERCISES**

For each of the parabolas in Exercises 1 to 12 determine the coordinates of the focus and vertex, and the equation of the directrix.

**1.** $y^2 = 8x$

**2.** $y^2 = -16x$

**3.** $y = \frac{1}{4}x^2$

**4.** $x^2 = -12y$

**5.** $x^2 = 24y$

**6.** $(y - 2)^2 = -8(x + 1)$

**7.** $y^2 = 16x - 32$

**8.** $x^2 = 12 - 4y$

**9.** $y^2 + 6y - 2x + 11 = 0$

**10.** $y^2 + 2y - 5x - 14 = 0$

**11.** $x^2 - 6x + 4y + 17 = 0$

**12.** $x^2 - 2x - 7y - 34 = 0$

In Exercises 13 to 24 determine the equations of the parabolas satisfying the given conditions.

**13.** Vertex at the origin, directrix $x = 3$

**14.** Vertex at the origin, focus (0,2)

**15.** Vertex at the origin, focus (3,0)

**16.** Vertex at the origin, axis of symmetry on the $X$ axis, open to the right, length of latus rectum equal to 12

**17.** Focus at $(-5,0)$, directrix $x = 5$

**18.** Focus at (0,4), directrix $y = -4$

**19.** Vertex at (3,4), focus at (3,6)

**20.** Vertex at $(-2,5)$, directrix $x = -5$

**21.** Focus at (0,6), directrix $y = 0$

**22.** Vertex at $(-1,-3)$, focus at $(3,-3)$

**23.** Focus at (4,2), directrix $x + 2 = 0$

**24.** Vertex at (8,2), focus at $(8,-4)$

## 12-4 THE ELLIPSE

An ellipse is defined as a locus of points in the plane such that the sum of their distances from two fixed points in the plane is a constant which is greater than the distance between the two fixed points. The fixed points are called the *foci* of the ellipse, and each one of them is called a *focus* of the ellipse. The line passing through the foci is called the *principal axis* of the ellipse, and the point midway between the foci is called the *center* of the ellipse.

We shall first consider an ellipse with center at the origin and foci lying on the $X$ axis. Let $F$ and $F'$ represent the foci and let $2c$ represent the distance between the foci. Let $P(x,y)$ be any point satisfying the definition of the ellipse and let $2a$ represent the sum of the distances from $P(x,y)$ to the foci. Since the center of the ellipse is at the origin and the foci $F$ and $F'$ are on the $X$ axis with $\overline{FF'} = 2c$, we can take the coordinates of $F$ as $(c,0)$ and those of $F'$ as $(-c,0)$ (Fig. 12-14).

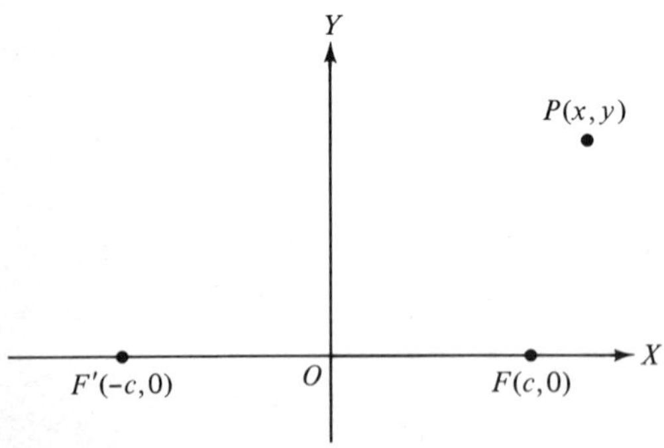


**Fig. 12-14**

From the definition of the ellipse,

$$\overline{PF} + \overline{PF'} = 2a$$

By the distance formula

$$\begin{aligned}
\sqrt{(x-c)^2 + y^2} + \sqrt{(x+c)^2 + y^2} &= 2a \\
\sqrt{(x-c)^2 + y^2} &= 2a - \sqrt{(x+c)^2 + y^2} \\
(x-c)^2 + y^2 &= 4a^2 - 4a\sqrt{(x+c)^2 + y^2} \\
&\quad + (x+c)^2 + y^2 \\
x^2 - 2cx + c^2 + y^2 &= 4a^2 - 4a\sqrt{(x+c)^2 + y^2} \\
&\quad + x^2 + 2cx + c^2 + y^2 \\
4a\sqrt{(x+c)^2 + y^2} &= 4a^2 + 4cx \\
a\sqrt{(x+c)^2 + y^2} &= a^2 + cx \\
a^2[(x+c)^2 + y^2] &= a^4 + 2a^2cx + c^2x^2 \\
a^2x^2 + 2a^2cx + a^2c^2 + a^2y^2 &= a^4 + 2a^2cx + c^2x^2 \\
a^2x^2 + a^2c^2 + a^2y^2 &= a^4 + c^2x^2 \\
a^2x^2 - c^2x^2 + a^2y^2 &= a^4 - a^2c^2 \\
(a^2 - c^2)x^2 + a^2y^2 &= a^2(a^2 - c^2)
\end{aligned}$$

From the definition of the ellipse

$$2a > 2c$$

and thus $a^2 > c^2$,

$$a^2 - c^2 > 0$$

We shall make the substitution

$$b^2 = a^2 - c^2$$

Our equation then becomes

$$b^2x^2 + a^2y^2 = a^2b^2$$

An alternate form of the equation can be obtained by dividing each term by $a^2b^2$.

$$\boxed{\frac{x^2}{a^2} + \frac{y^2}{b^2} = 1}$$

The equation that has been derived is the equation for an ellipse with center at the origin, principal axis on the $X$ axis, distance between the foci equal to $2c$, sum of the distances from any point on the ellipse to the foci equal to $2a$, and $b^2 = a^2 - c^2$.

To aid us in sketching the curve we note that the $x$ intercepts are $(a,0)$ and $(-a,0)$, the $y$ intercepts are $(0,b)$ and $(0,-b)$, and the curve is symmetric with respect to both axes and the origin (Fig. 12-15).

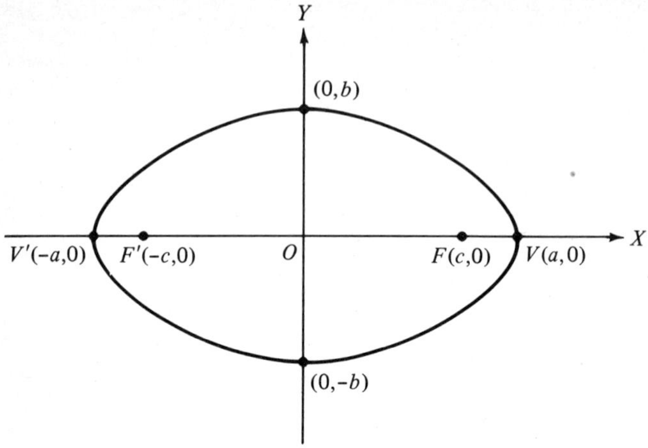


**Fig. 12-15**

The line segment connecting the points $(a,0)$ and $(-a,0)$ is sometimes referred to as the *major axis*, and the line segment connecting the points $(0,b)$ and $(0,-b)$ is called the *minor axis*. The endpoints of the major axis are called the *vertices* of the ellipse.

**Example 1** Determine an equation for the ellipse with foci at (3,0) and (−3,0), and vertices at (5,0) and (−5,0).

*Solution* Our equation has the form

$$\frac{x^2}{a^2} + \frac{y^2}{b^2} = 1$$

Since the vertices are at (5,0) and (−5,0) and the foci are at (3,0) and (−3,0), we have

$$\begin{aligned} a &= 5 \\ c &= 3 \\ b^2 &= a^2 - c^2 \\ &= 25 - 9 \\ &= 16 \end{aligned}$$

Hence our equation is

$$\frac{x^2}{25} + \frac{y^2}{16} = 1$$

**Example 2** Determine an equation for the ellipse with center at the origin, major axis equal to 8, and minor axis equal to 6.

*Solution* The major axis has a length of $2a$ and the minor axis has a length of $2b$. Thus,

$$\begin{aligned} 2a &= 8 \\ 2b &= 6 \\ a &= 4 \qquad b = 3 \end{aligned}$$

Our equation is then

$$\frac{x^2}{16} + \frac{y^2}{9} = 1$$

Thus far we have only considered an ellipse with center at the origin, major axis on the $X$ axis, and minor axis on the $Y$ axis. For an ellipse with center at the origin, major axis of length $2a$ on the $Y$ axis, and minor axis of length $2b$ on the $X$ axis, the roles of $x$ and $y$ are reversed in the equation and the equation of the ellipse becomes

$$\frac{y^2}{a^2} + \frac{x^2}{b^2} = 1$$

It is important to note that since $b^2 = a^2 - c^2$, the value of $a$ is always greater than the value of $b$ and therefore the length of the major axis is always greater than the length of the minor axis.

**Example 3** Determine an equation for the ellipse with center at the origin, foci at (0,2) and (0, −2), and major axis of length 10.

*Solution* The foci are on the $Y$ axis, and thus our equation will have the form

$$\frac{y^2}{a^2} + \frac{x^2}{b^2} = 1$$

$$c = 2$$

$$2a = 10$$

$$a = 5$$

To find $b^2$ we use the equation

$$\begin{aligned} b^2 &= a^2 - c^2 \\ &= 25 - 4 \\ &= 21 \end{aligned}$$

Hence, our equation is

$$\frac{y^2}{25} + \frac{x^2}{21} = 1$$

**Example 4** Determine the equation of the ellipse with center at the origin and which intersects the coordinate axes at (8,0), (−8,0), (0,5), and (0, −5).

*Solution*

$$a = 8 \qquad b = 5$$

The major axis will be on the $X$ axis and thus our equation will have the form

$$\frac{x^2}{a^2} + \frac{y^2}{b^2} = 1$$

Hence, our equation is

$$\frac{x^2}{64} + \frac{y^2}{25} = 1$$

Having considered the equation of an ellipse with center at the origin, we can now consider the more general case, the ellipse with center at the point $(h,k)$ and major axis parallel to one of the coordinate axes (Fig. 12-16).

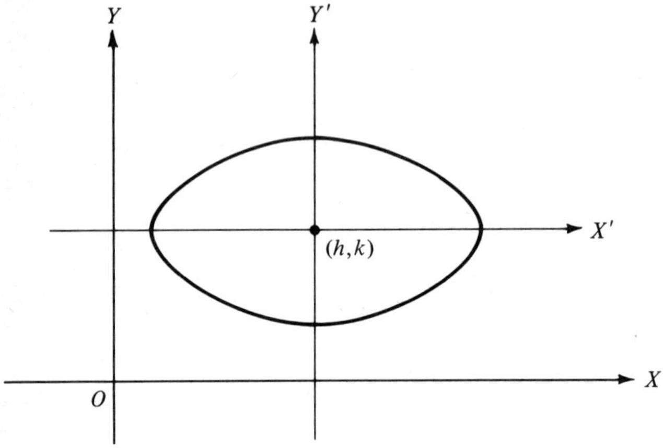


**Fig. 12-16**

As we did with the parabola, we can consider a new set of axes, $X'$ and $Y'$, such that the point $(h,k)$ is the origin of the new coordinate system. We can then proceed with a transformation of coordinates to again establish the following relationship between the two sets of axes:

$$x' = x - h$$
$$y' = y - k$$

If the major axis is parallel to the $X$ axis, our equation becomes

$$\frac{(x-h)^2}{a^2} + \frac{(y-k)^2}{b^2} = 1$$

If the major axis is parallel to the $Y$ axis, our equation becomes

$$\frac{(y-k)^2}{a^2} + \frac{(x-h)^2}{b^2} = 1$$

**Example 5** Determine an equation for the ellipse with foci (0,4) and (6,4) and vertices (−2,4) and (8,4) (Fig. 12-17).

*Solution* The major axis is parallel to the $X$ axis, and therefore our equation will have the form

$$\frac{(x-h)^2}{a^2} + \frac{(y-k)^2}{b^2} = 1$$

The center of the ellipse is midway between the foci. Hence, the center of the ellipse is at the point (3,4). Also,

$$a = 5$$
$$c = 3$$
$$b^2 = a^2 - c^2 = 25 - 9 = 16$$

Our equation is thus

$$\frac{(x-3)^2}{25} + \frac{(y-4)^2}{16} = 1$$

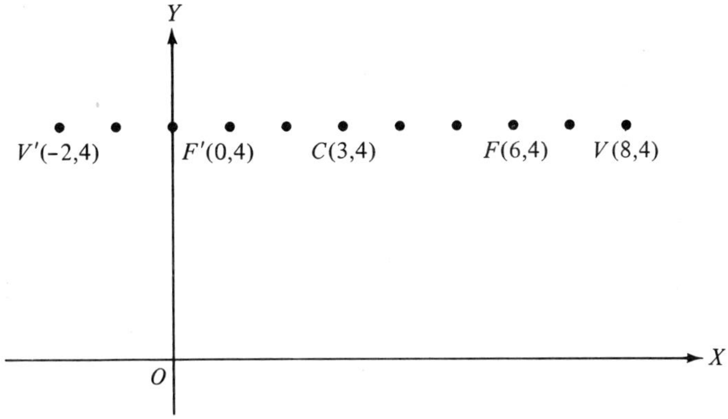


**Fig. 12-17**

**Example 6** For the ellipse $4x^2 + 9y^2 + 8x - 54y + 49 = 0$, determine the coordinates of the center, the foci, and the vertices.

*Solution* Using the method of completing the square,

$$\begin{aligned} 4x^2 + 9y^2 + 8x - 54y + 49 &= 0 \\ 4x^2 + 8x + 9y^2 - 54y &= -49 \\ 4(x^2 + 2x) + 9(y^2 - 6y) &= -49 \\ 4(x^2 + 2x + 1) + 9(y^2 - 6y + 9) &= -49 + 4 + 81 \\ 4(x+1)^2 + 9(y-3)^2 &= 36 \\ \frac{(x+1)^2}{9} + \frac{(y-3)^2}{4} &= 1 \end{aligned}$$

Thus, $\quad h = -1 \qquad k = 3 \qquad a = 3 \qquad b = 2$

$$\begin{aligned} b^2 &= a^2 - c^2 \\ c^2 &= a^2 - b^2 \\ &= 9 - 4 \\ c &= \pm\sqrt{5} \end{aligned}$$

Hence,

$$\begin{aligned} \text{Center} &= (-1,3) \\ \text{Foci} &= (-1 + \sqrt{5}, 3), (-1 - \sqrt{5}, 3) \\ \text{Vertices} &= (-4,3), (2,3) \end{aligned}$$

An indication of the shape of the ellipse is often obtained from the value of the *eccentricity* $e$, which is defined as the ratio

$$e = \frac{c}{a}$$

Since $a$ is always greater than $c$, the eccentricity is always a number between 0 and 1. Using the relationship

$$c^2 = a^2 - b^2$$

when $c$ is very small, then $a$ and $b$ are almost equal and the curve is nearly circular. When $c$ is very large, then $a$ is much greater than $b$ and the curve is narrow and elongated. Thus, when $e = \frac{c}{a}$ is close to 0, then $c$ is close to 0 and the curve is nearly circular. As the value of $e$ gets larger, the value of $c$ gets larger. When the value of $e$ gets closer to 1, the values of $c$ and $a$ are almost equal and $b^2 = a^2 - c^2$ is very small, thus producing an ellipse which is long and narrow.

## EXERCISES

For each of the ellipses in Exercises 1 to 8 find the coordinates of the foci and the vertices.

**1.** $\frac{x^2}{4} + \frac{y^2}{1} = 1$

**2.** $\frac{y^2}{9} + \frac{x^2}{1} = 1$

**3.** $\frac{y^2}{25} + \frac{x^2}{16} = 1$

**4.** $\frac{x^2}{81} + \frac{y^2}{25} = 1$

**5.** $\frac{x^2}{169} + \frac{y^2}{25} = 1$

**6.** $\frac{x^2}{144} + \frac{y^2}{169} = 1$

**7.** $4y^2 + 9x^2 = 36$

**8.** $16x^2 + 9y^2 = 144$

In Exercises 9 to 14 determine an equation for the ellipses satisfying the given conditions.

**9.** Foci $(\pm 3,0)$, vertices $(\pm 5,0)$

**10.** Center at the origin, axis of the ellipse on the $X$ axis, length of major axis 8, length of minor axis 4

**11.** Vertices $(0, \pm 8)$, eccentricity $\frac{3}{4}$

**12.** Foci $(\pm 7,0)$, length of minor axis 4

**13.** Foci $(0, \pm 5)$, one vertex at $(0,8)$

**14.** Vertices $(0, \pm 13)$, length of minor axis 24.

For each of the ellipses in Exercises 15 to 22 find the coordinates of the center, the foci, and the vertices.

**15.** $\frac{(x-3)^2}{16} + \frac{(y-2)^2}{9} = 1$

**16.** $\frac{(y-4)^2}{9} + \frac{(x+1)^2}{4} = 1$

**17.** $\frac{(x+3)^2}{169} + \frac{(y-3)^2}{144} = 1$

**18.** $\frac{(x-5)^2}{16} + \frac{(y-5)^2}{25} = 1$

**19.** $9x^2 + 4y^2 - 36x + 24y + 36 = 0$

**20.** $16x^2 + 9y^2 + 32x - 36y - 92 = 0$

**21.** $9x^2 + 25y^2 + 54x - 200y + 256 = 0$

**22.** $16x^2 + 4y^2 - 160x + 336 = 0$

In Exercises 23 to 28 determine the equations of the ellipses satisfying the given conditions.

**23.** Vertices (13,4) and (−7,4), length of minor axis 12

**24.** Foci (2,5) and (2,−1), vertices (2,7) and (2,−3)

**25.** Vertices (5,8) and (5,2), eccentricity $\frac{2}{3}$

**26.** Center at (4,6), major axis parallel to the $X$ axis and of length 10, minor axis of length 8

**27.** Vertices (4,7) and (4,1), foci (4,6) and (4,2)

**28.** Center at (6,6), one vertex at (6,10), minor axis of length 4

## 12-5 THE HYPERBOLA

A hyperbola is defined as a locus of points in the plane such that the difference of their distances from two fixed points in the plane is a constant. As before, the fixed points are called the foci of the hyperbola, the point midway between the foci is called the center of the hyperbola, and the line passing through the foci is called the principal axis of the hyperbola.

As we did with the ellipse, we shall first consider a hyperbola with center at the origin and principal axis on the $X$ axis. Let $F$ and $F'$ represent the foci and let $2c$ represent the distance between the foci. Let $P(x,y)$ be any point satisfying the definition of the hyperbola and let $\pm 2a$ represent the difference of the distances from $P(x,y)$ to the foci. (Why $\pm$?) Since the center of the hyperbola is at the origin and the foci $F$ and $F'$ are on the $X$ axis with $FF' = 2c$, we shall label the points $F$ and $F'$ as $(c,0)$ and $(-c,0)$, respectively (Fig. 12-18).

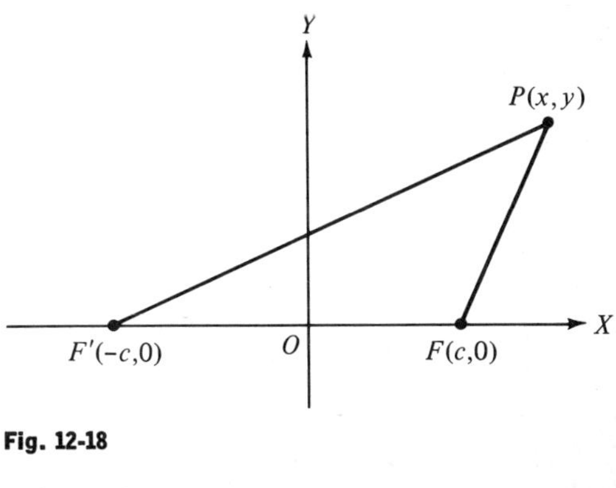


**Fig. 12-18**

From Fig. 12-18 we note that in triangle $PFF'$,

*Case* 1 If $\overline{PF} > \overline{PF'}$,

$$\overline{FF'} + \overline{PF'} > \overline{PF} \qquad \text{(Why?)}$$
$$\overline{FF'} > \overline{PF} - \overline{PF'}$$

*Case* 2 If $\overline{PF'} > \overline{PF}$,

$$\overline{FF'} + \overline{PF} > \overline{PF'} \qquad \text{(Why?)}$$
$$\overline{FF'} > \overline{PF'} - \overline{PF}$$

In either case,

$$2c > 2a \qquad c > a$$

We now proceed to the task of deriving an equation for the hyperbola. From the definition of the hyperbola

$$\begin{aligned} \overline{PF'} - \overline{PF} &= \pm 2a \\ \sqrt{(x+c)^2 + y^2} - \sqrt{(x-c)^2 + y^2} &= \pm 2a \\ \sqrt{(x+c)^2 + y^2} &= \pm 2a + \sqrt{(x-c)^2 + y^2} \\ x^2 + 2cx + c^2 + y^2 &= 4a^2 \pm 4a\sqrt{(x-c)^2 + y^2} + x^2 \\ &\quad - 2cx + c^2 + y^2 \\ 4cx - 4a^2 &= \pm 4a\sqrt{(x-c)^2 + y^2} \\ cx - a^2 &= \pm a\sqrt{(x-c)^2 + y^2} \end{aligned}$$

Squaring both sides,

$$\begin{aligned} c^2x^2 - 2a^2cx + a^4 &= a^2(x^2 - 2cx + c^2 + y^2) \\ c^2x^2 - 2a^2cx + a^4 &= a^2x^2 - 2a^2cx + a^2c^2 + a^2y^2 \\ c^2x^2 + a^4 &= a^2x^2 + a^2c^2 + a^2y^2 \\ c^2x^2 - a^2x^2 - a^2y^2 &= a^2c^2 - a^4 \\ (c^2 - a^2)x^2 - a^2y^2 &= a^2(c^2 - a^2) \end{aligned}$$

Since $c > a$, $c^2 - a^2 > 0$. Let $b^2 = c^2 - a^2$. Then our equation becomes

$$b^2x^2 - a^2y^2 = a^2b^2$$

or

$$\boxed{\frac{x^2}{a^2} - \frac{y^2}{b^2} = 1}$$

The equation that we have derived is the equation for a hyperbola with center at the origin, principal axis on the $X$ axis, distance between the foci equal to $2c$, difference of the distances from any point on the hyperbola to the foci equal to $\pm 2a$, and $b^2 = c^2 - a^2$.

A discussion of the equation yields the following results:

1. $x$ intercepts $\pm a$, no $y$ intercepts.
2. Symmetry with respect to both axes and the origin.
3. The only permissible $x$ values are $x > a$ and $x < -a$. All $y$ values are permissible.
4. There are no vertical or horizontal asymptotes.

The intercepts $(\pm a,0)$ on the principal axis of the hyperbola are called the vertices of the hyperbola, and the segment between the vertices is called the transverse axis. The segment from $(0,-b)$ to $(0,b)$ is called the conjugate axis.

From our equation $\frac{x^2}{a^2} - \frac{y^2}{b^2} = 1$, if we solve for $y$ in terms of $x$, we get

$$y = \pm \frac{b}{a} x \sqrt{1 - \frac{a^2}{x^2}}$$

As the value of $x$ gets infinitely large, $\frac{a^2}{x^2}$ approaches 0 and the $y$ value approaches $\pm \frac{b}{a} x$. Thus, the lines $y = \frac{b}{a} x$ and $y = -\frac{b}{a} x$ are asymptotes of the hyperbola. With the above information the hyperbola $\frac{x^2}{a^2} - \frac{y^2}{b^2} = 1$ can easily be graphed (Fig. 12-19).

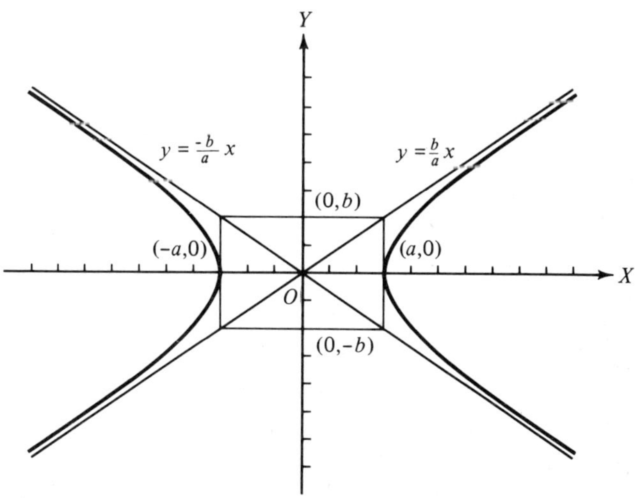


Fig. 12-19

As an aid to sketching we note that if one constructs a rectangle with "center" at the origin (Fig. 12-19) and sides equal to the lengths of the transverse and conjugate axes, respectively, then the diagonals of the rectangle are the asymptotes of the hyperbola $\left(\text{that is, } y = \frac{b}{a}x \text{ and } y = -\frac{b}{a}x\right)$.

**Example 1** Find an equation for the hyperbola with transverse axis equal to 8 and foci at $(\pm 5,0)$.

*Solution* The transverse axis equals 8. Therefore,

$$2a = 8 \qquad a = 4$$

The foci are at $(\pm 5,0)$, and hence

$$\begin{aligned} c &= 5 \\ b^2 &= c^2 - a^2 \\ &= 25 - 16 \\ &= 9 \end{aligned}$$

Our equation is thus $\frac{x^2}{16} - \frac{y^2}{9} = 1$.

**Example 2** Determine an equation for the hyperbola with vertices at $(\pm 7,0)$ and conjugate axis equal to 10.

*Solution* The vertices are at $(\pm 7,0)$, and thus

$$a = 7$$

The conjugate axis equals 10, and thus

$$2b = 10 \qquad b = 5$$

Our equation is then $\frac{x^2}{49} - \frac{y^2}{25} = 1$

It can easily be shown that if the foci of the hyperbola are on the $Y$ axis at $(0, \pm c)$, with transverse and conjugate axes equal to $2a$ and $2b$, respectively, the equation of the hyperbola becomes

$$\frac{y^2}{a^2} - \frac{x^2}{b^2} = 1$$

**Example 3** Determine the coordinates of the vertices and the foci for the hyperbola $4x^2 - 9y^2 = 36$.

*Solution*

$$4x^2 - 9y^2 = 36$$

$$\frac{x^2}{9} - \frac{y^2}{4} = 1$$

Our equation is in the form

$$\frac{x^2}{a^2} - \frac{y^2}{b^2} = 1$$

with $a = 3 \qquad b = 2$

and

$$\begin{aligned} c^2 &= a^2 + b^2 \\ &= 9 + 4 \\ &= 13 \end{aligned}$$

The vertices are at $(\pm 3,0)$. The foci are at $(\pm \sqrt{13},0)$.

**Example 4** Determine the coordinates of the vertices and the foci for the hyperbola $16y^2 - 9x^2 = 144$.

*Solution*

$$16y^2 - 9x^2 = 144$$

$$\frac{y^2}{9} - \frac{x^2}{16} = 1$$

Our equation is in the form

$$\frac{y^2}{a^2} - \frac{x^2}{b^2} = 1$$

with $a = 3 \qquad b = 4$

and

$$\begin{aligned} c^2 &= a^2 + b^2 \\ &= 9 + 16 \\ &= 25 \\ c &= \pm 5 \end{aligned}$$

The vertices are at $(0, \pm 3)$.
The foci are at $(0, \pm 5)$.

As with the ellipse, the eccentricity $e$ of the hyperbola is defined as the ratio

$$e = \frac{c}{a}$$

For the hyperbola, $c > a$ and thus $\frac{c}{a} > 1$. If $e$ is very close to 1, then the value of $a$ is very close to the value of $c$ and $b = \sqrt{c^2 - a^2}$ is very small. Thus, the asymptotes make a small angle with the transverse axis. As $e$ gets larger and larger, the angle between the asymptotes and the transverse axis gets larger and larger, approaching a right angle.

For a hyperbola with center at $(h,k)$ and principal axis parallel to one of the coordinate axes we can perform the familiar transformation of coordinates so that the $x$ and $y$ terms are replaced by $(x - h)$ and $(y - k)$, respectively. If the principal axis is parallel to the $X$ axis, our equation becomes

$$\frac{(x-h)^2}{a^2} - \frac{(y-k)^2}{b^2} = 1$$

If the principal axis is parallel to the $Y$ axis, the equation becomes

$$\frac{(y-k)^2}{a^2} - \frac{(x-h)^2}{b^2} = 1$$

**Example 5** Find an equation for the hyperbola with center at $(3,-2)$, vertices at $(3,4)$ and $(3,-8)$, and conjugate axis of length 6.

*Solution* The vertices are at $(3,4)$ and $(3,-8)$. The transverse axis is parallel to the $Y$ axis and has a length of 12. The equation of the hyperbola is then in the form

$$\frac{(y-k)^2}{a^2} - \frac{(x-h)^2}{b^2} = 1$$

with

$$h = 3 \qquad k = -2 \qquad a = 6 \qquad b = 3$$

Hence the equation is

$$\frac{(y+2)^2}{36} - \frac{(x-3)^2}{9} = 1$$

**Example 6** For the hyperbola $9x^2 - 16y^2 - 18x + 64y - 199 = 0$ find the coordinates of the center, the vertices, and the foci (Fig. 12-20).

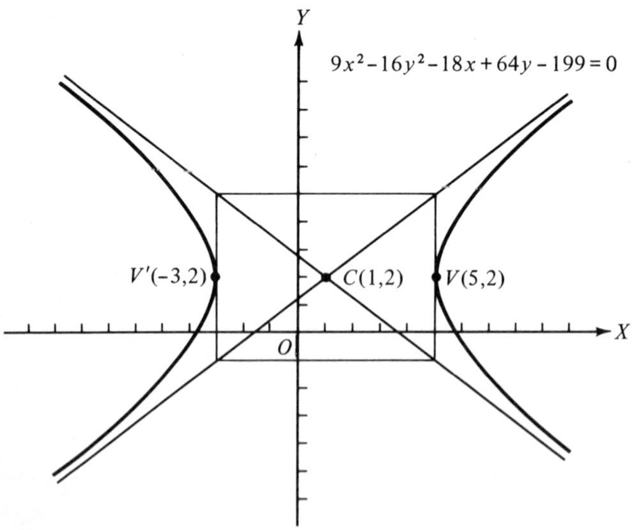


**Fig. 12-20**

*Solution* We shall complete the square. We have

$$\begin{aligned}
9x^2 - 16y^2 - 18x + 64y - 199 &= 0 \\
9x^2 - 18x - 16y^2 + 64y &= 199 \\
9(x^2 - 2x \quad) - 16(y^2 - 4y \quad) &= 199 \\
9(x^2 - 2x + 1) - 16(y^2 - 4y + 4) &= 199 + 9 - 64 \\
9(x - 1)^2 - 16(y - 2)^2 &= 144 \\
\frac{(x-1)^2}{16} - \frac{(y-2)^2}{9} &= 1
\end{aligned}$$

Now $\quad h = 1 \quad k = 2 \quad a = 4 \quad b = 3$

and
$$\begin{aligned}
c^2 &= a^2 + b^2 \\
&= 16 + 9 \\
&= 25 \\
c &= 5
\end{aligned}$$

The center is at (1,2).
The vertices are at (5,2) and (−3,2).
The foci are at (6,2) and (−4,2).

## EXERCISES

For each of the hyperbolas in Exercises 1 to 8 find the coordinates of the foci and the vertices.

**1.** $\dfrac{x^2}{4} - \dfrac{y^2}{1} = 1$

**2.** $\dfrac{x^2}{144} - \dfrac{y^2}{25} = 1$

**3.** $\dfrac{y^2}{9} - \dfrac{x^2}{4} = 1$

**4.** $\dfrac{x^2}{144} - \dfrac{y^2}{81} = 1$

**5.** $\dfrac{x^2}{64} - \dfrac{y^2}{36} = 1$

**6.** $\dfrac{y^2}{49} - \dfrac{x^2}{1} = 1$

**7.** $\dfrac{y^2}{16} - \dfrac{x^2}{9} = 1$

**8.** $9x^2 - 4y^2 = 1$

In Exercises 9 to 14 determine an equation for the hyperbolas satisfying the given conditions.

**9.** Vertices $(0, \pm 3)$, foci $(0, \pm 5)$

**10.** Vertices $(\pm 12, 0)$, foci $(\pm 13, 0)$

**11.** Vertices $(\pm 1, 0)$, conjugate axis of length 6

**12.** Center at the origin, one focus at $(\sqrt{17}, 0)$, transverse axis of length 8

**13.** Vertices $(\pm 6, 0)$, eccentricity $\frac{4}{3}$

**14.** Center at the origin, transverse axis on the $Y$ axis and of length 10, conjugate axis of length 12

For each of the hyperbolas in Exercises 15 to 20 find the coordinates of the center, the foci, and the vertices.

**15.** $\dfrac{(x+2)^2}{4} - \dfrac{(y+3)^2}{9} = 1$

**16.** $\dfrac{(y-2)^2}{25} - \dfrac{(x+1)^2}{9} = 1$

**17.** $\dfrac{(y-4)^2}{16} - \dfrac{(x+2)^2}{9} = 1$

**18.** $\dfrac{(x+3)^2}{144} - \dfrac{(y+7)^2}{25} = 1$

**19.** $4x^2 - 9y^2 - 16x - 72y - 129 = 0$

**20.** $9y^2 - 25x^2 + 180y + 200x + 275 = 0$

In Exercises 21 to 24 determine the equations of the hyperbolas satisfying the given conditions.

**21.** Vertices (0,4) and (6,4), foci $(-2,4)$ and (8,4)

**22.** Center at $(-3,-3)$, one focus at $(-3,9)$, conjugate axis of length 10

**23.** Vertices at (1,5) and (9,5), eccentricity $\frac{5}{4}$

**24.** Vertices at $(-2,8)$ and $(-2,-8)$, conjugate axis of length 12

**25.** When the transverse and conjugate axes of one hyperbola are, respectively, the conjugate and transverse axes of a second hyperbola, the two hyperbolas are referred to as conjugate hyperbolas. Show that $9x^2 - 4y^2 = 36$ and $4y^2 - 9x^2 = 36$ are conjugate hyperbolas and sketch them on one set of axes.

**26.** Write an equation for the conjugate hyperbola of the hyperbola $9x^2 - 16y^2 = 144$. Sketch the pair of conjugate hyperbolas.

# 13
# Polar Coordinates

## 13-1 POLAR-COORDINATE SYSTEM

In a rectangular-coordinate system the position of a point in the plane is specified by its directed distances from two perpendicular axes. We will now consider a coordinate system in which the position of a point in the plane is specified by means of its distance and direction from a fixed point in the plane called the *pole* (or *origin*), and a fixed half line in the plane called the *polar axis*. The polar axis is generally constructed horizontally and to the right of the pole.

Let $O$ (the pole) be a fixed point in the plane and let $OA$ (the polar axis) be a fixed half line in the plane. Let $P$ be any point in the plane. We shall denote the distance from point $O$ to point $P$ by the letter $r$ and we shall denote the angle from $OA$ to $OP$ by the letter $\theta$ (Fig. 13-1). The counterclockwise direction is considered the positive direction and the clockwise direction is considered negative. The position of the point $P$ in a polar coordinate system is specified by the directed distance $r$ and the directed angle $\theta$. The coordinates of $P$ are represented as $(r,\theta)$.

The directed distance $r$ ($\overline{OP}$) is called the *radius vector* of the point $P$, and the directed angle $\theta$ is called the *vectorial angle* of the point $P$. The radius vector is always listed first and the vectorial angle second.

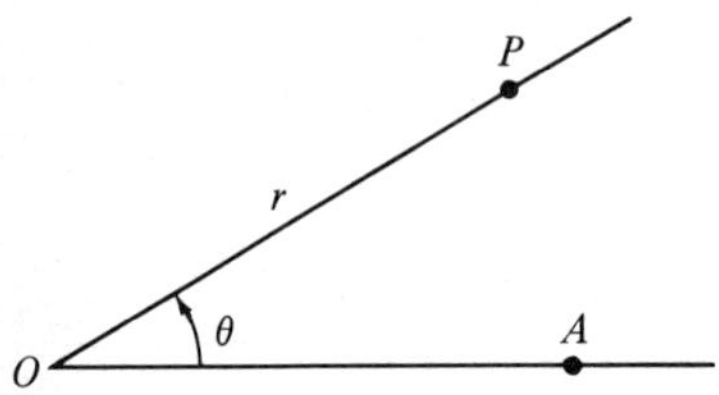


**Fig. 13-1**

Both $r$ and $\theta$ can be either positive or negative. As stated before, the positive direction for $\theta$ is counterclockwise and the negative direction is clockwise. When the radius vector $r$ is measured along the terminal side of $\theta$, then $r$ is positive. To plot a negative radius vector we must measure $r$ in the opposite direction of the terminal side of $\theta$.

**Example**

(*a*) Plot the point $\left(2, \frac{\pi}{4}\right)$.

(*b*) Plot the point $\left(-2, \frac{\pi}{4}\right)$.

(*c*) Plot the point $\left(2, -\frac{\pi}{4}\right)$.

*Solution* See Fig. 13-2.

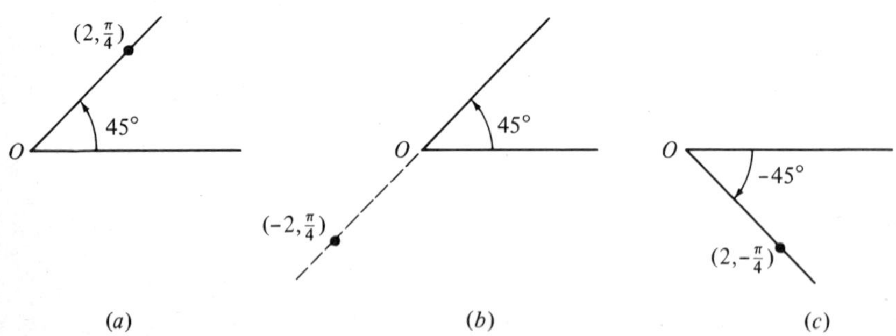


**Fig. 13-2** Solutions for (*a*) Example 1*a*, (*b*) Example 1*b*, and (*c*) Example 1*c*.

We note that the points $(r,\theta)$, $(r, \theta + 2\pi)$, and $(r, \theta + 4\pi)$ are all identical because $\theta$, $\theta + 2\pi$, and $\theta + 4\pi$ are angles with the same initial and terminal sides. In fact, any point in the plane represented by the coordinates $(r,\theta)$ can also be represented by $(r, \theta \pm 2n\pi)$, $n = 1, 2, 3, 4, \ldots$ . A careful analysis of the coordinate system will also indicate that any point that can be represented by $(r,\theta)$ can also be represented by $(-r, \theta \pm n\pi)$ where $n$ is an odd integer. Hence, unlike our result in a rectangular-coordinate system, in a polar-coordinate system we do not have a one-to-one correspondence between points in the plane and sets of coordinates. For every point in the plane there are an infinite number of pairs of coordinates corresponding to that point.

If one wishes to construct the coordinate system so that there is exactly one pair of coordinates for each point in the plane, he must set certain restrictions. The values of $\theta$ can be limited from 0 to $2\pi$ $(0 \leq \theta \leq 2\pi)$, and the radius vector $r$ can be taken as positive. When this is done, the pair of coordinates of the point are referred to as the principal set of polar coordinates of the point.

Plotting in polar coordinates is often facilitated by using polar-coordinate paper, which consists of equally spaced concentric circles and equally spaced radial lines.

### EXERCISES

**1.** Plot the following points whose polar coordinates are given:

(*a*) $(2,\pi)$ (*b*) $(-3,2\pi)$ (*c*) $(\frac{1}{2}, 30°)$

(*d*) $\left(3, -\frac{\pi}{2}\right)$ (*e*) $\left(4, \frac{3\pi}{2}\right)$ (*f*) $(-2, -90°)$

(*g*) $(3, -\pi)$ (*h*) $(2, 135°)$ (*i*) $\left(1, \frac{5\pi}{4}\right)$

**2.** Describe the set of points for which the radius vector is 5.

**3.** Describe the set of points for which the vectorial angle is $\frac{\pi}{4}$.

**4.** List three other pairs of coordinates for the point $\left(3, -\frac{2\pi}{3}\right)$.

## 13-2 THE RELATIONSHIP BETWEEN POLAR AND RECTANGULAR COORDINATES

We shall now establish a relationship between the polar and rectangular coordinates of a point in the plane. We shall place the two systems so that the pole and the origin coincide and the polar axis lies along the positive $X$ axis (Fig. 13-3).

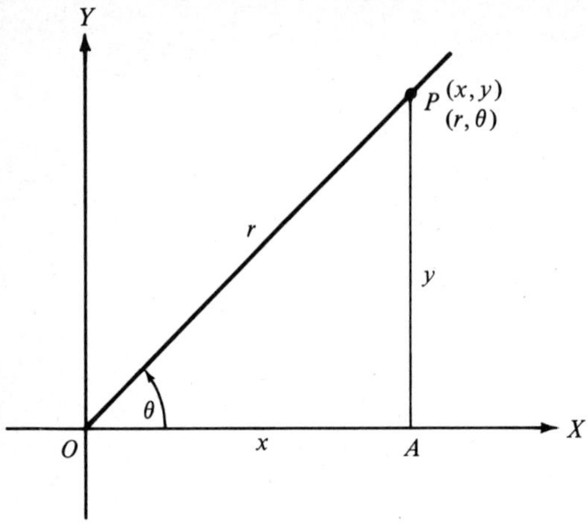


**Fig. 13-3**

Let $P$ be a point in the plane with coordinates $(x,y)$ and $(r,\theta)$. From triangle $OAP$,

$$\cos\theta = \frac{x}{r} \qquad \sin\theta = \frac{y}{r}$$

$$\boxed{x = r\cos\theta}$$

$$\boxed{y = r\sin\theta}$$

$$x^2 + y^2 = r^2 \qquad \tan\theta = \frac{y}{x}$$

$$\boxed{r = \pm\sqrt{x^2 + y^2}}$$

$$\boxed{\theta = \arctan\frac{y}{x}}$$

Also, $\qquad \theta = \arcsin\dfrac{y}{r} \qquad$ and $\qquad \theta = \arccos\dfrac{x}{r}$

**Example 1** Determine the rectangular coordinates of the point whose polar coordinates are $\left(2, \dfrac{\pi}{3}\right)$.

*Solution*

$$
\begin{aligned}
x &= r\cos\theta \\
y &= r\sin\theta \\
x &= 2\cos\frac{\pi}{3} \\
&= 2(\tfrac{1}{2}) = 1 \\
y &= 2\sin\frac{\pi}{3} \\
&= 2\left(\frac{\sqrt{3}}{2}\right) = \sqrt{3}
\end{aligned}
$$

The rectangular coordinates are $(1,\sqrt{3})$.

**Example 2** Express the rectangular coordinates $(-3,3)$ in terms of polar coordinates.

*Solution*

$$
\begin{aligned}
x^2 + y^2 &= r^2 \\
9 + 9 &= r^2 \\
r^2 &= 18 \\
r &= \pm 3\sqrt{2} \\
\theta &= \arctan\frac{3}{-3} \\
&= \arctan -1
\end{aligned}
$$

The point $(-3,3)$ is in the second quadrant. If we choose $r = 3\sqrt{2}$, we must choose $\theta$ as an angle in the second quadrant whose tangent equals $-1$. Thus, $\theta = \frac{3\pi}{4}$. Our coordinates are then $\left(3\sqrt{2}, \frac{3\pi}{4}\right)$. As an alternate solution we can choose $\left(-3\sqrt{2}, \frac{7\pi}{4}\right)$ as the polar coordinates of the point $(-3,3)$. In general, we will choose our polar coordinates with $r$ positive unless we have reason to do otherwise.

**Example 3** Find a rectangular equation for the curve whose polar equation is $r = 2\cos\theta$.

*Solution*

$$x = r\cos\theta \qquad r^2 = x^2 + y^2$$

Since $r = 2\cos\theta$,

$$
\begin{aligned}
r^2 &= 2r\cos\theta \\
x^2 + y^2 &= 2x
\end{aligned}
$$

**Example 4** Transform the equation $r = 5$ to a rectangular-coordinate equation.

*Solution*

$$
\begin{aligned}
r &= 5 \\
r^2 &= 25 \\
x^2 + y^2 &= r^2 \\
x^2 + y^2 &= 25
\end{aligned}
$$

Hence, $r = 5$ is the equation of a circle with center at the origin and radius equal to 5.

**EXERCISES**

Find the nonnegative polar coordinates of the following points.

**1.** (2,2) **2.** (4,0) **3.** (4, −3)
**4.** (5,12) **5.** (0,3) **6.** (−2,0)
**7.** (−12,5) **8.** (−5, −5) **9.** $(3\sqrt{3}, -3)$
**10.** (−4,4) **11.** (0, −6) **12.** (−3,4)
**13.** $(-\sqrt{5}, -\sqrt{5})$ **14.** (0,0) **15.** (3, −3)

Find the rectangular coordinates of the following points.

**16.** (2, 45°) **17.** (5, 120°) **18.** $\left(-2, \frac{3\pi}{4}\right)$
**19.** (2, 60°) **20.** (5, 180°) **21.** $\left(2, \frac{2\pi}{3}\right)$
**22.** $\left(\sqrt{2}, \frac{5\pi}{4}\right)$ **23.** (−1, −30°) **24.** $(2\sqrt{3}, 240°)$
**25.** $\left(-3\sqrt{2}, \frac{7\pi}{4}\right)$ **26.** (4, 270°) **27.** $(4,\pi)$

Change the following polar coordinate equations to rectangular coordinate equations.

**28.** $r = 4 \sin \theta$ **29.** $r^2 - r \cos \theta + 4 = 0$
**30.** $3r \sin \theta = 5$ **31.** $r = 2 \sin \theta + 3 \cos \theta$
**32.** $\theta = \frac{\pi}{4}$ **33.** $r^2 \sin \theta \cos \theta = 1$

Change the following rectangular coordinate equations to polar coordinate equations.

**34.** $x + y = 5$ **35.** $y = 2$
**36.** $x^2 + y^2 = 36$ **37.** $y = x$
**38.** $y = 4x^2$ **39.** $x^2 - y^2 = 4$

## 13-3 CURVE SKETCHING IN POLAR COORDINATES

The methods used to sketch a curve in polar coordinates are essentially the same as the methods used in rectangular coordinates. In considering a polar equation in the form $r = f(\theta)$ we can construct a table of values for $r$ and $\theta$ choosing a convenient interval for $\theta$. As in the case of rectangular coordinates, curve sketching will be greatly facilitated by an algebraic analysis of the equation. Our analysis will consist of a discussion of *intercepts, symmetry,* and *extent* (excluded values).

### INTERCEPTS

In our discussion of curve sketching in rectangular coordinates we considered intercepts with reference to the $X$ and $Y$ axes and the origin. Analogous to these in polar coordinates we shall consider intercepts with reference to the polar axis, the 90° axis, and the pole. For the equation

$r = f(\theta)$ the intercepts on the polar axis can be found by setting $\theta = 0$, $\pm\pi$, $\pm 2\pi$, or, in general, $\pm n\pi$. To find the intercepts on the 90° axis we set $\theta = \frac{\pi}{2}, \frac{3\pi}{2}$, or, in general, $\frac{n\pi}{2}$ where $n$ is an odd integer. Any value of $\theta$ for which $r = 0$ will give us an intercept at the pole.

**Example 1** Determine the intercepts for the curve $r = 1 - \cos\theta$.

*Solution*

Polar axis: When $\theta = 0$, $r = 0$.
When $\theta = \pi$, $r = 2$.

90° axis: When $\theta = \frac{\pi}{2}$, $r = 1$.
When $\theta = \frac{3\pi}{2}$, $r = 1$.

Pole: $r = 0$ when $\theta = 0, 2\pi, \ldots, 2\pi n$, $n$ an integer.

### SYMMETRY

In a polar-coordinate system there are many different representations for the same point and therefore it will not be possible to establish a foolproof test for symmetry. However, when any of the tests that we shall introduce is satisfied in a given equation, then we are certain of that particular symmetry. On the other hand, if the test fails that does not disprove the symmetry in question. We shall list two tests for each symmetry.

#### Symmetry with respect to the polar axis (Fig. 13-4)

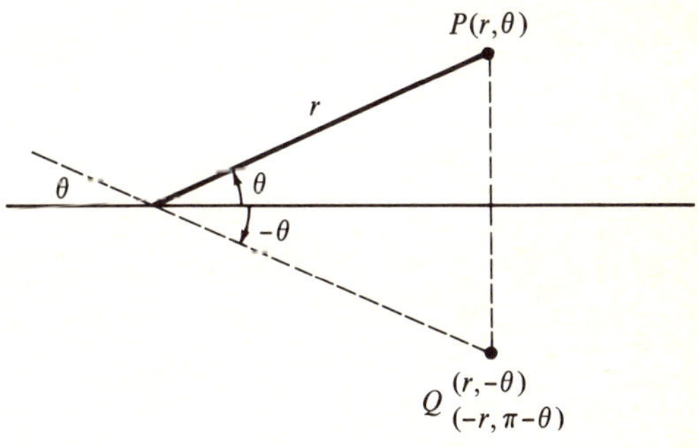


**Fig. 13-4**

| | |
|---|---|
| Test 1 | Replace $\theta$ by $-\theta$; $r$ remains unchanged. |
| Test 2 | Replace $r$ by $-r$ and $\theta$ by $\pi - \theta$. |

If the equation remains unchanged when either test 1 or test 2 is applied, then we have symmetry with respect to the polar axis.

### Symmetry with respect to the 90° axis (Fig. 13-5)

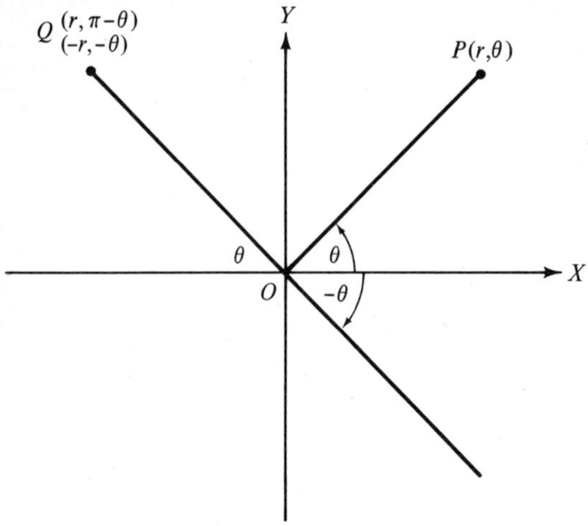


Fig. 13-5

| | |
|---|---|
| Test 1 | Replace $\theta$ by $\pi - \theta$; $r$ remains unchanged. |
| Test 2 | Replace $r$ by $-r$ and $\theta$ by $-\theta$. |

If the equation remains unchanged when either test 1 or test 2 is applied, then we have symmetry with respect to the 90° axis.

### Symmetry with respect to the pole (Fig. 13-6)

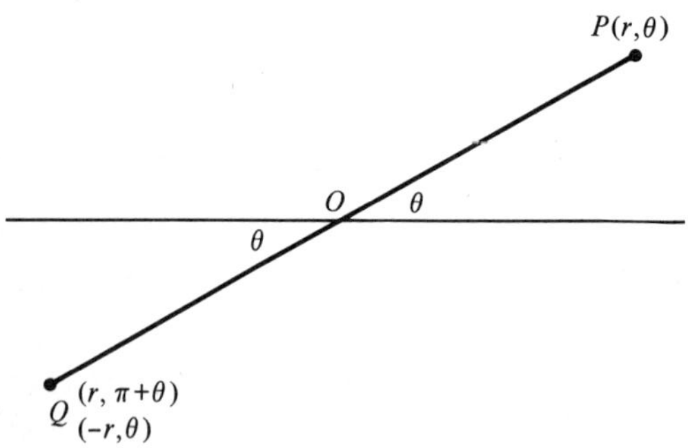


Fig. 13-6

| | |
|---|---|
| Test 1 | Replace $\theta$ by $\pi + \theta$; $r$ remains unchanged. |
| Test 2 | Replace $r$ by $-r$; $\theta$ remains unchanged. |

If the equation remains unchanged when either test 1 or test 2 is applied, then we have symmetry with respect to the pole.

**Example 2** Test the curve $r = 1 - \cos\theta$ for symmetry with respect to the polar axis, the 90° axis, and the pole.

*Solution* To test for symmetry with respect to the polar axis we shall replace $\theta$ by $-\theta$.

$$r = 1 - \cos(-\theta)$$

But, $\cos(-\theta) = \cos\theta$

Hence, our equation becomes

$$r = 1 - \cos\theta$$

The test for symmetry is satisfied and we have symmetry with respect to the polar axis.

To test for symmetry with respect to the 90° axis we shall replace $\theta$ by $\pi - \theta$.

$$\begin{aligned} r &= 1 - \cos(\pi - \theta) \\ \cos(\pi - \theta) &= -\cos\theta \\ r &= 1 + \cos\theta \end{aligned}$$

Test 1 fails.

We shall now apply test 2 and replace $r$ by $-r$ and $\theta$ by $-\theta$:

$$\begin{aligned} -r &= 1 - \cos(-\theta) \\ &= 1 - \cos\theta \end{aligned}$$

Test 2 fails.

Finally, we shall test for symmetry with respect to the pole. We first replace $\theta$ by $\pi + \theta$.

$$\begin{aligned} r &= 1 - \cos(\pi + \theta) \\ \cos(\pi + \theta) &= -\cos\theta \\ r &= 1 + \cos\theta \end{aligned}$$

Test 1 fails.

We now apply test 2 and replace $r$ by $-r$:

$$-r = 1 - \cos\theta$$

Test 2 fails.

As a result of our tests for symmetry we can conclude that there is symmetry with respect to the polar axis and there does not appear to be symmetry with respect to the 90° axis or the pole.

## Extent (excluded values)

With some equations it will sometimes occur that for certain values of $\theta$ the corresponding values of $r$ will be imaginary. Such values of $\theta$ must be excluded, and this will sometimes cause us to exclude entire regions in the plane. It will also be helpful for us to determine whether the curve is a closed curve or not, and what the maximum and minimum values of $r$ are.

**Example 3** Find the extent for the curve $r = 1 - \cos\theta$ (Fig. 13-7).

*Solution* First we note that $r$ is defined and is real for all values of $\theta$. Since $-1 \le \cos\theta \le 1$, we can conclude that the curve is a closed curve for which the maximum value of $r$ is 2 and the minimum value is 0.

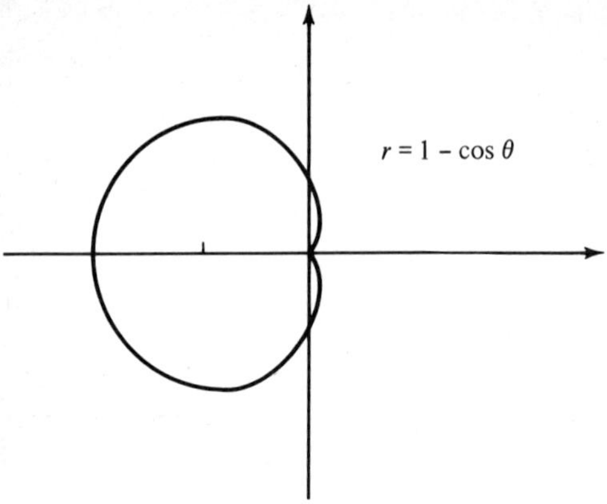


**Fig. 13-7**

**Example 4** Sketch the curve $r = 2 \cos 2\theta$.

*Solution* We shall consider intercepts, symmetry, and extent, and then we shall construct a table of values.

(*a*) *Intercepts*

$$r = 2 \cos 2\theta$$

Set:

$$\theta = 0 \qquad r = 2$$
$$\theta = \frac{\pi}{2} \qquad r = -2$$
$$\theta = \pi \qquad r = 2$$
$$\theta = \frac{3\pi}{2} \qquad r = -2$$
$$\theta = 2\pi \qquad r = 2$$

$r$ will equal 0 when $\cos 2\theta = 0$. This will occur when $2\theta = \frac{n\pi}{2}$, $n$ an odd integer, or equivalently, $\theta = \frac{n\pi}{4}$, $n$ an odd integer.

(*b*) *Symmetry*

Polar axis: Replace $\theta$ by $-\theta$.

$$\begin{aligned} r &= 2 \cos 2(-\theta) \\ &= 2 \cos 2\theta \qquad \text{Symmetric with respect to the polar axis} \end{aligned}$$

90° axis: Replace $\theta$ by $\pi - \theta$.

$$\begin{aligned} r &= 2 \cos 2(\pi - \theta) \\ &= 2 \cos (2\pi - 2\theta) \\ &= 2 \cos (-2\theta) \\ &= 2 \cos 2\theta \qquad \text{Symmetric with respect to the 90° axis} \end{aligned}$$

Pole: Replace $\theta$ by $\pi + \theta$.

$$\begin{aligned} r &= 2 \cos 2(\pi + \theta) \\ &= 2 \cos (2\pi + 2\theta) \\ &= 2 \cos 2\theta \qquad \text{Symmetric with respect to the pole} \end{aligned}$$

*Note:* The test for symmetry with respect to the pole was really not necessary in this example because a curve that is symmetric with respect to both the polar axis and the 90° axis must be symmetric with respect to the pole.

(*c*) *Extent*

$$r = 2 \cos 2\theta$$

$r$ is defined and real for all values of $\theta$. Since $-1 \le \cos 2\theta \le 1$, the curve is a closed curve for which the maximum value of $r$ is 2 and the minimum value is $-2$.

We now make a table of values for $0 \le \theta \le \frac{\pi}{2}$. Our knowledge of intercepts, symmetry, and extent will enable us to sketch the remainder of the curve.

| $\theta$ | $2\theta$ | $\cos 2\theta$ | $r = 2 \cos 2\theta$ |
|---|---|---|---|
| 0 | 0 | 1 | 2 |
| $\frac{\pi}{6}$ | $\frac{\pi}{3}$ | 0.5 | 1 |
| $\frac{\pi}{4}$ | $\frac{\pi}{2}$ | 0 | 0 |
| $\frac{\pi}{3}$ | $\frac{2\pi}{3}$ | $-0.5$ | $-1$ |
| $\frac{\pi}{2}$ | $\pi$ | $-1$ | $-2$ |

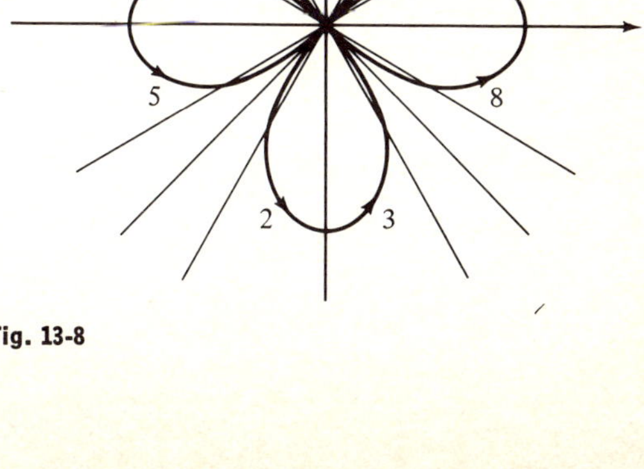


Fig. 13-8

The curve is known as a *four-leafed rose.* In general, the equations $r = a \sin n\theta$ and $r = b \cos n\theta$ will represent closed curves which are known as *rose curves.* If $n$ is odd, the curve will consist of $n$ equally spaced closed loops. If $n$ is even, there will be $2n$ equally spaced loops.

**Example 5** Sketch the curve $r = 1 + 2 \cos \theta$.

*Solution*

(*a*) *Intercepts*

$$r = 1 + 2 \cos \theta$$

Set:

$$\begin{array}{ll} \theta = 0 & r = 3 \\ \theta = \dfrac{\pi}{2} & r = 1 \\ \theta = \pi & r = -1 \\ \theta = \dfrac{3\pi}{2} & r = 1 \\ \theta = 2\pi & r = 3 \end{array}$$

$r$ will equal 0 when $2 \cos \theta = -1$. This will occur when $\cos \theta = -\frac{1}{2}$ (that is, $\theta = \dfrac{2\pi}{3}, \dfrac{4\pi}{3}$, etc.).

(*b*) *Symmetry*

Polar axis: Replace $\theta$ by $-\theta$.

$$\begin{aligned} r &= 1 + 2 \cos(-\theta) \\ &= 1 + 2 \cos \theta \qquad \text{Symmetric with respect to the polar axis} \end{aligned}$$

90° axis: Replace $\theta$ by $\pi - \theta$.

$$\begin{aligned} r &= 1 + 2 \cos(\pi - \theta) \\ &= 1 - 2 \cos \theta \qquad \text{Test fails} \end{aligned}$$

We now apply the second test and we replace $r$ by $-r$ and $\theta$ by $-\theta$.

$$\begin{aligned} -r &= 1 + 2 \cos(-\theta) \\ &= 1 + 2 \cos \theta \qquad \text{Test fails} \end{aligned}$$

Pole: Replace $\theta$ by $\pi + \theta$.

$$\begin{aligned} r &= 1 + 2 \cos(\pi + \theta) \\ &= 1 - 2 \cos \theta \qquad \text{Test fails} \end{aligned}$$

Replace $r$ by $-r$.

$$-r = 1 + 2 \cos \theta \qquad \text{Test fails}$$

Hence, we have symmetry with respect to the polar axis.

(*c*) *Extent*

$$r = 1 + 2 \cos \theta$$

$r$ is defined and real for all values of $\theta$. Since $-1 \le \cos \theta \le 1$, the curve is a closed curve for which the maximum value of $r$ is 3 and the minimum value is $-1$.

| $\theta$ | $\cos\theta$ | $2\cos\theta$ | $r = 1 + 2\cos\theta$ |
|---|---|---|---|
| $0$ | $1$ | $2$ | $3$ |
| $\frac{\pi}{6}$ | $\frac{\sqrt{3}}{2}$ | $\sqrt{3}$ | $1+\sqrt{3}$ |
| $\frac{\pi}{3}$ | $\frac{1}{2}$ | $1$ | $2$ |
| $\frac{\pi}{2}$ | $0$ | $0$ | $1$ |
| $\frac{2\pi}{3}$ | $-\frac{1}{2}$ | $-1$ | $0$ |
| $\frac{5\pi}{6}$ | $-\frac{\sqrt{3}}{2}$ | $-\sqrt{3}$ | $1-\sqrt{3}$ |
| $\pi$ | $-1$ | $-2$ | $-1$ |

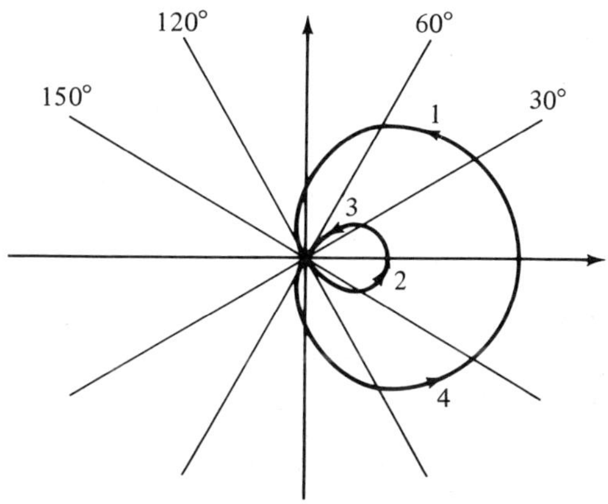


Fig. 13-9

The curve is known as a *limaçon.*

**EXERCISES**

Sketch the graph of the following equations after first discussing intercepts, symmetry, and extent.

**1.** $r = 5$

**2.** $r = 1 + \cos\theta$

**3.** $r = \sin 2\theta$

**4.** $\theta = \frac{\pi}{4}$

**5.** $r = \cos 3\theta$

6. $r = 2 + 3 \sin \theta$ (limaçon)
7. $r = 2 + 2 \cos \theta$ (cardioid)
8. $r = 24 \cos \theta$
9. $r = 2(1 - \sin \theta)$
10. $r = a(1 + \cos \theta)$ (where $a$ is a positive constant)
11. $r = 2 + \cos \theta$
12. $r = 2 \sin 3\theta$
13. $r^2 = 4 \sin 2\theta$ (lemniscate)
14. $r^2 = a^2 \cos 2\theta$ (where $a$ is a positive constant)
15. $r = 3 \cos 5\theta$
16. $r = 1 - 2 \cos \theta$
17. $r = 12 \sin \theta$
18. $r = 2 \sin 4\theta$
19. $r = 3 + \sin \theta$
20. $r = 2\theta,\ \theta \geq 0$ (spiral of Archimedes)

## 13-4 AREAS USING POLAR COORDINATES

We shall now consider the problem of finding areas of regions bounded by curves defined by polar-coordinate equations. Let $r = f(\theta)$ be continuous for $\alpha \leq \theta \leq \beta$. We shall consider the region bounded by the curve $r = f(\theta)$ and the lines $\theta = \alpha$ and $\theta = \beta$ (Fig. 13-10).

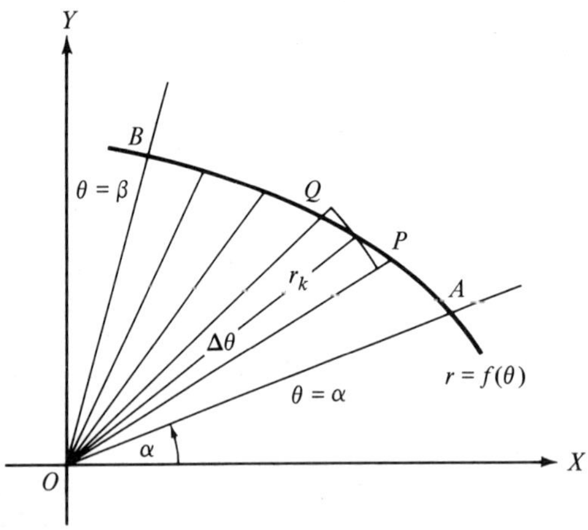


Fig. 13-10

We shall divide the region bounded by $r = f(\theta)$, $\theta = \alpha$, and $\theta = \beta$ into $n$ subregions by dividing the angle $\beta - \alpha$ into $n$ equal parts with $\Delta\theta = \dfrac{\beta - \alpha}{n}\cdot$ Let $POQ$ be a typical subregion and let $r_k$ be any radius vector in this subregion. We shall approximate the area of subregion $POQ$ by the area of the sector of a circle with radius $r$ and central angle $\Delta\theta$. The area of the sector of the circle is $\frac{1}{2}r_k^2\,\Delta\theta$. We shall repeat this procedure for all $n$ subregions. Thus, the area of region $AOB$ is approximately

$$\sum_{k=1}^{n} \tfrac{1}{2}r_k^2\,\Delta\theta$$

or, equivalently,

$$\sum_{k=1}^{n} \tfrac{1}{2}[f(\theta_k)]^2\,\Delta\theta$$

To find the actual area we consider the limit as $n \to \infty$. Hence,

$$A = \lim_{n\to\infty} \sum_{k=1}^{n} \tfrac{1}{2}[f(\theta_k)]^2\,\Delta\theta$$

Since $f(\theta)$ is continuous, this limit exists and is, in fact, the definite integral.

$$A = \lim_{n\to\infty} \sum_{k=1}^{n} \tfrac{1}{2}[f(\theta_k)]^2\,\Delta\theta = \int_\alpha^\beta \tfrac{1}{2}[f(\theta)]^2\,d\theta$$

The area of the region bounded by the continuous function $r = f(\theta)$ and the lines $\theta = \alpha$ and $\theta = \beta$ is given by

$$A = \tfrac{1}{2}\int_\alpha^\beta [f(\theta)]^2\,d\theta$$

**Example 1** Find the area of the region included within the cardioid $r = 1 + \cos\theta$ (Fig. 13-11).

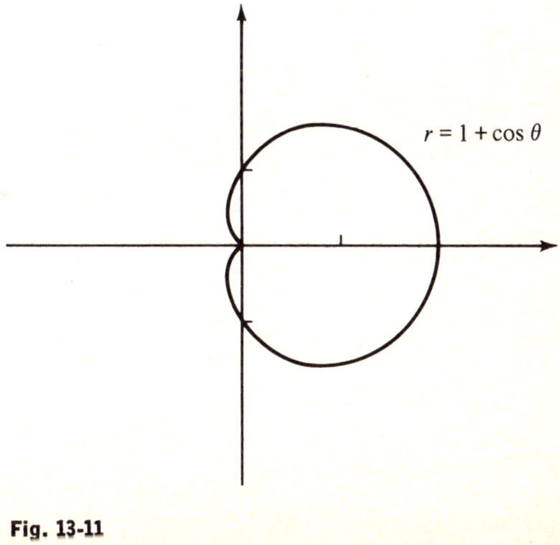


**Fig. 13-11**

*Solution*

$$A = \tfrac{1}{2}\int_{\alpha}^{\beta} [f(\theta)]^2\, d\theta$$

Since the curve is symmetric with respect to the polar axis, we shall trace the curve from 0 to $\pi$ and then multiply by 2.

$$\begin{aligned} A &= 2(\tfrac{1}{2})\int_0^{\pi} (1+\cos\theta)^2\, d\theta \\ &= \int_0^{\pi} (1 + 2\cos\theta + \cos^2\theta)\, d\theta \\ &= \int_0^{\pi} d\theta + 2\int_0^{\pi} \cos\theta\, d\theta + \int_0^{\pi} \cos^2\theta\, d\theta \end{aligned}$$

To evaluate $\int_0^{\pi} \cos^2\theta\, d\theta$ we shall make use of the trigonometric identity $\cos^2\theta = \dfrac{1+\cos 2\theta}{2}$.

$$\begin{aligned} A &= \int_0^{\pi} d\theta + 2\int_0^{\pi} \cos\theta\, d\theta + \int_0^{\pi}\left(\frac{1}{2} + \frac{\cos 2\theta}{2}\right) d\theta \\ &= \theta\Big|_0^{\pi} + 2\sin\theta\Big|_0^{\pi} + \left(\frac{\theta}{2} + \frac{\sin 2\theta}{4}\right)\Big|_0^{\pi} \\ &= \pi + 0 + \frac{\pi}{2} \\ &= \frac{3\pi}{2} \end{aligned}$$

**Example 2** Find the area of the region inside the curve $r = 4\sin\theta$ and outside the curve $r = 2$ (Fig. 13-12).

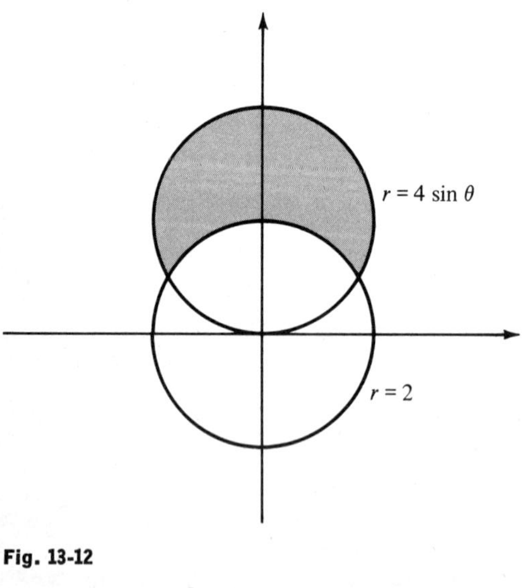


**Fig. 13-12**

*Solution* An analysis of the curve $r = 4 \sin \theta$ reveals that it is a circle with center at $\left(2, \frac{\pi}{2}\right)$ and radius of length 2. The curve $r = 2$ is a circle with center at the origin and radius of length 2. To find the points of intersection of the two curves, we set

$$4 \sin \theta = 2$$
$$\sin \theta = \tfrac{1}{2}$$
$$\theta = \frac{\pi}{6} \qquad \theta = \frac{5\pi}{6}$$

The area of the region is

$$A = \tfrac{1}{2} \int_{\pi/6}^{5\pi/6} (4 \sin \theta)^2 \, d\theta - \tfrac{1}{2} \int_{\pi/6}^{5\pi/6} (2)^2 \, d\theta$$

Because we have symmetry with respect to the 90° axis, we can also express the area as

$$\begin{aligned} A &= 2(\tfrac{1}{2}) \int_{\pi/6}^{\pi/2} (4 \sin \theta)^2 \, d\theta - 2(\tfrac{1}{2}) \int_{\pi/6}^{\pi/2} (2)^2 \, d\theta \\ &= \int_{\pi/6}^{\pi/2} 16 \sin^2 \theta \, d\theta - \int_{\pi/6}^{\pi/2} 4 \, d\theta \\ &= \int_{\pi/6}^{\pi/2} (16 \sin^2 \theta - 4) \, d\theta \end{aligned}$$

Now, since $\sin^2 \theta = \dfrac{1 - \cos 2\theta}{2}$,

$$\begin{aligned} A &= \int_{\pi/6}^{\pi/2} [8(1 - \cos 2\theta) - 4] \, d\theta \\ &= \int_{\pi/6}^{\pi/2} (4 - 8 \cos 2\theta) \, d\theta \\ &= 4\theta - 4 \sin 2\theta \Big|_{\pi/6}^{\pi/2} \\ &= (2\pi - 0) - \left(\frac{2\pi}{3} - 2\sqrt{3}\right) \\ &= \frac{4\pi}{3} + 2\sqrt{3} \end{aligned}$$

## EXERCISES

In each of the following exercises draw the given curves and shade the area to be found.

1. Find the area of the region bounded by $r = \sin \theta$, $\theta = 0$, and $\theta = \frac{\pi}{2}$.
2. Find the area of the region bounded by $r = 1 - \cos \theta$, $\theta = 0$, and $\theta = \pi$.
3. Find the area of the region included within the curve $r = 8 \cos \theta$.
4. Find the area of one loop of the lemniscate $r^2 = 4 \cos 2\theta$.
5. Find the area of one loop of the four-leafed rose $r = a \cos 2\theta$.
6. Find the area of one loop of the three-leafed rose $r = 2 \sin 3\theta$.
7. Find the area of the region bounded by $r = e^\theta$, $\theta = 0$, and $\theta = \pi$.

**8.** Find the area of the region included within the curve $r = 2 + 2 \sin \theta$.

**9.** Find the area of the region which lies inside the cardioid $r = 1 + \cos \theta$ and outside the circle $r = 1$.

**10.** Find the area of the region which lies inside the lemniscate $r^2 = 8 \cos 2\theta$ and outside the circle $r = 2$.

**11.** Find the area common to the circles $r = 2 \sin \theta$ and $r = 2 \cos \theta$.

**12.** Find the area of the region which lies within the smaller loop of the limaçon $r = 1 + 2 \cos \theta$.

# Appendix

## REFERENCE FORMULAS

A. *Algebra*

1. Laws of exponents

$$b^m b^n = b^{m+n}$$

$$\frac{b^m}{b^n} = b^{m-n}$$

$$(b^m)^n = b^{mn}$$

$$(ab)^m = a^m b^m$$

$$\left(\frac{a}{b}\right)^m = \frac{a^m}{b^m}$$

$$a^0 = 1$$

$$a^{-m} = \frac{1}{a^m}$$

2. Laws of logarithms

$$\log_b AB = \log_b A + \log_b B$$

$$\log_b \frac{A}{B} = \log_b A - \log_b B$$

$$\log_b A^n = n \log_b A$$

$$\log_b \sqrt[n]{A} = \frac{1}{n} \log_b A$$

3. Binomial theorem for $n$ = positive integer

$$(a + b)^n = a^n + na^{n-1}b + \frac{n(n-1)}{1(2)} a^{n-2}b^2 + \frac{n(n-1)(n-2)}{1(2)(3)} a^{n-3}b^3 + \cdots + nab^{n-1} + b^n$$

4. Quadratic formula: If $a \neq 0$, then the roots of the equation $ax^2 + bx + c = 0$ are given by the formula

$$x = \frac{-b \pm \sqrt{b^2 - 4ac}}{2a}$$

B. *Geometry* Let $r$ denote radius, $h$ altitude, $b$ length of base, $s$ slant height, and $\theta$ central angle measured in radians.

1. Triangle: area $= \frac{1}{2}bh$
2. Trapezoid: area $= \frac{1}{2}h(b_1 + b_2)$
3. Circle: area $= \pi r^2$, circumference $= 2\pi r$
4. Circular sector: area $= \frac{1}{2}r^2\theta$
5. Sphere: volume $= \frac{4}{3}\pi r^3$, surface area $= 4\pi r^2$
6. Right circular cylinder: volume $= \pi r^2 h$, lateral surface area $= 2\pi rh$
7. Right circular cone: volume $= \frac{1}{3}\pi r^2 h$, lateral surface area $= \pi rs$

C. *Trigonometry*

1. Definitions of the trigonometric functions

$$\sin \theta = \frac{y}{r}$$

$$\cos \theta = \frac{x}{r}$$

$$\tan \theta = \frac{y}{x}$$

$$\cot \theta = \frac{x}{y}$$

$$\sec \theta = \frac{r}{x}$$

$$\csc \theta = \frac{r}{y}$$

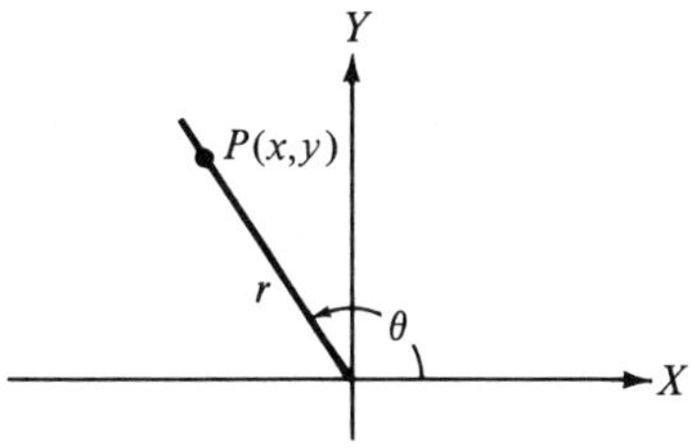


2. Reciprocal relations

$$\sin\theta = \frac{1}{\csc\theta} \qquad \csc\theta = \frac{1}{\sin\theta}$$

$$\cos\theta = \frac{1}{\sec\theta} \qquad \sec\theta = \frac{1}{\cos\theta}$$

$$\tan\theta = \frac{1}{\cot\theta} \qquad \cot\theta = \frac{1}{\tan\theta}$$

3. Fundamental identities

$$\tan\theta = \frac{\sin\theta}{\cos\theta}$$

$$\cot\theta = \frac{\cos\theta}{\sin\theta}$$

$$\sin^2\theta + \cos^2\theta = 1$$
$$1 + \tan^2\theta = \sec^2\theta$$
$$1 + \cot^2\theta = \csc^2\theta$$
$$\sin(-\theta) = -\sin\theta$$
$$\cos(-\theta) = \cos\theta$$
$$\sin(A \pm B) = \sin A\cos B \pm \cos A\sin B$$
$$\cos(A \pm B) = \cos A\cos B \mp \sin A\sin B$$

$$\tan(A \pm B) = \frac{\tan A \pm \tan B}{1 \mp \tan A\tan B}$$

$$\sin A + \sin B = 2\sin\frac{A+B}{2}\cos\frac{A-B}{2}$$

$$\sin A - \sin B = 2\cos\frac{A+B}{2}\sin\frac{A-B}{2}$$

$$\cos A + \cos B = 2\cos\frac{A+B}{2}\cos\frac{A-B}{2}$$

$$\cos A - \cos B = -2\sin\frac{A+B}{2}\sin\frac{A-B}{2}$$

$$\sin 2A = 2\sin A\cos A$$
$$\cos 2A = \cos^2 A - \sin^2 A = 2\cos^2 A - 1 = 1 - 2\sin^2 A$$

$$\tan 2A = \frac{2 \tan A}{1 - \tan^2 A}$$

$$\sin \tfrac{1}{2}A = \pm \sqrt{\frac{1 - \cos A}{2}}$$

$$\cos \tfrac{1}{2}A = \pm \sqrt{\frac{1 + \cos A}{2}}$$

$$\tan \tfrac{1}{2}A = \pm \sqrt{\frac{1 - \cos A}{1 + \cos A}} = \frac{\sin A}{1 + \cos A} = \frac{1 - \cos A}{\sin A}$$

$$\sin \left(\frac{\pi}{2} - A\right) = \cos A$$

$$\cos \left(\frac{\pi}{2} - A\right) = \sin A$$

4. Triangle relationships

Law of cosines: $c^2 = a^2 + b^2 - 2ab \cos C$

Law of sines: $\dfrac{\sin A}{a} = \dfrac{\sin B}{b} = \dfrac{\sin C}{c}$

Area $= \tfrac{1}{2}bc \sin A = \tfrac{1}{2}ac \sin B = \tfrac{1}{2}ab \sin C$

**Table A-1 Some integrals**

1 $\displaystyle\int \frac{x\,dx}{ax+b} = \frac{x}{a} - \frac{b}{a^2}\ln|ax+b|$

2 $\displaystyle\int \frac{x\,dx}{(ax+b)^2} = \frac{b}{a^2(ax+b)} + \frac{1}{a^2}\ln|ax+b|$

3 $\displaystyle\int \frac{x^2\,dx}{ax+b} = \frac{x^2}{2a} - \frac{bx}{a^2} + \frac{b^2}{a^3}\ln|ax+b|$

4 $\displaystyle\int \frac{x^2\,dx}{(ax+b)^2} = \frac{x}{a^2} - \frac{b^2}{a^3(ax+b)} - \frac{2b}{a^3}\ln|ax+b|$

5 $\displaystyle\int \frac{dx}{x(ax+b)} = \frac{1}{b}\ln\left|\frac{x}{ax+b}\right|$

6 $\displaystyle\int x(ax+b)^n\,dx = \frac{x(ax+b)^{n+1}}{a(n+1)} - \frac{(ax+b)^{n+2}}{a^2(n+1)(n+2)} \qquad n \neq -1, -2$

7 $\displaystyle\int \frac{dx}{(ax+b)(cx+d)} = \frac{1}{ad-bc}\ln\left|\frac{ax+b}{cx+d}\right|$

8 $\displaystyle\int \frac{x\,dx}{(ax+b)(cx+d)} = \frac{1}{ad-bc}\left[\frac{d}{c}\ln|cx+d| - \frac{b}{a}\ln|ax+b|\right]$

9 $\displaystyle\int \frac{dx}{x(ax+b)^2} = \frac{1}{b(ax+b)} + \frac{1}{b^2}\ln\left|\frac{x}{ax+b}\right|$

10 $\displaystyle\int x\sqrt{ax+b}\,dx = \frac{2}{a^2}\left[\frac{(ax+b)^{5/2}}{5} - \frac{b(ax+b)^{3/2}}{3}\right]$

11 $\displaystyle\int x^2\sqrt{ax+b}\,dx = \frac{2}{a^3}\left[\frac{(ax+b)^{7/2}}{7} - \frac{2b(ax+b)^{5/2}}{5} + \frac{b^2(ax+b)^{3/2}}{3}\right]$

12 $\displaystyle\int \frac{dx}{x\sqrt{ax+b}} = \frac{1}{\sqrt{b}}\ln\left|\frac{\sqrt{ax+b}-\sqrt{b}}{\sqrt{ax+b}+\sqrt{b}}\right| \qquad b > 0$

$\displaystyle\qquad = \frac{2}{\sqrt{-b}}\operatorname{Arctan}\frac{\sqrt{ax+b}}{\sqrt{-b}} \qquad b < 0$

13 $\displaystyle\int \frac{x\,dx}{\sqrt{ax+b}} = \frac{2(ax-2b)}{3a^2}\sqrt{ax+b}$

14 $\displaystyle\int \frac{\sqrt{ax+b}}{x}\,dx = 2\sqrt{ax+b} + b\int\frac{dx}{x\sqrt{ax+b}}$

15 $\displaystyle\int \sqrt{a^2-x^2}\,dx = \frac{x}{2}\sqrt{a^2-x^2} + \frac{a^2}{2}\operatorname{Arcsin}\frac{x}{a}$

16 $\displaystyle\int x^2\sqrt{a^2-x^2}\,dx = -\frac{x}{4}(a^2-x^2)^{3/2} + \frac{a^2}{4}\int\sqrt{a^2-x^2}\,dx$

17 $\displaystyle\int (a^2-x^2)^{3/2}\,dx = \frac{x}{4}(a^2-x^2)^{3/2} + \frac{3a^2}{4}\int\sqrt{a^2-x^2}\,dx$

18 $\displaystyle\int \frac{dx}{(a^2-x^2)^{3/2}} = \frac{x}{a^2\sqrt{a^2-x^2}}$

19 $\displaystyle\int \frac{dx}{x\sqrt{a^2-x^2}} = -\frac{1}{a}\ln\left|\frac{a+\sqrt{a^2-x^2}}{x}\right|$

*Source:* Rees and Sparks, "Calculus with Analytic Geometry," © 1969 by McGraw-Hill, Inc. Used with permission of McGraw-Hill Book Company.

**Table A-1 Some integrals** *(continued)*

20 $$\int \frac{dx}{x^2\sqrt{a^2-x^2}} = -\frac{\sqrt{a^2-x^2}}{a^2x}$$

21 $$\int \frac{\sqrt{a^2-x^2}}{x}\,dx = \sqrt{a^2-x^2} - a\ln\left|\frac{a+\sqrt{a^2-x^2}}{x}\right|$$

22 $$\int \sqrt{x^2 \pm a^2}\,dx = \frac{x}{2}\sqrt{x^2 \pm a^2} \pm \frac{a^2}{2}\ln|x+\sqrt{x^2 \pm a^2}|$$

23 $$\int \frac{dx}{\sqrt{x^2 \pm a^2}} = \ln|x+\sqrt{x^2 \pm a^2}|$$

24 $$\int \frac{dx}{(x^2 \pm a^2)^{3/2}} = \frac{\pm x}{a^2\sqrt{x^2 \pm a^2}}$$

25 $$\int \frac{\sqrt{x^2 \pm a^2}}{x^2}\,dx = -\frac{\sqrt{x^2 \pm a^2}}{x} + \ln|x+\sqrt{x^2 \pm a^2}|$$

26 $$\int \frac{dx}{x\sqrt{x^2+a^2}} = -\frac{1}{a}\ln\frac{a+\sqrt{x^2+a^2}}{|x|}$$

27 $$\int \frac{dx}{x\sqrt{x^2-a^2}} = \frac{1}{a}\,\text{Arctan}\,\frac{\sqrt{x^2-a^2}}{a}$$

28 $$\int \frac{\sqrt{x^2 \pm a^2}}{x}\,dx = \sqrt{x^2 \pm a^2} \pm a^2\int \frac{dx}{x\sqrt{x^2 \pm a^2}}$$

29 $$\int \sin^2 x\,dx = \frac{x}{2} - \frac{\sin 2x}{4}$$

30 $$\int x\sin x\,dx = \sin x - x\cos x$$

31 $$\int x\cos x\,dx = \cos x + x\sin x$$

32 $$\int x^2\sin x\,dx = 2x\sin x - (x^2-2)\cos x$$

33 $$\int x^2\cos x\,dx = 2x\cos x + (x^2-2)\sin x$$

34 $$\int x\sin^2 x\,dx = \frac{1}{4}[x^2 + \sin^2 x - x\sin 2x]$$

35 $$\int \sin^3 x\,dx = \frac{\cos^3 x}{3} - \cos x$$

36 $$\int \sin^4 x\,dx = \frac{3x}{8} - \frac{3\sin 2x}{16} - \frac{\sin^3 x\cos x}{4}$$

37 $$\int \frac{dx}{1+k\sin x} = \frac{-1}{\sqrt{k^2-1}}\ln\left|\frac{k+\sin x+\sqrt{k^2-1}\cos x}{1+k\sin x}\right| \qquad k^2>1$$

$$= \frac{1}{\sqrt{1-k^2}}\,\text{Arcsin}\,\frac{k+\sin x}{1+k\sin x} \qquad k^2<1$$

38 $$\int \sin ax\sin bx\,dx = \frac{\sin(a-b)x}{2(a-b)} - \frac{\sin(a+b)x}{2(a+b)} \qquad a^2 \neq b^2$$

**Table A-1 Some integrals** *(continued)*

39 $\int xe^{ax}\,dx = \frac{1}{a^2}(ax - 1)e^{ax}$

40 $\int x^2e^{ax}\,dx = \frac{1}{a^3}(a^2x^2 - 2ax + 2)e^{ax}$

41 $\int e^{ax}\sin bx\,dx = \frac{1}{a^2 + b^2}(a \sin bx - b \cos bx)e^{ax}$

42 $\int e^{ax}\cos bx\,dx = \frac{1}{a^2 + b^2}(a \cos bx + b \sin bx)e^{ax}$

43 $\int \ln |x|\,dx = x \ln |x| - x$

44 $\int x^n \ln |x|\,dx = \frac{x^{n+1}}{n + 1}\left(\ln |x| - \frac{1}{n + 1}\right) \qquad n \neq -1$

**Table A-2 Trigonometric functions**

| Angles | Sines | | Cosines | | Tangents | | Cotangents | | Angles |
|---|---|---|---|---|---|---|---|---|---|
| | Nat. | Log. | Nat. | Log. | Nat. | Log. | Nat. | Log. | |
| 0° 00′ | .0000 | ∞ | 1.0000 | **0**.0000 | .0000 | ∞ | ∞ | ∞ | 90° 00′ |
| 10 | .0029 | **7**.4637 | 1.0000 | 0000 | .0029 | **7**.4637 | 343.77 | **2**.5363 | 50 |
| 20 | .0058 | 7648 | 1.0000 | 0000 | .0058 | 7648 | 171.89 | 2352 | 40 |
| 30 | .0087 | 9408 | 1.0000 | 0000 | .0087 | 9409 | 114.59 | 0591 | 30 |
| 40 | .0116 | **8**.0658 | .9999 | 0000 | .0116 | **8**.0658 | 85.940 | **1**.9342 | 20 |
| 50 | .0145 | 1627 | .9999 | 0000 | .0145 | 1627 | 68.750 | 8373 | 10 |
| 1° 00′ | .0175 | **8**.2419 | .9998 | **9**.9999 | .0175 | **8**.2419 | 57.290 | **1**.7581 | 89° 00′ |
| 10 | .0204 | 3088 | .9998 | 9999 | .0204 | 3089 | 49.104 | 6911 | 50 |
| 20 | .0233 | 3668 | .9997 | 9999 | .0233 | 3669 | 42.964 | 6331 | 40 |
| 30 | .0262 | 4179 | .9997 | 9999 | .0262 | 4181 | 38.188 | 5819 | 30 |
| 40 | .0291 | 4637 | .9996 | 9998 | .0291 | 4638 | 34.368 | 5362 | 20 |
| 50 | .0320 | 5050 | .9995 | 9998 | .0320 | 5053 | 31.242 | 4947 | 10 |
| 2° 00′ | .0349 | **8**.5428 | .9994 | **9**.9997 | .0349 | **8**.5431 | 28.636 | **1**.4569 | 88° 00′ |
| 10 | .0378 | 5776 | .9993 | 9997 | .0378 | 5779 | 26.432 | 4221 | 50 |
| 20 | .0407 | 6097 | .9992 | 9996 | .0407 | 6101 | 24.542 | 3899 | 40 |
| 30 | .0436 | 6397 | .9990 | 9996 | .0437 | 6401 | 22.904 | 3599 | 30 |
| 40 | .0465 | 6677 | .9989 | 9995 | .0466 | 6682 | 21.470 | 3318 | 20 |
| 50 | .0494 | 6940 | .9988 | 9995 | .0495 | 6945 | 20.206 | 3055 | 10 |
| 3° 00′ | .0523 | **8**.7188 | .9986 | **9**.9994 | .0524 | **8**.7194 | 19.081 | **1**.2806 | 87° 00′ |
| 10 | .0552 | 7423 | .9985 | 9993 | .0553 | 7429 | 18.075 | 2571 | 50 |
| 20 | .0581 | 7645 | .9983 | 9993 | .0582 | 7652 | 17.169 | 2348 | 40 |
| 30 | .0610 | 7857 | .9981 | 9992 | .0612 | 7865 | 16.350 | 2135 | 30 |
| 40 | .0640 | 8059 | .9980 | 9991 | .0641 | 8067 | 15.605 | 1933 | 20 |
| 50 | .0669 | 8251 | .9978 | 9990 | .0670 | 8261 | 14.924 | 1739 | 10 |
| 4° 00′ | .0698 | **8**.8436 | .9976 | **9**.9989 | .0699 | **8**.8446 | 14.301 | **1**.1554 | 86° 00′ |
| 10 | .0727 | 8613 | .9974 | 9989 | .0729 | 8624 | 13.727 | 1376 | 50 |
| 20 | .0756 | 8783 | .9971 | 9988 | .0758 | 8795 | 13.197 | 1205 | 40 |
| 30 | .0785 | 8946 | .9969 | 9987 | .0787 | 8960 | 12.706 | 1040 | 30 |
| 40 | .0814 | 9104 | .9967 | 9986 | .0816 | 9118 | 12.251 | 0882 | 20 |
| 50 | .0843 | 9256 | .9964 | 9985 | .0846 | 9272 | 11.826 | 0728 | 10 |
| 5° 00′ | .0872 | **8**.9403 | .9962 | **9**.9983 | .0875 | **8**.9420 | 11.430 | **1**.0580 | 85° 00′ |
| 10 | .0901 | 9545 | .9959 | 9982 | .0904 | 9563 | 11.059 | 0437 | 50 |
| 20 | .0929 | 9682 | .9957 | 9981 | .0934 | 9701 | 10.712 | 0299 | 40 |
| 30 | .0958 | 9816 | .9954 | 9980 | .0963 | 9836 | 10.385 | 0164 | 30 |
| 40 | .0987 | 9945 | .9951 | 9979 | .0992 | 9966 | 10.078 | 0034 | 20 |
| 50 | .1016 | **9**.0070 | .9948 | 9977 | .1022 | **9**.0093 | 9.7882 | **0**.9907 | 10 |
| 6° 00′ | .1045 | **9**.0192 | .9945 | **9**.9976 | .1051 | **9**.0216 | 9.5144 | **0**.9784 | 84° 00 |
| 10 | .1074 | 0311 | .9942 | 9975 | .1080 | 0336 | 9.2553 | 9664 | 50 |
| 20 | .1103 | 0426 | .9939 | 9973 | .1110 | 0453 | 9.0098 | 9547 | 40 |
| 30 | .1132 | 0539 | .9936 | 9972 | .1139 | 0567 | 8.7769 | 9433 | 30 |
| 40 | .1161 | 0648 | .9932 | 9971 | .1169 | 0678 | 8.5555 | 9322 | 20 |
| 50 | .1190 | 0755 | .9920 | 9969 | .1198 | 0786 | 8.3450 | 9214 | 10 |
| 7° 00′ | .1219 | **9**.0859 | .9925 | **9**.9968 | .1228 | **9**.0891 | 8.1443 | **0**.9109 | 83° 00′ |
| 10 | .1248 | 0961 | .9922 | 9966 | .1257 | 0995 | 7.9530 | 9005 | 50 |
| 20 | .1276 | 1060 | .9918 | 9964 | .1287 | 1096 | 7.7704 | 8904 | 40 |
| 30 | .1305 | 1157 | .9914 | 9963 | .1317 | 1194 | 7.5958 | 8806 | 30 |
| 40 | .1334 | 1252 | .9911 | 9961 | .1346 | 1291 | 7.4287 | 8709 | 20 |
| 50 | .1363 | 1345 | .9907 | 9959 | .1376 | 1385 | 7.2687 | 8615 | 10 |
| 8° 00′ | .1392 | **9**.1436 | .9903 | **9**.9958 | .1405 | **9**.1478 | 7.1154 | **0**.8522 | 82° 00′ |
| 10 | .1421 | 1525 | .9899 | 9956 | .1435 | 1569 | 6.9682 | 8431 | 50 |
| 20 | .1449 | 1612 | .9894 | 9954 | .1465 | 1658 | 6.8269 | 8342 | 40 |
| 30 | .1478 | 1697 | .9890 | 9952 | .1495 | 1745 | 6.6912 | 8255 | 30 |
| 40 | .1507 | 1781 | .9886 | 9950 | .1524 | 1831 | 6.5606 | 8169 | 20 |
| 50 | .1536 | 1863 | .9881 | 9948 | .1554 | 1915 | 6.4348 | 8085 | 10 |
| 9° 00′ | .1564 | **9**.1943 | .9877 | **9**.9946 | .1584 | **9**.1997 | 6.3138 | **0**.8003 | 81° 00′ |
| | Nat. | Log. | Nat. | Log. | Nat. | Log. | Nat. | Log. | |
| Angles | Cosines | | Sines | | Cotangents | | Tangents | | Angles |

*Source:* Middlemiss, Marks, and Smart, "Analytic Geometry," 3d ed., © 1968 by McGraw-Hill, Inc. Used with permission of McGraw-Hill Book Company.

**Table A-2 Trigonometric functions** *(continued)*

| Angles | Sines | | Cosines | | Tangents | | Cotangents | | Angles |
|---|---|---|---|---|---|---|---|---|---|
| | Nat. | Log. | Nat. | Log. | Nat. | Log. | Nat. | Log. | |
| 9° 00′ | .1564 | 9.1943 | .9877 | 9.9946 | .1584 | 9.1997 | 6.3138 | 0.8003 | 81° 00′ |
| 10 | .1593 | 2022 | .9872 | 9944 | .1614 | 2078 | 6.1970 | 7922 | 50 |
| 20 | .1622 | 2100 | .9868 | 9942 | .1644 | 2158 | 6.0844 | 7842 | 40 |
| 30 | .1650 | 2176 | .9863 | 9940 | .1673 | 2236 | 5.9758 | 7764 | 30 |
| 40 | .1679 | 2251 | .9858 | 9938 | .1703 | 2313 | 5.8708 | 7687 | 20 |
| 50 | .1708 | 2324 | .9853 | 9936 | .1733 | 2389 | 5.7694 | 7611 | 10 |
| 10° 00′ | .1736 | 9.2397 | .9848 | 9.9934 | .1763 | 9.2463 | 5.6713 | 0.7537 | 80° 00′ |
| 10 | .1765 | 2468 | .9843 | 9931 | .1793 | 2536 | 5.5764 | 7464 | 50 |
| 20 | .1794 | 2538 | .9838 | 9929 | .1823 | 2609 | 5.4845 | 7391 | 40 |
| 30 | .1822 | 2606 | .9833 | 9927 | .1853 | 2680 | 5.3955 | 7320 | 30 |
| 40 | .1851 | 2674 | .9827 | 9924 | .1883 | 2750 | 5.3093 | 7250 | 20 |
| 50 | .1880 | 2740 | .9822 | 9922 | .1914 | 2819 | 5.2257 | 7181 | 10 |
| 11° 00′ | .1908 | 9.2806 | .9816 | 9.9919 | .1944 | 9.2887 | 5.1446 | 0.7113 | 79° 00′ |
| 10 | .1937 | 2870 | .9811 | 9917 | .1974 | 2953 | 5.0658 | 7047 | 50 |
| 20 | .1965 | 2934 | .9805 | 9914 | .2004 | 3020 | 4.9894 | 6980 | 40 |
| 30 | .1994 | 2997 | .9799 | 9912 | .2035 | 3085 | 4.9152 | 6915 | 30 |
| 40 | .2022 | 3058 | .9793 | 9909 | .2065 | 3149 | 4.8430 | 6851 | 20 |
| 50 | .2051 | 3119 | .9787 | 9907 | .2095 | 3212 | 4.7729 | 6788 | 10 |
| 12° 00′ | .2079 | 9.3179 | .9781 | 9.9904 | .2126 | 9.3275 | 4.7046 | 0.6725 | 78° 00′ |
| 10 | .2108 | 3238 | .9775 | 9901 | .2156 | 3336 | 4.6382 | 6664 | 50 |
| 20 | .2136 | 3296 | .9769 | 9899 | .2186 | 3397 | 4.5736 | 6603 | 40 |
| 30 | .2164 | 3353 | .9763 | 9896 | .2217 | 3458 | 4.5107 | 6542 | 30 |
| 40 | .2193 | 3410 | .9757 | 9893 | .2247 | 3517 | 4.4494 | 6483 | 20 |
| 50 | .2221 | 3466 | .9750 | 9890 | .2278 | 3576 | 4.3897 | 6424 | 10 |
| 13° 00′ | .2250 | 9.3521 | .9744 | 9.9887 | .2309 | 9.3634 | 4.3315 | 0.6366 | 77° 00′ |
| 10 | .2278 | 3575 | .9737 | 9884 | .2339 | 3691 | 4.2747 | 6309 | 50 |
| 20 | .2306 | 3629 | .9730 | 9881 | .2370 | 3748 | 4.2193 | 6252 | 40 |
| 30 | .2334 | 3682 | .9724 | 9878 | .2401 | 3804 | 4.1653 | 6196 | 30 |
| 40 | .2363 | 3734 | .9717 | 9875 | .2432 | 3859 | 4.1126 | 6141 | 20 |
| 50 | .2391 | 3786 | .9710 | 9872 | .2462 | 3914 | 4.0611 | 6086 | 10 |
| 14° 00′ | .2419 | 9.3837 | .9703 | 9.9869 | .2493 | 9.3968 | 4.0108 | 0.6032 | 76° 00′ |
| 10 | .2447 | 3887 | .9696 | 9866 | .2524 | 4021 | 3.9617 | 5979 | 50 |
| 20 | .2476 | 3937 | .9689 | 9863 | .2555 | 4074 | 3.9136 | 5926 | 40 |
| 30 | .2504 | 3986 | .9681 | 9859 | .2586 | 4127 | 3.8667 | 5873 | 30 |
| 40 | .2532 | 4035 | .9674 | 9856 | .2617 | 4178 | 3.8208 | 5822 | 20 |
| 50 | .2560 | 4083 | .9667 | 9853 | .2648 | 4230 | 3.7760 | 5770 | 10 |
| 15° 00′ | .2588 | 9.4130 | .9659 | 9.9849 | .2679 | 9.4281 | 3.7321 | 0.5719 | 75° 00′ |
| 10 | .2616 | 4177 | .9652 | 9846 | .2711 | 4331 | 3.6891 | 5669 | 50 |
| 20 | .2644 | 4223 | .9644 | 9843 | .2742 | 4381 | 3.6470 | 5619 | 40 |
| 30 | .2672 | 4269 | .9636 | 9839 | .2773 | 4430 | 3.6059 | 5570 | 30 |
| 40 | .2700 | 4314 | .9628 | 9836 | .2805 | 4479 | 3.5656 | 5521 | 20 |
| 50 | .2728 | 4359 | .9621 | 9832 | .2836 | 4527 | 3.5261 | 5473 | 10 |
| 16° 00′ | .2756 | 9.4403 | .0613 | 9.9828 | .2867 | 9.4575 | 3.4874 | 0.5425 | 74° 00′ |
| 10 | .2784 | 4447 | .9605 | 9825 | .2899 | 4622 | 3.4495 | 5378 | 50 |
| 20 | .2812 | 4491 | .9596 | 9821 | .2931 | 4669 | 3.4124 | 5331 | 40 |
| 30 | .2840 | 4533 | .9588 | 9817 | .2962 | 4716 | 3.3759 | 5284 | 30 |
| 40 | .2868 | 4576 | .9580 | 9814 | .2994 | 4762 | 3.3402 | 5238 | 20 |
| 50 | .2896 | 4618 | .9572 | 9810 | .3026 | 4808 | 3.3052 | 5192 | 10 |
| 17° 00′ | .2924 | 9.4659 | .9563 | 9.9806 | .3057 | 9.4853 | 3.2709 | 0.5147 | 73° 00′ |
| 10 | .2952 | 4700 | .9555 | 9802 | .3089 | 4898 | 3.2371 | 5102 | 50 |
| 20 | .2979 | 4741 | .9546 | 9798 | .3121 | 4943 | 3.2041 | 5057 | 40 |
| 30 | .3007 | 4781 | .9537 | 9794 | .3153 | 4987 | 3.1716 | 5013 | 30 |
| 40 | .3035 | 4821 | .9528 | 9790 | .3185 | 5031 | 3.1397 | 4969 | 20 |
| 50 | .3062 | 4861 | .9520 | 9786 | .3217 | 5075 | 3.1084 | 4925 | 10 |
| 18° 00′ | .3090 | 9.4900 | .9511 | 9.9782 | .3249 | 9.5118 | 3.0777 | 0.4882 | 72° 00′ |
| | Nat. | Log. | Nat. | Log. | Nat. | Log. | Nat. | Log. | |
| Angles | Cosines | | Sines | | Cotangents | | Tangents | | Angles |

**Table A-2 Trigonometric functions** *(continued)*

| Angles | Sines | | Cosines | | Tangents | | Cotangents | | Angles |
|---|---|---|---|---|---|---|---|---|---|
| | Nat. | Log. | Nat. | Log. | Nat. | Log. | Nat. | Log. | |
| 18° 00′ | .3090 | 9.4900 | .9511 | 9.9782 | .3249 | 9.5118 | 3.0777 | 0.4882 | 72° 00′ |
| 10 | .3118 | 4939 | .9502 | 9778 | .3281 | 5161 | 3.0475 | 4839 | 50 |
| 20 | .3145 | 4977 | .9492 | 9774 | .3314 | 5203 | 3.0178 | 4797 | 40 |
| 30 | .3173 | 5015 | .9483 | 9770 | .3346 | 5245 | 2.9887 | 4755 | 30 |
| 40 | .3201 | 5052 | .9474 | 9765 | .3378 | 5287 | 2.9600 | 4713 | 20 |
| 50 | .3228 | 5090 | .9465 | 9761 | .3411 | 5329 | 2.9319 | 4671 | 10 |
| 19° 00′ | .3256 | 9.5126 | .9455 | 9.9757 | .3443 | 9.5370 | 2.9042 | 0.4630 | 71° 00′ |
| 10 | .3283 | 5163 | .9446 | 9752 | .3476 | 5411 | 2.8770 | 4589 | 50 |
| 20 | .3311 | 5199 | .9436 | 9748 | .3508 | 5451 | 2.8502 | 4549 | 40 |
| 30 | .3338 | 5235 | .9426 | 9743 | .3541 | 5491 | 2.8239 | 4509 | 30 |
| 40 | .3365 | 5270 | .9417 | 9739 | .3574 | 5531 | 2.7980 | 4469 | 20 |
| 50 | .3393 | 5306 | .9407 | 9734 | .3607 | 5571 | 2.7725 | 4429 | 10 |
| 20° 00′ | .3420 | 9.5341 | .9397 | 9.9730 | .3640 | 9.5611 | 2.7475 | 0.4389 | 70° 00′ |
| 10 | .3448 | 5375 | .9387 | 9725 | .3673 | 5650 | 2.7228 | 4350 | 50 |
| 20 | .3475 | 5409 | .9377 | 9721 | .3706 | 5689 | 2.6985 | 4311 | 40 |
| 30 | .3502 | 5443 | .9367 | 9716 | .3739 | 5727 | 2.6746 | 4273 | 30 |
| 40 | .3529 | 5477 | .9356 | 9711 | .3772 | 5766 | 2.6511 | 4234 | 20 |
| 50 | .3557 | 5510 | .9346 | 9706 | .3805 | 5804 | 2.6279 | 4196 | 10 |
| 21° 00′ | .3584 | 9.5543 | .9336 | 9.9702 | .3839 | 9.5842 | 2.6051 | 0.4158 | 69° 00′ |
| 10 | .3611 | 5576 | .9325 | 9697 | .3872 | 5879 | 2.5826 | 4121 | 50 |
| 20 | .3638 | 5609 | .9315 | 9692 | .3906 | 5917 | 2.5605 | 4083 | 40 |
| 30 | .3665 | 5641 | .9304 | 9687 | .3939 | 5954 | 2.5386 | 4046 | 30 |
| 40 | .3692 | 5673 | .9293 | 9682 | .3973 | 5991 | 2.5172 | 4009 | 20 |
| 50 | .3719 | 5704 | .9283 | 9677 | .4006 | 6028 | 2.4960 | 3972 | 10 |
| 22° 00′ | .3746 | 9.5736 | .9272 | 9.9672 | .4040 | 9.6064 | 2.4751 | 0.3936 | 68° 00′ |
| 10 | .3773 | 5767 | .9261 | 9667 | .4074 | 6100 | 2.4545 | 3900 | 50 |
| 20 | .3800 | 5798 | .9250 | 9661 | .4108 | 6136 | 2.4342 | 3864 | 40 |
| 30 | .3827 | 5828 | .9239 | 9656 | .4142 | 6172 | 2.4142 | 3828 | 30 |
| 40 | .3854 | 5859 | .9228 | 9651 | .4176 | 6208 | 2.3945 | 3792 | 20 |
| 50 | .3881 | 5889 | .9216 | 9646 | .4210 | 6243 | 2.3750 | 3757 | 10 |
| 23° 00′ | .3907 | 9.5919 | .9205 | 9.9640 | .4245 | 9.6279 | 2.3559 | 0.3721 | 67° 00′ |
| 10 | .3934 | 5948 | .9194 | 9635 | .4279 | 6314 | 2.3369 | 3686 | 50 |
| 20 | .3961 | 5978 | .9182 | 9629 | .4314 | 6348 | 2.3183 | 3652 | 40 |
| 30 | .3987 | 6007 | .9171 | 9624 | .4348 | 6383 | 2.2998 | 3617 | 30 |
| 40 | .4014 | 6036 | .9159 | 9618 | .4383 | 6417 | 2.2817 | 3583 | 20 |
| 50 | .4041 | 6065 | .9147 | 9613 | .4417 | 6452 | 2.2637 | 3548 | 10 |
| 24° 00′ | .4067 | 9.6093 | .9135 | 9.9607 | .4452 | 9.6486 | 2.2460 | 0.3514 | 66° 00′ |
| 10 | .4094 | 6121 | .9124 | 9602 | .4487 | 6520 | 2.2286 | 3480 | 50 |
| 20 | .4120 | 6149 | .9112 | 9596 | .4522 | 6553 | 2.2113 | 3447 | 40 |
| 30 | .4147 | 6177 | .9100 | 9590 | .4557 | 6587 | 2.1943 | 3413 | 30 |
| 40 | .4173 | 6205 | .9088 | 9584 | .4592 | 6620 | 2.1775 | 3380 | 20 |
| 50 | .4200 | 6232 | .9075 | 9579 | .4628 | 6654 | 2.1609 | 3346 | 10 |
| 25° 00′ | .4226 | 9.6259 | .9063 | 9.9573 | .4663 | 9.6687 | 2.1445 | 0.3313 | 65° 00′ |
| 10 | .4253 | 6286 | .9051 | 9567 | .4699 | 6720 | 2.1283 | 3280 | 50 |
| 20 | .4279 | 6313 | .9038 | 9561 | .4734 | 6752 | 2.1123 | 3248 | 40 |
| 30 | .4305 | 6340 | .9026 | 9555 | .4770 | 6785 | 2.0965 | 3215 | 30 |
| 40 | .4331 | 6366 | .9013 | 9549 | .4806 | 6817 | 2.0809 | 3183 | 20 |
| 50 | .4358 | 6392 | .9001 | 9543 | .4841 | 6850 | 2.0655 | 3150 | 10 |
| 26° 00′ | .4384 | 9.6418 | .8988 | 9.9537 | .4877 | 9.6882 | 2.0503 | 0.3118 | 64° 00′ |
| 10 | .4410 | 6444 | .8975 | 9530 | .4913 | 6914 | 2.0353 | 3086 | 50 |
| 20 | .4436 | 6470 | .8962 | 9524 | .4950 | 6946 | 2.0204 | 3054 | 40 |
| 30 | .4462 | 6495 | .8949 | 9518 | .4986 | 6977 | 2.0057 | 3023 | 30 |
| 40 | .4488 | 6521 | .8936 | 9512 | .5022 | 7009 | 1.9912 | 2991 | 20 |
| 50 | .4514 | 6546 | .8923 | 9505 | .5059 | 7040 | 1.9768 | 2960 | 10 |
| 27° 00′ | .4540 | 9.6570 | .8910 | 9.9499 | .5095 | 9.7072 | 1.9626 | 0.2928 | 63° 00′ |
| | Nat. | Log. | Nat. | Log. | Nat. | Log. | Nat. | Log. | |
| Angles | Cosines | | Sines | | Cotangents | | Tangents | | Angles |

**Table A-2 Trigonometric functions** *(continued)*

| Angles | Sines | | Cosines | | Tangents | | Cotangents | | Angles |
|---|---|---|---|---|---|---|---|---|---|
| | Nat. | Log. | Nat. | Log. | Nat. | Log. | Nat. | Log. | |
| 27° 00′ | .4540 | 9.6570 | .8910 | 9.9499 | .5095 | 9.7072 | 1.9626 | 0.2928 | 63° 00′ |
| 10 | .4566 | 6595 | .8897 | 9492 | .5132 | 7103 | 1.9486 | 2897 | 50 |
| 20 | .4592 | 6620 | .8884 | 9486 | .5169 | 7134 | 1.9347 | 2866 | 40 |
| 30 | .4617 | 6644 | .8870 | 9479 | .5206 | 7165 | 1.9210 | 2835 | 30 |
| 40 | .4643 | 6668 | .8857 | 9473 | .5243 | 7196 | 1.9074 | 2804 | 20 |
| 50 | .4669 | 6692 | .8843 | 9466 | .5280 | 7226 | 1.8940 | 2774 | 10 |
| 28° 00′ | .4695 | 9.6716 | .8829 | 9.9459 | .5317 | 9.7257 | 1.8807 | 0.2743 | 62° 00′ |
| 10 | .4720 | 6740 | .8816 | 9453 | .5354 | 7287 | 1.8676 | 2713 | 50 |
| 20 | .4746 | 6763 | .8802 | 9446 | .5392 | 7317 | 1.8546 | 2683 | 40 |
| 30 | .4772 | 6787 | .8788 | 9439 | .5430 | 7348 | 1.8418 | 2652 | 30 |
| 40 | .4797 | 6810 | .8774 | 9432 | .5467 | 7378 | 1.8291 | 2622 | 20 |
| 50 | .4823 | 6833 | .8760 | 9425 | .5505 | 7408 | 1.8165 | 2592 | 10 |
| 29° 00′ | .4848 | 9.6856 | .8746 | 9.9418 | .5543 | 9.7438 | 1.8040 | 0.2562 | 61° 00′ |
| 10 | .4874 | 6878 | .8732 | 9411 | .5581 | 7467 | 1.7917 | 2533 | 50 |
| 20 | .4899 | 6901 | .8718 | 9404 | .5619 | 7497 | 1.7796 | 2503 | 40 |
| 30 | .4924 | 6923 | .8704 | 9397 | .5658 | 7526 | 1.7675 | 2474 | 30 |
| 40 | .4950 | 6946 | .8689 | 9390 | .5696 | 7556 | 1.7556 | 2444 | 20 |
| 50 | .4975 | 6968 | .8675 | 9383 | .5735 | 7585 | 1.7437 | 2415 | 10 |
| 30° 00′ | .5000 | 9.6990 | .8660 | 9.9375 | .5774 | 9.7614 | 1.7321 | 0.2386 | 60° 00′ |
| 10 | .5025 | 7012 | .8646 | 9368 | .5812 | 7644 | 1.7205 | 2356 | 50 |
| 20 | .5050 | 7033 | .8631 | 9361 | .5851 | 7673 | 1.7090 | 2327 | 40 |
| 30 | .5075 | 7055 | .8616 | 9353 | .5890 | 7701 | 1.6977 | 2299 | 30 |
| 40 | .5100 | 7076 | .8601 | 9346 | .5930 | 7730 | 1.6864 | 2270 | 20 |
| 50 | .5125 | 7097 | .8587 | 9338 | .5969 | 7759 | 1.6753 | 2241 | 10 |
| 31° 00′ | .5150 | 9.7118 | .8572 | 9.9331 | .6009 | 9.7788 | 1.6643 | 0.2212 | 59° 00′ |
| 10 | .5175 | 7139 | .8557 | 9323 | .6048 | 7816 | 1.6534 | 2184 | 50 |
| 20 | .5200 | 7160 | .8542 | 9315 | .6088 | 7845 | 1.6426 | 2155 | 40 |
| 30 | .5225 | 7181 | .8526 | 9308 | .6128 | 7873 | 1.6319 | 2127 | 30 |
| 40 | .5250 | 7201 | .8511 | 9300 | .6168 | 7902 | 1.6212 | 2098 | 20 |
| 50 | .5275 | 7222 | .8496 | 9292 | .6208 | 7930 | 1.6107 | 2070 | 10 |
| 32° 00′ | .5299 | 9.7242 | .8480 | 9.9284 | .6249 | 9.7958 | 1.6003 | 0.2042 | 58° 00′ |
| 10 | .5324 | 7262 | .8465 | 9276 | .6289 | 7986 | 1.5900 | 2014 | 50 |
| 20 | .5348 | 7282 | .8450 | 9268 | .6330 | 8014 | 1.5798 | 1986 | 40 |
| 30 | .5373 | 7302 | .8434 | 9260 | .6371 | 8042 | 1.5697 | 1958 | 30 |
| 40 | .5398 | 7322 | .8418 | 9252 | .6412 | 8070 | 1.5597 | 1930 | 20 |
| 50 | .5422 | 7342 | .8403 | 9244 | .6453 | 8097 | 1.5497 | 1903 | 10 |
| 33° 00′ | .5446 | 9.7361 | .8387 | 9.9236 | .6494 | 9.8125 | 1.5399 | 0.1875 | 57° 00′ |
| 10 | .5471 | 7380 | .8371 | 9228 | .6536 | 8153 | 1.5301 | 1847 | 50 |
| 20 | .5495 | 7400 | .8355 | 9219 | .6577 | 8180 | 1.5204 | 1820 | 40 |
| 30 | .5519 | 7419 | .8339 | 9211 | .6619 | 8208 | 1.5108 | 1792 | 30 |
| 40 | .5544 | 7438 | .8323 | 9203 | .6661 | 8235 | 1.5013 | 1765 | 20 |
| 50 | .5568 | 7457 | .8307 | 9194 | .6703 | 8263 | 1.4919 | 1737 | 10 |
| 34° 00′ | .5592 | 9.7476 | .8290 | 9.9186 | .6745 | 9.8290 | 1.4826 | 0.1710 | 56° 00′ |
| 10 | .5616 | 7494 | .8274 | 9177 | .6787 | 8317 | 1.4733 | 1683 | 50 |
| 20 | .5640 | 7513 | .8258 | 9169 | .6830 | 8344 | 1.4641 | 1656 | 40 |
| 30 | .5664 | 7531 | .8241 | 9160 | .6873 | 8371 | 1.4550 | 1629 | 30 |
| 40 | .5688 | 7550 | .8225 | 9151 | .6916 | 8398 | 1.4460 | 1602 | 20 |
| 50 | .5712 | 7568 | .8208 | 9142 | .6959 | 8425 | 1.4370 | 1575 | 10 |
| 35° 00′ | .5736 | 9.7586 | .8192 | 9.9134 | .7002 | 9.8452 | 1.4281 | 0.1548 | 55° 00′ |
| 10 | .5760 | 7604 | .8175 | 9125 | .7046 | 8479 | 1.4193 | 1521 | 50 |
| 20 | .5783 | 7622 | .8158 | 9116 | .7089 | 8506 | 1.4106 | 1494 | 40 |
| 30 | .5807 | 7640 | .8141 | 9107 | .7133 | 8533 | 1.4019 | 1467 | 30 |
| 40 | .5831 | 7657 | .8124 | 9098 | .7177 | 8559 | 1.3934 | 1441 | 20 |
| 50 | .5854 | 7675 | .8107 | 9089 | .7221 | 8586 | 1.3848 | 1414 | 10 |
| 36° 00′ | .5878 | 9.7692 | .8090 | 9.9080 | .7265 | 9.8613 | 1.3764 | 0.1387 | 54° 00′ |
| | Nat. | Log. | Nat. | Log. | Nat. | Log. | Nat. | Log. | |
| Angles | Cosines | | Sines | | Cotangents | | Tangents | | Angles |

**Table A-2 Trigonometric functions** *(continued)*

| Angles | Sines | | Cosines | | Tangents | | Cotangents | | Angles |
|---|---|---|---|---|---|---|---|---|---|
| | Nat. | Log. | Nat. | Log. | Nat. | Log. | Nat. | Log. | |
| **36°** 00′ | .5878 | **9**.7692 | .8090 | **9**.9080 | .7265 | **9**.8613 | 1.3764 | **0**.1387 | 54° 00′ |
| 10 | .5901 | 7710 | .8073 | 9070 | .7310 | 8639 | 1.3680 | 1361 | 50 |
| 20 | .5925 | 7727 | .8056 | 9061 | .7355 | 8666 | 1.3597 | 1334 | 40 |
| 30 | .5948 | 7744 | .8039 | 9052 | .7400 | 8692 | 1.3514 | 1308 | 30 |
| 40 | .5972 | 7761 | .8021 | 9042 | .7445 | 8718 | 1.3432 | 1282 | 20 |
| 50 | .5995 | 7778 | .8004 | 9033 | .7490 | 8745 | 1.3351 | 1255 | 10 |
| **37°** 00′ | .6018 | **9**.7795 | .7986 | **9**.9023 | .7536 | **9**.8771 | 1.3270 | **0**.1229 | 53° 00′ |
| 10 | .6041 | 7811 | .7969 | 9014 | .7581 | 8797 | 1.3190 | 1203 | 50 |
| 20 | .6065 | 7828 | .7951 | 9004 | .7627 | 8824 | 1.3111 | 1176 | 40 |
| 30 | .6088 | 7844 | .7934 | 8995 | .7673 | 8850 | 1.3032 | 1150 | 30 |
| 40 | .6111 | 7861 | .7916 | 8985 | .7720 | 8876 | 1.2954 | 1124 | 20 |
| 50 | .6134 | 7877 | .7898 | 8975 | .7766 | 8902 | 1.2876 | 1098 | 10 |
| **38°** 00′ | .6157 | **9**.7893 | .7880 | **9**.8965 | .7813 | **9**.8928 | 1.2790 | **0**.1072 | 52° 00′ |
| 10 | .6180 | 7910 | .7862 | 8955 | .7860 | 8954 | 1.2723 | 1046 | 50 |
| 20 | .6202 | 7926 | .7844 | 8945 | .7907 | 8980 | 1.2647 | 1020 | 40 |
| 30 | .6225 | 7941 | .7826 | 8935 | .7954 | 9006 | 1.2572 | 0994 | 30 |
| 40 | .6248 | 7957 | .7808 | 8925 | .8002 | 9032 | 1.2497 | 0968 | 20 |
| 50 | .6271 | 7973 | .7790 | 8915 | .8050 | 9058 | 1.2423 | 0942 | 10 |
| **39°** 00′ | .6293 | **9**.7989 | .7771 | **9**.8905 | .8098 | **9**.9084 | 1.2349 | **0**.0916 | 51° 00′ |
| 10 | .6316 | 8004 | .7753 | 8895 | .8146 | 9110 | 1.2276 | 0890 | 50 |
| 20 | .6338 | 8020 | .7735 | 8884 | .8195 | 9135 | 1.2203 | 0865 | 40 |
| 30 | .6361 | 8035 | .7716 | 8874 | .8243 | 9161 | 1.2131 | 0839 | 30 |
| 40 | .6383 | 8050 | .7698 | 8864 | .8292 | 9187 | 1.2059 | 0813 | 20 |
| 50 | .6406 | 8066 | .7679 | 8853 | .8342 | 9212 | 1.1988 | 0788 | 10 |
| **40°** 00′ | .6428 | **9**.8081 | .7660 | **9**.8843 | .8391 | **9**.9238 | 1.1918 | **0**.0762 | 50° 00′ |
| 10 | .6450 | 8096 | .7642 | 8832 | .8441 | 9264 | 1.1847 | 0736 | 50 |
| 20 | .6472 | 8111 | .7623 | 8821 | .8491 | 9289 | 1.1778 | 0711 | 40 |
| 30 | .6494 | 8125 | .7604 | 8810 | .8541 | 9315 | 1.1708 | 0685 | 30 |
| 40 | .6517 | 8140 | .7585 | 8800 | .8591 | 9341 | 1.1640 | 0659 | 20 |
| 50 | .6539 | 8155 | .7566 | 8789 | .8642 | 9366 | 1.1571 | 0634 | 10 |
| **41°** 00′ | .6561 | **9**.8169 | .7547 | **9**.8778 | .8693 | **9**.9392 | 1.1504 | **0**.0608 | 49° 00′ |
| 10 | .6583 | 8184 | .7528 | 8767 | .8744 | 9417 | 1.1436 | 0583 | 50 |
| 20 | .6604 | 8198 | .7509 | 8756 | .8796 | 9443 | 1.1369 | 0557 | 40 |
| 30 | .6626 | 8213 | .7490 | 8745 | .8847 | 9468 | 1.1303 | 0532 | 30 |
| 40 | .6648 | 8227 | .7470 | 8733 | .8899 | 9494 | 1.1237 | 0506 | 20 |
| 50 | .6670 | 8241 | .7451 | 8722 | .8952 | 9519 | 1.1171 | 0481 | 10 |
| **42°** 00′ | .6691 | **9**.8255 | .7431 | **9**.8711 | .9004 | **9**.9544 | 1.1106 | **0**.0456 | 48° 00′ |
| 10 | .6713 | 8269 | .7412 | 8699 | .9057 | 9570 | 1.1041 | 0430 | 50 |
| 20 | .6734 | 8283 | .7392 | 8688 | .9110 | 9595 | 1.0977 | 0405 | 40 |
| 30 | .6756 | 8297 | .7373 | 8676 | .9163 | 9621 | 1.0913 | 0379 | 30 |
| 40 | .6777 | 8311 | .7353 | 8665 | .9217 | 9646 | 1.0850 | 0354 | 20 |
| 50 | .6799 | 8324 | .7333 | 8653 | .9271 | 9671 | 1.0786 | 0329 | 10 |
| **43°** 00′ | .6820 | **9**.8338 | .7314 | **9**.8641 | .9325 | **9**.9697 | 1.0724 | **0**.0303 | 47° 00′ |
| 10 | .6841 | 8351 | .7294 | 8629 | .9380 | 9722 | 1.0661 | 0278 | 50 |
| 20 | .6862 | 8365 | .7274 | 8618 | .9435 | 9747 | 1.0599 | 0253 | 40 |
| 30 | .6884 | 8378 | .7254 | 8606 | .9490 | 9772 | 1.0538 | 0228 | 30 |
| 40 | .6905 | 8391 | .7234 | 8594 | .9545 | 9798 | 1.0477 | 0202 | 20 |
| 50 | .6926 | 8405 | .7214 | 8582 | .9601 | 9823 | 1.0416 | 0177 | 10 |
| **44°** 00′ | .6947 | **9**.8418 | .7193 | **9**.8569 | .9657 | **9**.9848 | 1.0355 | **0**.0152 | 46° 00′ |
| 10 | .6967 | 8431 | .7173 | 8557 | .9713 | 9874 | 1.0295 | 0126 | 50 |
| 20 | .6988 | 8444 | .7153 | 8545 | .9770 | 9899 | 1.0235 | 0101 | 40 |
| 30 | .7009 | 8457 | .7133 | 8532 | .9827 | 9924 | 1.0176 | 0076 | 30 |
| 40 | .7030 | 8469 | .7112 | 8520 | .9884 | 9949 | 1.0117 | 0051 | 20 |
| 50 | .7050 | 8482 | .7092 | 8507 | .9942 | 9975 | 1.0058 | 0025 | 10 |
| **45°** 00′ | .7071 | **9**.8495 | .7071 | **9**.8495 | 1.0000 | **0**.0000 | 1.0000 | **0**.0000 | 45° 00′ |
| | Nat. | Log. | Nat. | Log. | Nat. | Log. | Nat. | Log. | |
| Angles | Cosines | | Sines | | Cotangents | | Tangents | | Angles |

**Table A-3 Natural logarithms***

| N | 0 | 1 | 2 | 3 | 4 | 5 | 6 | 7 | 8 | 9 |
|---|---|---|---|---|---|---|---|---|---|---|
| **1.0** | 0.0 000 | 100 | 198 | 296 | 392 | 488 | 583 | 677 | 770 | 862 |
| 1.1 | 953 | *044 | *133 | *222 | *310 | *398 | *484 | *570 | *655 | *740 |
| 1.2 | 0.1 823 | 906 | 989 | *070 | *151 | *231 | *311 | *390 | *469 | *546 |
| 1.3 | 0.2 624 | 700 | 776 | 852 | 927 | *001 | *075 | *148 | *221 | *293 |
| 1.4 | 0.3 365 | 436 | 507 | 577 | 646 | 716 | 784 | 853 | 920 | 988 |
| 1.5 | 0.4 055 | 121 | 187 | 253 | 318 | 383 | 447 | 511 | 574 | 637 |
| 1.6 | 700 | 762 | 824 | 886 | 947 | *008 | *068 | *128 | *188 | *247 |
| 1.7 | 0.5 306 | 365 | 423 | 481 | 539 | 596 | 653 | 710 | 766 | 822 |
| 1.8 | 878 | 933 | 988 | *043 | *098 | *152 | *206 | *259 | *313 | *366 |
| 1.9 | 0.6 419 | 471 | 523 | 575 | 627 | 678 | 729 | 780 | 831 | 881 |
| **2.0** | 931 | 981 | *031 | *080 | *129 | *178 | *227 | *275 | *324 | *372 |
| 2.1 | 0.7 419 | 467 | 514 | 561 | 608 | 655 | 701 | 747 | 793 | 839 |
| 2.2 | 885 | 930 | 975 | *020 | *065 | *109 | *154 | *198 | *242 | *286 |
| 2.3 | 0.8 329 | 372 | 416 | 459 | 502 | 544 | 587 | 629 | 671 | 713 |
| 2.4 | 755 | 796 | 838 | 879 | 920 | 961 | *002 | *042 | *083 | *123 |
| 2.5 | 0.9 163 | 203 | 243 | 282 | 322 | 361 | 400 | 439 | 478 | 517 |
| 2.6 | 555 | 594 | 632 | 670 | 708 | 746 | 783 | 821 | 858 | 895 |
| 2.7 | 933 | 969 | *006 | *043 | *080 | *116 | *152 | *188 | *225 | *260 |
| 2.8 | 1.0 296 | 332 | 367 | 403 | 438 | 473 | 508 | 543 | 578 | 613 |
| 2.9 | 647 | 682 | 716 | 750 | 784 | 818 | 852 | 886 | 919 | 953 |
| **3.0** | 986 | *019 | *053 | *086 | *119 | *151 | *184 | *217 | *249 | *282 |
| 3.1 | 1.1 314 | 346 | 378 | 410 | 442 | 474 | 506 | 537 | 569 | 600 |
| 3.2 | 632 | 663 | 694 | 725 | 756 | 787 | 817 | 848 | 878 | 909 |
| 3.3 | 939 | 969 | *000 | *030 | *060 | *090 | *119 | *149 | *179 | *208 |
| 3.4 | 1.2 238 | 267 | 296 | 326 | 355 | 384 | 413 | 442 | 470 | 499 |
| 3.5 | 528 | 556 | 585 | 613 | 641 | 669 | 698 | 726 | 754 | 782 |
| 3.6 | 809 | 837 | 865 | 892 | 920 | 947 | 975 | *002 | *029 | *056 |
| 3.7 | 1.3 083 | 110 | 137 | 164 | 191 | 218 | 244 | 271 | 297 | 324 |
| 3.8 | 350 | 376 | 402 | 429 | 455 | 481 | 507 | 533 | 558 | 584 |
| 3.9 | 610 | 635 | 661 | 686 | 712 | 737 | 762 | 788 | 813 | 838 |
| **4.0** | 863 | 888 | 913 | 938 | 962 | 987 | *012 | *036 | *061 | *085 |
| 4.1 | 1.4 110 | 134 | 159 | 183 | 207 | 231 | 255 | 279 | 303 | 327 |
| 4.2 | 351 | 375 | 398 | 422 | 446 | 469 | 493 | 516 | 540 | 563 |
| 4.3 | 586 | 609 | 633 | 656 | 679 | 702 | 725 | 748 | 770 | 793 |
| 4.4 | 816 | 839 | 861 | 884 | 907 | 929 | 951 | 974 | 996 | *019 |
| 4.5 | 1.5 041 | 063 | 085 | 107 | 129 | 151 | 173 | 195 | 217 | 239 |
| 4.6 | 261 | 282 | 304 | 326 | 347 | 369 | 390 | 412 | 433 | 454 |
| 4.7 | 476 | 497 | 518 | 539 | 560 | 581 | 602 | 623 | 644 | 665 |
| 4.8 | 686 | 707 | 728 | 748 | 769 | 790 | 810 | 831 | 851 | 872 |
| 4.9 | 892 | 913 | 933 | 953 | 974 | 994 | *014 | *034 | *054 | *074 |
| **5.0** | 1.6 094 | 114 | 134 | 154 | 174 | 194 | 214 | 233 | 253 | 273 |

* If given number $n = N \times 10^m$, then $\log_e n = \log_e N + m \log_e 10$. Find $m \log_e 10$ from the following table:

*Multiples of* $\log_e 10$

| | |
|---|---|
| $\log_e 10 = 2.3026$ | $-\log_e 10 = 7.6974 - 10$ |
| $2 \log_e 10 = 4.6052$ | $-2 \log_e 10 = 5.3948 - 10$ |
| $3 \log_e 10 = 6.9078$ | $-3 \log_e 10 = 3.0922 - 10$ |
| $4 \log_e 10 = 9.2103$ | $-4 \log_e 10 = 0.7897 - 10$ |
| $5 \log_e 10 = 11.5129$ | $-5 \log_e 10 = 8.4871 - 20$ |

*Source:* Middlemiss, Marks, and Smart, "Analytic Geometry," 3d ed., © 1968 by McGraw-Hill, Inc. Used with permission of McGraw-Hill Book Company.

**Table A-3 Natural logarithms** *(continued)*

| N | 0 | 1 | 2 | 3 | 4 | 5 | 6 | 7 | 8 | 9 |
|---|---|---|---|---|---|---|---|---|---|---|
| **5.0** | 1.6 094 | 114 | 134 | 154 | 174 | 194 | 214 | 233 | 253 | 273 |
| 5.1 | 292 | 312 | 332 | 351 | 371 | 390 | 409 | 429 | 448 | 467 |
| 5.2 | 487 | 506 | 525 | 544 | 563 | 582 | 601 | 620 | 639 | 658 |
| 5.3 | 677 | 696 | 715 | 734 | 752 | 771 | 790 | 808 | 827 | 845 |
| 5.4 | 864 | 882 | 901 | 919 | 938 | 956 | 974 | 993 | *011 | *029 |
| 5.5 | 1.7 047 | 066 | 084 | 102 | 120 | 138 | 156 | 174 | 192 | 210 |
| 5.6 | 228 | 246 | 263 | 281 | 299 | 317 | 334 | 352 | 370 | 387 |
| 5.7 | 405 | 422 | 440 | 457 | 475 | 492 | 509 | 527 | 544 | 561 |
| 5.8 | 579 | 596 | 613 | 630 | 647 | 664 | 681 | 699 | 716 | 733 |
| 5.9 | 750 | 766 | 783 | 800 | 817 | 834 | 851 | 867 | 884 | 901 |
| **6.0** | 918 | 934 | 951 | 967 | 984 | *001 | *017 | *034 | *050 | *066 |
| 6.1 | 1.8 083 | 099 | 116 | 132 | 148 | 165 | 181 | 197 | 213 | 229 |
| 6.2 | 245 | 262 | 278 | 294 | 310 | 326 | 342 | 358 | 374 | 390 |
| 6.3 | 405 | 421 | 437 | 453 | 469 | 485 | 500 | 516 | 532 | 547 |
| 6.4 | 563 | 579 | 594 | 610 | 625 | 641 | 656 | 672 | 687 | 703 |
| 6.5 | 718 | 733 | 749 | 764 | 779 | 795 | 810 | 825 | 840 | 856 |
| 6.6 | 871 | 886 | 901 | 916 | 931 | 946 | 961 | 976 | 991 | *006 |
| 6.7 | 1.9 021 | 036 | 051 | 066 | 081 | 095 | 110 | 125 | 140 | 155 |
| 6.8 | 169 | 184 | 199 | 213 | 228 | 242 | 257 | 272 | 286 | 301 |
| 6.9 | 315 | 330 | 344 | 359 | 373 | 387 | 402 | 416 | 430 | 445 |
| **7.0** | 459 | 473 | 488 | 502 | 516 | 530 | 544 | 559 | 573 | 587 |
| 7.1 | 601 | 615 | 629 | 643 | 657 | 671 | 685 | 699 | 713 | 727 |
| 7.2 | 741 | 755 | 769 | 782 | 796 | 810 | 824 | 838 | 851 | 865 |
| 7.3 | 879 | 892 | 906 | 920 | 933 | 947 | 961 | 974 | 988 | *001 |
| 7.4 | 2.0 015 | 028 | 042 | 055 | 069 | 082 | 096 | 109 | 122 | 136 |
| 7.5 | 149 | 162 | 176 | 189 | 202 | 215 | 229 | 242 | 255 | 268 |
| 7.6 | 281 | 295 | 308 | 321 | 334 | 347 | 360 | 373 | 386 | 399 |
| 7.7 | 412 | 425 | 438 | 451 | 464 | 477 | 490 | 503 | 516 | 528 |
| 7.8 | 541 | 554 | 567 | 580 | 592 | 605 | 618 | 631 | 643 | 656 |
| 7.9 | 669 | 681 | 694 | 707 | 719 | 732 | 744 | 757 | 769 | 782 |
| **8.0** | 794 | 807 | 819 | 832 | 844 | 857 | 869 | 882 | 894 | 906 |
| 8.1 | 919 | 931 | 943 | 956 | 968 | 980 | 992 | *005 | *017 | *029 |
| 8.2 | 2.1 041 | 054 | 066 | 080 | 090 | 102 | 114 | 126 | 138 | 150 |
| 8.3 | 163 | 175 | 187 | 199 | 211 | 223 | 235 | 247 | 258 | 270 |
| 8.4 | 282 | 294 | 306 | 318 | 330 | 342 | 353 | 365 | 377 | 389 |
| 8.5 | 401 | 412 | 424 | 436 | 448 | 460 | 471 | 483 | 494 | 506 |
| 8.6 | 518 | 529 | 541 | 552 | 564 | 576 | 587 | 599 | 610 | 622 |
| 8.7 | 633 | 645 | 656 | 668 | 679 | 691 | 702 | 713 | 725 | 736 |
| 8.8 | 748 | 759 | 770 | 782 | 793 | 804 | 815 | 827 | 838 | 849 |
| 8.9 | 861 | 872 | 883 | 894 | 905 | 917 | 928 | 939 | 950 | 961 |
| **9.0** | 972 | 983 | 994 | *006 | *017 | *028 | *039 | *050 | *061 | *072 |
| 9.1 | 2.2 083 | 094 | 105 | 116 | 127 | 137 | 148 | 159 | 170 | 181 |
| 9.2 | 192 | 203 | 214 | 225 | 235 | 246 | 257 | 268 | 279 | 289 |
| 9.3 | 300 | 311 | 322 | 332 | 343 | 354 | 364 | 375 | 386 | 396 |
| 9.4 | 407 | 418 | 428 | 439 | 450 | 460 | 471 | 481 | 492 | 502 |
| 9.5 | 513 | 523 | 534 | 544 | 555 | 565 | 576 | 586 | 597 | 607 |
| 9.6 | 618 | 628 | 638 | 649 | 659 | 670 | 680 | 690 | 701 | 711 |
| 9.7 | 721 | 732 | 742 | 752 | 762 | 773 | 783 | 793 | 803 | 814 |
| 9.8 | 824 | 834 | 844 | 854 | 865 | 875 | 885 | 895 | 905 | 915 |
| 9.9 | 925 | 935 | 946 | 956 | 966 | 976 | 986 | 996 | *006 | *016 |
| **10.** | 2.3 026 | 036 | 046 | 056 | 066 | 076 | 086 | 096 | 106 | 115 |

**Table A-4 Powers and roots**

| No. | Sq. | Sq. Root | Cube | Cube Root | No. | Sq. | Sq. Root | Cube | Cube Root |
|---|---|---|---|---|---|---|---|---|---|
| 1 | 1 | 1.000 | 1 | 1.000 | 51 | 2,601 | 7.141 | 132,651 | 3.708 |
| 2 | 4 | 1.414 | 8 | 1.260 | 52 | 2,704 | 7.211 | 140,608 | 3.733 |
| 3 | 9 | 1.732 | 27 | 1.442 | 53 | 2,809 | 7.280 | 148,877 | 3.756 |
| 4 | 16 | 2.000 | 64 | 1.587 | 54 | 2,916 | 7.348 | 157,464 | 3.780 |
| 5 | 25 | 2.236 | 125 | 1.710 | 55 | 3,025 | 7.416 | 166,375 | 3.803 |
| 6 | 36 | 2.449 | 216 | 1.817 | 56 | 3,136 | 7.483 | 175,616 | 3.826 |
| 7 | 49 | 2.646 | 343 | 1.913 | 57 | 3,249 | 7.550 | 185,193 | 3.849 |
| 8 | 64 | 2.828 | 512 | 2.000 | 58 | 3,364 | 7.616 | 195,112 | 3.871 |
| 9 | 81 | 3.000 | 729 | 2.080 | 59 | 3,481 | 7.681 | 205,379 | 3.893 |
| 10 | 100 | 3.162 | 1,000 | 2.154 | 60 | 3,600 | 7.746 | 216,000 | 3.915 |
| 11 | 121 | 3.317 | 1,331 | 2.224 | 61 | 3,721 | 7.810 | 226,981 | 3.936 |
| 12 | 144 | 3.464 | 1,728 | 2.289 | 62 | 3,844 | 7.874 | 238,328 | 3.958 |
| 13 | 169 | 3.606 | 2,197 | 2.351 | 63 | 3,969 | 7.937 | 250,047 | 3.979 |
| 14 | 196 | 3.742 | 2,744 | 2.410 | 64 | 4,096 | 8.000 | 262,144 | 4.000 |
| 15 | 225 | 3.873 | 3,375 | 2.466 | 65 | 4,225 | 8.062 | 274,625 | 4.021 |
| 16 | 256 | 4.000 | 4,096 | 2.520 | 66 | 4,356 | 8.124 | 287,496 | 4.041 |
| 17 | 289 | 4.123 | 4,913 | 2.571 | 67 | 4,489 | 8.185 | 300,763 | 4.062 |
| 18 | 324 | 4.243 | 5,832 | 2.621 | 68 | 4,624 | 8.246 | 314,432 | 4.082 |
| 19 | 361 | 4.359 | 6,859 | 2.668 | 69 | 4,761 | 8.307 | 328,509 | 4.102 |
| 20 | 400 | 4.472 | 8,000 | 2.714 | 70 | 4,900 | 8.367 | 343,000 | 4.121 |
| 21 | 441 | 4.583 | 9,261 | 2.759 | 71 | 5,041 | 8.426 | 357,911 | 4.141 |
| 22 | 484 | 4.690 | 10,648 | 2.802 | 72 | 5,184 | 8.485 | 373,248 | 4.160 |
| 23 | 529 | 4.796 | 12,167 | 2.844 | 73 | 5,329 | 8.544 | 389,017 | 4.179 |
| 24 | 576 | 4.899 | 13,824 | 2.884 | 74 | 5,476 | 8.602 | 405,224 | 4.198 |
| 25 | 625 | 5.000 | 15,625 | 2.924 | 75 | 5,625 | 8.660 | 421,875 | 4.217 |
| 26 | 676 | 5.099 | 17,576 | 2.962 | 76 | 5,776 | 8.718 | 438,976 | 4.236 |
| 27 | 729 | 5.196 | 19,683 | 3.000 | 77 | 5,929 | 8.775 | 456,533 | 4.254 |
| 28 | 784 | 5.291 | 21,952 | 3.037 | 78 | 6,084 | 8.832 | 474,552 | 4.273 |
| 29 | 841 | 5.385 | 24,389 | 3.072 | 79 | 6,241 | 8.888 | 493,039 | 4.291 |
| 30 | 900 | 5.477 | 27,000 | 3.107 | 80 | 6,400 | 8.944 | 512,000 | 4.309 |
| 31 | 961 | 5.568 | 29,791 | 3.141 | 81 | 6,561 | 9.000 | 531,441 | 4.327 |
| 32 | 1,024 | 5.657 | 32,768 | 3.175 | 82 | 6,724 | 9.055 | 551,368 | 4.344 |
| 33 | 1,089 | 5.745 | 35,937 | 3.208 | 83 | 6,889 | 9.110 | 571,787 | 4.362 |
| 34 | 1,156 | 5.831 | 39,304 | 3.240 | 84 | 7,056 | 9.165 | 592,704 | 4.380 |
| 35 | 1,225 | 5.916 | 42,875 | 3.271 | 85 | 7,225 | 9.220 | 014,125 | 4.397 |
| 36 | 1,296 | 6.000 | 46,656 | 3.302 | 86 | 7,396 | 9.274 | 636,056 | 4.414 |
| 37 | 1,369 | 6.083 | 50,653 | 3.332 | 87 | 7,569 | 9.327 | 658,503 | 4.431 |
| 38 | 1,444 | 6.164 | 54,872 | 3.362 | 88 | 7,744 | 9.381 | 681,472 | 4.448 |
| 39 | 1,521 | 6.245 | 59,319 | 3.391 | 89 | 7,921 | 9.434 | 704,969 | 4.465 |
| 40 | 1,600 | 6.325 | 64,000 | 3.420 | 90 | 8,100 | 9.487 | 729,000 | 4.481 |
| 41 | 1,681 | 6.403 | 68,921 | 3.448 | 91 | 8,281 | 9.539 | 753,571 | 4.498 |
| 42 | 1,764 | 6.481 | 74,088 | 3.476 | 92 | 8,464 | 9.592 | 778,688 | 4.514 |
| 43 | 1,849 | 6.557 | 79,507 | 3.503 | 93 | 8,649 | 9.644 | 804,357 | 4.531 |
| 44 | 1,936 | 6.633 | 85,184 | 3.530 | 94 | 8,836 | 9.695 | 830,584 | 4.547 |
| 45 | 2,025 | 6.708 | 91,125 | 3.557 | 95 | 9,025 | 9.747 | 857,375 | 4.563 |
| 46 | 2,116 | 6.782 | 97,336 | 3.583 | 96 | 9,216 | 9.798 | 884,736 | 4.579 |
| 47 | 2,209 | 6.856 | 103,823 | 3.609 | 97 | 9,409 | 9.849 | 912,673 | 4.595 |
| 48 | 2,304 | 6.928 | 110,592 | 3.634 | 98 | 9,604 | 9.899 | 941,192 | 4.610 |
| 49 | 2,401 | 7.000 | 117,649 | 3.659 | 99 | 9,801 | 9.950 | 970,299 | 4.626 |
| 50 | 2,500 | 7.071 | 125,000 | 3.684 | 100 | 10,000 | 10.000 | 1,000,000 | 4.642 |

*Source:* Middlemiss, Marks, and Smart, "Analytic Geometry," 3d ed., © 1968 by McGraw-Hill, Inc. Used with permission of McGraw-Hill Book Company.

**Table A-5 Exponential and hyperbolic functions**

| $x$ | $e^x$ | $e^{-x}$ | sinh $x$ | cosh $x$ | tanh $x$ |
|---|---|---|---|---|---|
| .00 | 1.000 | 1.000 | .000 | 1.000 | .000 |
| .01 | 1.010 | .990 | .010 | 1.000 | .010 |
| .02 | 1.020 | .980 | .020 | 1.000 | .020 |
| .03 | 1.030 | .970 | .030 | 1.000 | .030 |
| .04 | 1.041 | .961 | .040 | 1.001 | .040 |
| .05 | 1.051 | .951 | .050 | 1.001 | .050 |
| .06 | 1.062 | .942 | .060 | 1.002 | .060 |
| .07 | 1.073 | .932 | .070 | 1.002 | .070 |
| .08 | 1.083 | .923 | .080 | 1.003 | .080 |
| .09 | 1.094 | .914 | .090 | 1.004 | .090 |
| .1 | 1.105 | .905 | .100 | 1.005 | .100 |
| .2 | 1.221 | .819 | .201 | 1.020 | .197 |
| .3 | 1.350 | .741 | .305 | 1.045 | .291 |
| .4 | 1.492 | .670 | .411 | 1.081 | .380 |
| .5 | 1.649 | .607 | .521 | 1.128 | .462 |
| .6 | 1.822 | .549 | .637 | 1.185 | .537 |
| .7 | 2.014 | .497 | .759 | 1.255 | .604 |
| .8 | 2.226 | .449 | .888 | 1.337 | .664 |
| .9 | 2.460 | .407 | 1.027 | 1.433 | .716 |
| 1.0 | 2.718 | .368 | 1.175 | 1.543 | .762 |
| 1.1 | 3.004 | .333 | 1.336 | 1.669 | .800 |
| 1.2 | 3.320 | .301 | 1.509 | 1.811 | .834 |
| 1.3 | 3.669 | .273 | 1.698 | 1.971 | .862 |
| 1.4 | 4.055 | .247 | 1.904 | 2.151 | .885 |
| 1.5 | 4.482 | .223 | 2.129 | 2.352 | .905 |
| 1.6 | 4.953 | .202 | 2.376 | 2.577 | .922 |
| 1.7 | 5.474 | .183 | 2.646 | 2.828 | .935 |
| 1.8 | 6.050 | .165 | 2.942 | 3.107 | .947 |
| 1.9 | 6.686 | .150 | 3.268 | 3.418 | .956 |
| 2.0 | 7.389 | .135 | 3.627 | 3.762 | .964 |
| 2.1 | 8.166 | .122 | 4.022 | 4.144 | .970 |
| 2.2 | 9.025 | .111 | 4.457 | 4.568 | .976 |
| 2.3 | 9.974 | .100 | 4.937 | 5.037 | .980 |
| 2.4 | 11.023 | .091 | 5.466 | 5.557 | .984 |
| 2.5 | 12.182 | .082 | 6.050 | 6.132 | .987 |
| 2.6 | 13.464 | .074 | 6.695 | 6.769 | .989 |
| 2.7 | 14.880 | .067 | 7.406 | 7.473 | .991 |
| 2.8 | 16.445 | .061 | 8.192 | 8.253 | .993 |
| 2.9 | 18.174 | .055 | 9.060 | 9.115 | .994 |
| 3.0 | 20.086 | .050 | 10.018 | 10.068 | .995 |
| 3.1 | 22.20 | .045 | 11.08 | 11.12 | .996 |
| 3.2 | 24.53 | .041 | 12.25 | 12.29 | .997 |
| 3.3 | 27.11 | .037 | 13.54 | 13.57 | .997 |
| 3.4 | 29.96 | .033 | 14.97 | 15.00 | .998 |
| 3.5 | 33.12 | .030 | 16.54 | 16.57 | .998 |
| 3.6 | 36.60 | .027 | 18.29 | 18.31 | .999 |
| 3.7 | 40.45 | .025 | 20.21 | 20.24 | .999 |
| 3.8 | 44.70 | .022 | 22.34 | 22.36 | .999 |
| 3.9 | 49.40 | .020 | 24.69 | 24.71 | .999 |
| 4.0 | 54.60 | .018 | 27.29 | 27.31 | .999 |
| 4.1 | 60.34 | .017 | 30.16 | 30.18 | .999 |
| 4.2 | 66.69 | .015 | 33.34 | 33.35 | 1.000 |
| 4.3 | 73.70 | .014 | 36.84 | 36.86 | 1.000 |
| 4.4 | 81.45 | .012 | 40.72 | 40.73 | 1.000 |
| 4.5 | 90.02 | .011 | 45.00 | 45.01 | 1.000 |
| 4.6 | 99.48 | .010 | 49.74 | 49.75 | 1.000 |
| 4.7 | 109.95 | .0090 | 54.97 | 54.98 | 1.000 |
| 4.8 | 121.51 | .0082 | 60.75 | 60.76 | 1.000 |
| 4.9 | 134.29 | .0074 | 67.14 | 67.15 | 1.000 |
| 5.0 | 148.41 | .0067 | 74.20 | 74.21 | 1.000 |
| 6.0 | 403.4 | .0025 | 201.7 | | 1.000 |
| 7.0 | 1096.6 | .00091 | 548.3 | | 1.000 |
| 8.0 | 2981.0 | .00034 | 1490.5 | | 1.000 |
| 9.0 | 8103.1 | .00012 | 4051.5 | | 1.000 |
| 10.0 | 22026.5 | .000045 | 11013.2 | | 1.000 |

*Source:* Middlemiss, Marks, and Smart, "Analytic Geometry," 3d ed., © 1968 by McGraw-Hill, Inc. Used with permission of McGraw-Hill Book Company.

Table A-6 Common logarithms

| N | 0 | 1 | 2 | 3 | 4 | 5 | 6 | 7 | 8 | 9 |
|---|---|---|---|---|---|---|---|---|---|---|
| 10 | 0000 | 0043 | 0086 | 0128 | 0170 | 0212 | 0253 | 0294 | 0334 | 0374 |
| 11 | 0414 | 0453 | 0492 | 0531 | 0569 | 0607 | 0645 | 0682 | 0719 | 0755 |
| 12 | 0792 | 0828 | 0864 | 0899 | 0934 | 0969 | 1004 | 1038 | 1072 | 1106 |
| 13 | 1139 | 1173 | 1206 | 1239 | 1271 | 1303 | 1335 | 1367 | 1399 | 1430 |
| 14 | 1461 | 1492 | 1523 | 1553 | 1584 | 1614 | 1644 | 1673 | 1703 | 1732 |
| 15 | 1761 | 1790 | 1818 | 1847 | 1875 | 1903 | 1931 | 1959 | 1987 | 2014 |
| 16 | 2041 | 2068 | 2095 | 2122 | 2148 | 2175 | 2201 | 2227 | 2253 | 2279 |
| 17 | 2304 | 2330 | 2355 | 2380 | 2405 | 2430 | 2455 | 2480 | 2504 | 2529 |
| 18 | 2553 | 2577 | 2601 | 2625 | 2648 | 2672 | 2695 | 2718 | 2742 | 2765 |
| 19 | 2788 | 2810 | 2833 | 2856 | 2878 | 2900 | 2923 | 2945 | 2967 | 2989 |
| 20 | 3010 | 3032 | 3054 | 3075 | 3096 | 3118 | 3139 | 3160 | 3181 | 3201 |
| 21 | 3222 | 3243 | 3263 | 3284 | 3304 | 3324 | 3345 | 3365 | 3385 | 3404 |
| 22 | 3424 | 3444 | 3464 | 3483 | 3502 | 3522 | 3541 | 3560 | 3579 | 3598 |
| 23 | 3617 | 3636 | 3655 | 3674 | 3692 | 3711 | 3729 | 3747 | 3766 | 3784 |
| 24 | 3802 | 3820 | 3838 | 3856 | 3874 | 3892 | 3909 | 3927 | 3945 | 3962 |
| 25 | 3979 | 3997 | 4014 | 4031 | 4048 | 4065 | 4082 | 4099 | 4116 | 4133 |
| 26 | 4150 | 4166 | 4183 | 4200 | 4216 | 4232 | 4249 | 4265 | 4281 | 4298 |
| 27 | 4314 | 4330 | 4346 | 4362 | 4378 | 4393 | 4409 | 4425 | 4440 | 4456 |
| 28 | 4472 | 4487 | 4502 | 4518 | 4533 | 4548 | 4564 | 4579 | 4594 | 4609 |
| 29 | 4624 | 4639 | 4654 | 4669 | 4683 | 4698 | 4713 | 4728 | 4742 | 4757 |
| 30 | 4771 | 4786 | 4800 | 4814 | 4829 | 4843 | 4857 | 4871 | 4886 | 4900 |
| 31 | 4914 | 4928 | 4942 | 4955 | 4969 | 4983 | 4997 | 5011 | 5024 | 5038 |
| 32 | 5051 | 5065 | 5079 | 5092 | 5105 | 5119 | 5132 | 5145 | 5159 | 5172 |
| 33 | 5185 | 5198 | 5211 | 5224 | 5237 | 5250 | 5263 | 5276 | 5289 | 5302 |
| 34 | 5315 | 5328 | 5340 | 5353 | 5366 | 5378 | 5391 | 5403 | 5416 | 5428 |
| 35 | 5441 | 5453 | 5465 | 5478 | 5490 | 5502 | 5514 | 5527 | 5539 | 5551 |
| 36 | 5563 | 5575 | 5587 | 5599 | 5611 | 5623 | 5635 | 5647 | 5658 | 5670 |
| 37 | 5682 | 5694 | 5705 | 5717 | 5729 | 5740 | 5752 | 5763 | 5775 | 5786 |
| 38 | 5798 | 5809 | 5821 | 5832 | 5843 | 5855 | 5866 | 5877 | 5888 | 5899 |
| 39 | 5911 | 5922 | 5933 | 5944 | 5955 | 5966 | 5977 | 5988 | 5999 | 6010 |
| 40 | 6021 | 6031 | 6042 | 6053 | 6064 | 6075 | 6085 | 6096 | 6107 | 6117 |
| 41 | 6128 | 6138 | 6149 | 6160 | 6170 | 6180 | 6191 | 6201 | 6212 | 6222 |
| 42 | 6232 | 6243 | 6253 | 6263 | 6274 | 6284 | 6294 | 6304 | 6314 | 6325 |
| 43 | 6335 | 6345 | 6355 | 6365 | 6375 | 6385 | 6395 | 6405 | 6415 | 6425 |
| 44 | 6435 | 6444 | 6454 | 6464 | 6474 | 6484 | 6493 | 6503 | 6513 | 6522 |
| 45 | 6532 | 6542 | 6551 | 6561 | 6571 | 6580 | 6590 | 6599 | 6609 | 6618 |
| 46 | 6628 | 6637 | 6646 | 6656 | 6665 | 6675 | 6684 | 6693 | 6702 | 6712 |
| 47 | 6721 | 6730 | 6739 | 6749 | 6758 | 6767 | 6776 | 6785 | 6794 | 6803 |
| 48 | 6812 | 6821 | 6830 | 6839 | 6848 | 6857 | 6866 | 6875 | 6884 | 6893 |
| 49 | 6902 | 6911 | 6920 | 6928 | 6937 | 6946 | 6955 | 6964 | 6972 | 6981 |
| 50 | 6990 | 6998 | 7007 | 7016 | 7024 | 7033 | 7042 | 7050 | 7059 | 7067 |
| 51 | 7076 | 7084 | 7093 | 7101 | 7110 | 7118 | 7126 | 7135 | 7143 | 7152 |
| 52 | 7160 | 7168 | 7177 | 7185 | 7193 | 7202 | 7210 | 7218 | 7226 | 7235 |
| 53 | 7243 | 7251 | 7259 | 7267 | 7275 | 7284 | 7292 | 7300 | 7308 | 7316 |
| 54 | 7324 | 7332 | 7340 | 7348 | 7356 | 7364 | 7372 | 7380 | 7388 | 7396 |
| N | 0 | 1 | 2 | 3 | 4 | 5 | 6 | 7 | 8 | 9 |

*Source:* Middlemiss, Marks, and Smart, "Analytic Geometry," 3d ed., © 1968 by McGraw-Hill, Inc. Used with permission of McGraw-Hill Book Company.

**Table A-6 Common logarithms** *(continued)*

| N | 0 | 1 | 2 | 3 | 4 | 5 | 6 | 7 | 8 | 9 |
|---|---|---|---|---|---|---|---|---|---|---|
| 55 | 7404 | 7412 | 7419 | 7427 | 7435 | 7443 | 7451 | 7459 | 7466 | 7474 |
| 56 | 7482 | 7490 | 7497 | 7505 | 7513 | 7520 | 7528 | 7536 | 7543 | 7551 |
| 57 | 7559 | 7566 | 7574 | 7582 | 7589 | 7597 | 7604 | 7612 | 7619 | 7627 |
| 58 | 7634 | 7642 | 7649 | 7657 | 7664 | 7672 | 7679 | 7686 | 7694 | 7701 |
| 59 | 7709 | 7716 | 7723 | 7731 | 7738 | 7745 | 7752 | 7760 | 7767 | 7774 |
| **60** | 7782 | 7789 | 7796 | 7803 | 7810 | 7818 | 7825 | 7832 | 7839 | 7846 |
| 61 | 7853 | 7860 | 7868 | 7875 | 7882 | 7889 | 7896 | 7903 | 7910 | 7917 |
| 62 | 7924 | 7931 | 7938 | 7945 | 7952 | 7959 | 7966 | 7973 | 7980 | 7987 |
| 63 | 7993 | 8000 | 8007 | 8014 | 8021 | 8028 | 8035 | 8041 | 8048 | 8055 |
| 64 | 8062 | 8069 | 8075 | 8082 | 8089 | 8096 | 8102 | 8109 | 8116 | 8122 |
| 65 | 8129 | 8136 | 8142 | 8149 | 8156 | 8162 | 8169 | 8176 | 8182 | 8189 |
| 66 | 8195 | 8202 | 8209 | 8215 | 8222 | 8228 | 8235 | 8241 | 8248 | 8254 |
| 67 | 8261 | 8267 | 8274 | 8280 | 8287 | 8293 | 8299 | 8306 | 8312 | 8319 |
| 68 | 8325 | 8331 | 8338 | 8344 | 8351 | 8357 | 8363 | 8370 | 8376 | 8382 |
| 69 | 8388 | 8395 | 8401 | 8407 | 8414 | 8420 | 8426 | 8432 | 8439 | 8445 |
| **70** | 8451 | 8457 | 8463 | 8470 | 8476 | 8482 | 8488 | 8494 | 8500 | 8506 |
| 71 | 8513 | 8519 | 8525 | 8531 | 8537 | 8543 | 8549 | 8555 | 8561 | 8567 |
| 72 | 8573 | 8579 | 8585 | 8591 | 8597 | 8603 | 8609 | 8615 | 8621 | 8627 |
| 73 | 8633 | 8639 | 8645 | 8651 | 8657 | 8663 | 8669 | 8675 | 8681 | 8686 |
| 74 | 8692 | 8698 | 8704 | 8710 | 8716 | 8722 | 8727 | 8733 | 8739 | 8745 |
| 75 | 8751 | 8756 | 8762 | 8768 | 8774 | 8779 | 8785 | 8791 | 8797 | 8802 |
| 76 | 8808 | 8814 | 8820 | 8825 | 8831 | 8837 | 8842 | 8848 | 8854 | 8859 |
| 77 | 8865 | 8871 | 8876 | 8882 | 8887 | 8893 | 8899 | 8904 | 8910 | 8915 |
| 78 | 8921 | 8927 | 8932 | 8938 | 8943 | 8949 | 8954 | 8960 | 8965 | 8971 |
| 79 | 8976 | 8982 | 8987 | 8993 | 8998 | 9004 | 9009 | 9015 | 9020 | 9025 |
| **80** | 9031 | 9036 | 9042 | 9047 | 9053 | 9058 | 9063 | 9069 | 9074 | 9079 |
| 81 | 9085 | 9090 | 9096 | 9101 | 9106 | 9112 | 9117 | 9122 | 9128 | 9133 |
| 82 | 9138 | 9143 | 9149 | 9154 | 9159 | 9165 | 9170 | 9175 | 9180 | 9186 |
| 83 | 9191 | 9196 | 9201 | 9206 | 9212 | 9217 | 9222 | 9227 | 9232 | 9238 |
| 84 | 9243 | 9248 | 9253 | 9258 | 9263 | 9269 | 9274 | 9279 | 9284 | 9289 |
| 85 | 9294 | 9299 | 9304 | 9309 | 9315 | 9320 | 9325 | 9330 | 9335 | 9340 |
| 86 | 9345 | 9350 | 9355 | 9360 | 9365 | 9370 | 9375 | 9380 | 9385 | 9390 |
| 87 | 9395 | 9400 | 9405 | 9410 | 9415 | 9420 | 9425 | 9430 | 9435 | 9440 |
| 88 | 9445 | 9450 | 9455 | 9460 | 9465 | 9469 | 9474 | 9479 | 9484 | 9489 |
| 89 | 9494 | 9499 | 9504 | 9509 | 9513 | 9518 | 9523 | 9528 | 9533 | 9538 |
| **90** | 9542 | 9547 | 9552 | 9557 | 9562 | 9566 | 9571 | 9576 | 9581 | 9586 |
| 91 | 9590 | 9595 | 9600 | 9605 | 9609 | 9614 | 9619 | 9624 | 9628 | 9633 |
| 92 | 9638 | 9643 | 9647 | 9652 | 9657 | 9661 | 9666 | 9671 | 9675 | 9680 |
| 93 | 9685 | 9689 | 9694 | 9699 | 9703 | 9708 | 9713 | 9717 | 9722 | 9727 |
| 94 | 9731 | 9736 | 9741 | 9745 | 9750 | 9754 | 9759 | 9763 | 9768 | 9773 |
| 95 | 9777 | 9782 | 9786 | 9791 | 9795 | 9800 | 9805 | 9809 | 9814 | 9818 |
| 96 | 9823 | 9827 | 9832 | 9836 | 9841 | 9845 | 9850 | 9854 | 9859 | 9863 |
| 97 | 9868 | 9872 | 9877 | 9881 | 9886 | 9890 | 9894 | 9899 | 9903 | 9908 |
| 98 | 9912 | 9917 | 9921 | 9926 | 9930 | 9934 | 9939 | 9943 | 9948 | 9952 |
| 99 | 9956 | 9961 | 9965 | 9969 | 9974 | 9978 | 9983 | 9987 | 9991 | 9996 |
| N | 0 | 1 | 2 | 3 | 4 | 5 | 6 | 7 | 8 | 9 |

# Answers to Odd-numbered Exercises

## CHAPTER 1

### SEC. 1-2

**1.** (*a*) 5; (*b*) 10; (*c*) 9; (*d*) $5\sqrt{2}$; (*e*) 6; (*f*) 5

**3.** $\overline{AB} = 4$, $\overline{BC} = 10$, $\overline{CD} = \sqrt{65}$, $\overline{AD} = 5$

### SEC. 1-3

**1.** (*a*) (3,5); (*b*) (1,2); (*c*) $(3,-4)$; (*d*) $(-\frac{1}{2},\frac{1}{2})$; (*e*) (2,1)

**3.** 5

**5.** Yes

### SEC. 1-4

**1.** (*a*) $-1$; (*b*) 2; (*c*) $-\frac{2}{9}$; (*d*) 1; (*e*) $-1$; (*f*) $-\frac{5}{2}$

**3.** $m\overline{AB} = m\overline{CD} = \frac{4}{9}$

**5.** Yes

**7.** Yes

**9.** Yes

### SEC. 1-5

**1.** (*a*) $2x + y = 0$; (*b*) $8x - 5y - 31 = 0$; (*c*) $6x - y + 4 = 0$
**3.** $m = \frac{7}{4}$
**5.** $2x - y + 1 = 0$
**7.** $2x - 5y = 0$
**9.** $3x + y - 19 = 0$

## CHAPTER 2

### SEC. 2-2

**1.** Intercepts: $x = \pm 6$, $y = \pm 6$; symmetric with respect to both axes and the origin; extent: $-6 \leq x \leq 6$, $-6 \leq y \leq 6$; no horizontal or vertical asymptotes

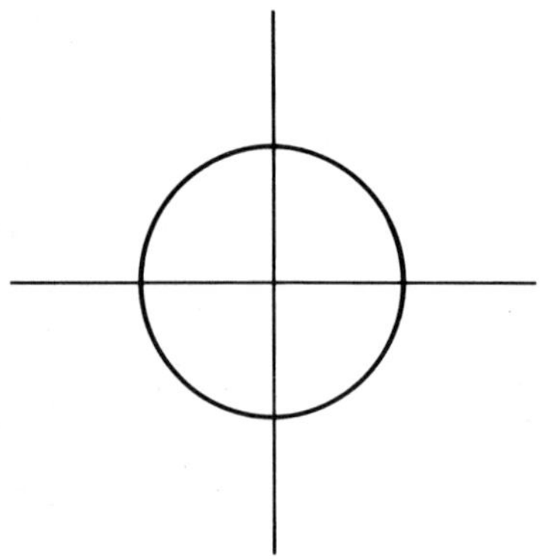

**3.** Intercepts: $x = 0$, $y = 0$; symmetric with respect to the $X$ axis; extent: $x \geq 0$, all values of $y$; no horizontal or vertical asymptotes

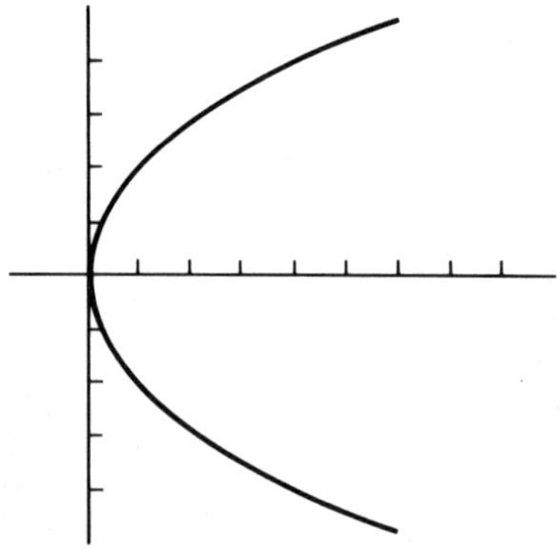

**5.** Intercepts: $y = -3$, no $x$ intercept; not symmetric with respect to the $X$ or $Y$ axes or the origin; extent: $x \neq 3$, $y \neq 0$; asymptotes: $x = 3$, $y = 0$

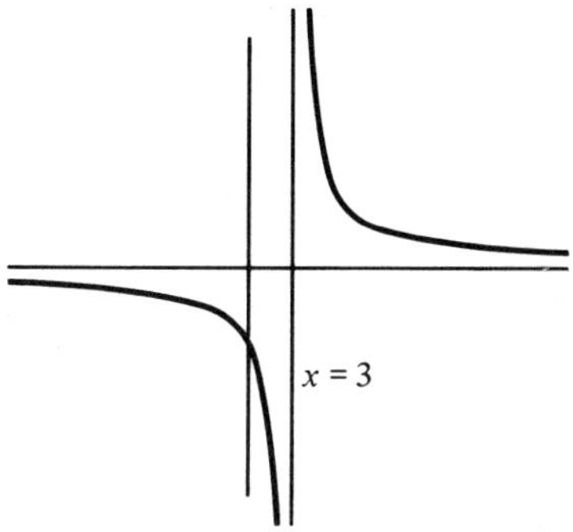


**7.** Intercepts: $y = -\frac{1}{2}$, no $x$ intercept; symmetric with respect to the $Y$ axis; extent: $x \neq \pm 2$, $y > 0$, $y \leq -\frac{1}{2}$; asymptotes: $x = \pm 2$, $y = 0$

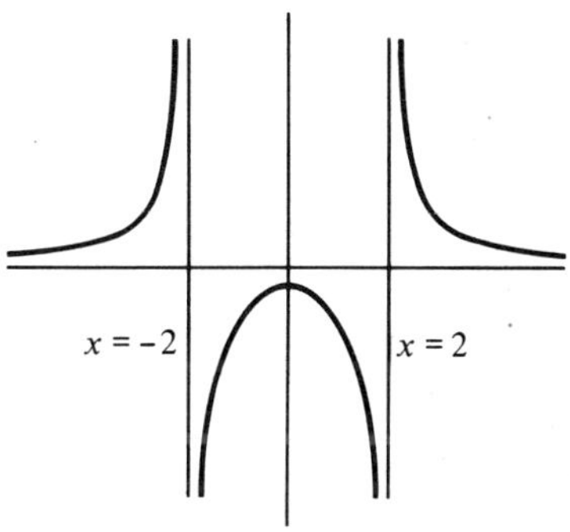


**9.** Intercepts: $x = 0$, $y = 0$; symmetric with respect to the origin; extent: all $x$, all $y$ ($x$ and $y$ must have the same sign); asymptotes: none

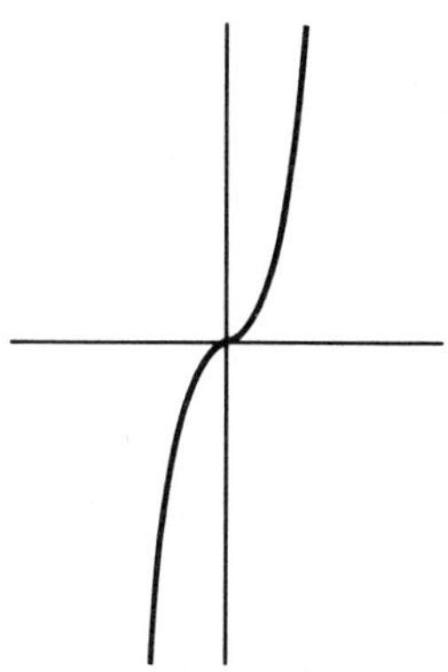

11. Intercepts: $x = 3$, $y = 9$; not symmetric with respect to the $X$ or $Y$ axes or the origin; extent: all $x$, $y \geq 0$; asymptotes: none

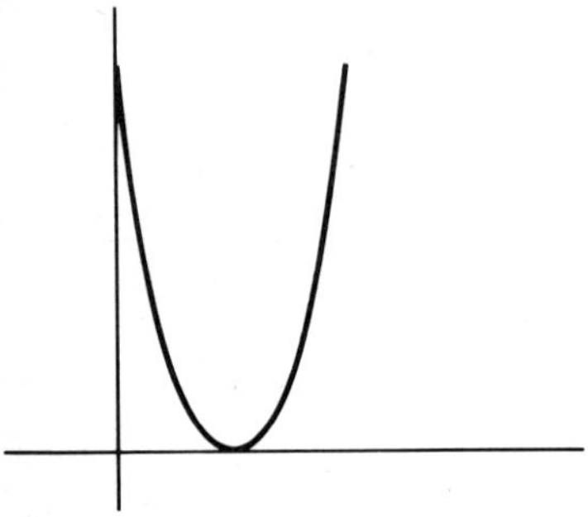

## CHAPTER 3

### SEC. 3-1

1. $D = \{1,2,3,5,6\}$, $R = \{3,4,5,7,9\}$
3. $D = \{1,3,5,7,9,11\}$, $R = \{2,4,6,8,10,12\}$
5. (*a*) $D = \{1,2,3,4,5\}$, $R = \{2,3,4,5\}$; (*b*) not a function; (*c*) $D = \{1,2,3,4,5\}$, $R = \{2\}$; (*d*) not a function
7.

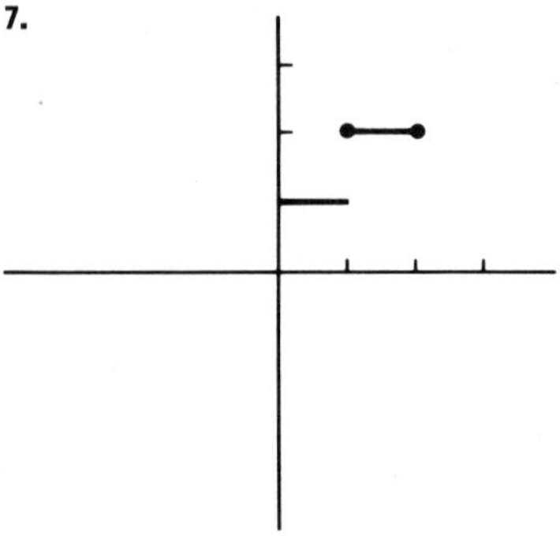

9. $f(2) = 4$, $f(3) = 9$, $f(h) = h^2$, $f(x + h) = x^2 + 2hx + h^2$
11. $f(0) = 1$, $f(\frac{1}{2}) = \frac{1}{2}$, $f(a) = 1 - 2a^2$, $f(a + 1) = -2a^2 - 2a - 1$

### SEC. 3-2

1. $A = \pi r^2$
3. $A = 20x - x^2$
5. $A = 40x - x^2$
7. $P = 2x + \dfrac{2{,}000}{x}$

### SEC. 3-3

1. $\Delta y = 0.51$

**3.** (*a*) $\Delta y = \Delta x - 2x\,\Delta x - \Delta x^2$, $\dfrac{\Delta y}{\Delta x} = 1 - 2x - \Delta x$; (*b*) $\Delta y = -0.31$, $\dfrac{\Delta y}{\Delta x} = -3.1$

**5.** (*a*) $\Delta y = -\dfrac{\Delta x}{x(x + \Delta x)}$, $\dfrac{\Delta y}{\Delta x} = -\dfrac{1}{x(x + \Delta x)}$; (*b*) $\Delta y = -0.024$, $\dfrac{\Delta y}{\Delta x} = -0.24$

**7.** (*a*) $\Delta y = \dfrac{\Delta x}{(x+1)(x + \Delta x + 1)}$, $\dfrac{\Delta y}{\Delta x} = \dfrac{1}{(x+1)(x + \Delta x + 1)}$; (*b*) $\Delta y = 0.0108$, $\dfrac{\Delta y}{\Delta x} = 0.108$

**9.** (*a*) $\Delta y = 4x\,\Delta x + 2\Delta x^2 + 4\Delta x$, $\dfrac{\Delta y}{\Delta x} = 4x + 2\Delta x + 4$; (*b*) $\Delta y = 1.22$, $\dfrac{\Delta y}{\Delta x} = 12.2$

**11.** (*a*) $\Delta y = 7\Delta x - 2x\,\Delta x - \Delta x^2$, $\dfrac{\Delta y}{\Delta x} = 7 - 2x - \Delta x$; (*b*) $\Delta y = 0.29$, $\dfrac{\Delta y}{\Delta x} = 2.9$

**13.** (*a*) $\Delta y = -6x^2\,\Delta x - 6x\,\Delta x^2 - 2\Delta x^3$, $\dfrac{\Delta y}{\Delta x} = -6x^2 - 6x\,\Delta x - 2\Delta x^2$; (*b*) $\Delta y = -2.522$, $\dfrac{\Delta y}{\Delta x} = -25.22$

**SEC. 3-4**

**1.** $\frac{1}{2}$
**3.** 3
**5.** $\dfrac{a}{a+1}$
**7.** 2
**9.** $\infty$
**11.** $\frac{11}{4}$

**SEC. 3-5**

**1.** 6
**3.** $\frac{2}{5}$
**5.** 0
**7.** $\infty$
**9.** 6
**11.** $-\frac{1}{8}$

**SEC. 3-6**

**1.** (*a*) yes; (*b*) no; (*c*) yes
**3.** (*a*) $x = \pm 1$; (*b*) $x = 2$; (*c*) $x = 0$; (*d*) $x = 1$, $x = -4$
**5.** Not continuous at $x = 1$

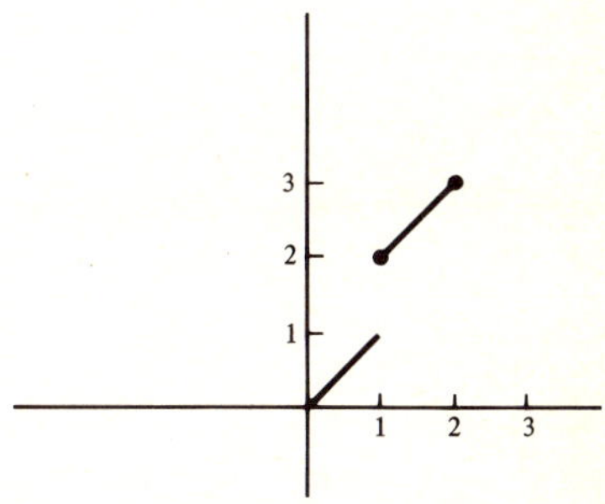


**SEC. 3-7**

**1.** $-3 < x < 13$
**3.** $-5 < x < 13$
**5.** $-5 < x < 15$

## CHAPTER 4

### SEC. 4-1

**1.** $(a)\ m = 2x_1 - 3$; $(b)\ m = -1$
**3.** $(a)\ m = -\dfrac{1}{x_1^2}$; $(b)\ m = -1$
**5.** $(a)\ m = 2 - 2x_1$; $(b)\ m = 0$
**7.** $y = 12x - 8$
**9.** $y = -\frac{1}{4}x + \frac{3}{4}$
**11.** $(\frac{9}{4}, -\frac{63}{8})$

### SEC. 4-2

**1.** $D_xy = 2x_1 + 2$
**3.** $D_xy = -\dfrac{2}{x_1^3}$
**5.** $D_xy = 3x_1^2$
**7.** $D_xy = 25$
**9.** $D_xy = \frac{8}{9}$
**11.** $(a)\ m = 6$; $(b)\ y = 6x - 4$

### SEC. 4-3

**1.** Average velocity $= 12$ ft/sec
**3.** $v = 64 - 32t$, at $t = 1$ $v = 32$
**5.** $(a)$ $-32$ ft/sec; $(b)$ $-160$ ft/sec; $(c)$ $t = 10$ sec

### SEC. 4-4

**1.** $D_qP = 4q$
**3.** At $V = 20$, $D_vP = -\frac{5}{4}$

### SEC. 4-5

**1.** $(a)$ Average current $= 2.12$ amperes; $(b)$ $I = 6t + 2$; $(c)$ $I = 2.06$ amperes
**3.** $I = 1.995$ amperes
**5.** $I = 0.000012$ ampere
**7.** $(a)$ $P = 10t + 2$ watts; $(b)$ $P = 32$ watts

## CHAPTER 5

### SEC. 5-5

**1.** $y' = 3x^2 - 4$
**3.** $y' = 16x^3 + x^2 + 2x$
**5.** $y' = -7 + 2x + 27x^2$
**7.** $y' = 40x^4 - 12x^2 + 1$
**9.** $y' = 16x^7 + 20x^4 - 21x^2$
**11.** $(a)\ m = 14$; $(b)\ m = 81$
**13.** $(a)\ y = -\frac{1}{8}x + \frac{17}{4}$; $(b)\ y = -\frac{1}{4}x$; $(c)\ y = \frac{1}{2}x + \frac{1}{2}$
**15.** $(a)$ $v = 160$ ft/sec; $(b)$ 5 sec, 400 ft
**17.** $I = 0.000075$ ampere

**SEC. 5-6**

**1.** $y' = 16x - 22$
**3.** $y' = 5x^4 + 9x^2 - 28$
**5.** $y' = 12x^3 - 40x$
**7.** $y = 31x - 92$

**SEC. 5-8**

**1.** $y' = -\dfrac{5}{x^6}$

**3.** $y' = -\dfrac{1}{x^2}$

**5.** $y' = -15x^{-6} - 8x^{-3}$

**7.** $y' = \dfrac{-2x^2 + 2x + 2}{(x^2 + 1)^2}$

**9.** $y' = \dfrac{-9}{(x - 1)^2}$

**11.** $m = 6$
**13.** $y = -5x + 10$
**15.** (a) $v = \dfrac{-t^2 - 4t - 1}{(t^2 - 1)^2}$; (b) $v = 2 - \dfrac{1}{t^2}$; (c) $v = \dfrac{-t - 2}{t^3}$

**SEC. 5-9**

**1.** $D_xy = 12x + 4x^3$

**3.** $D_xy = \dfrac{2x}{(x + 1)^3}$

**5.** $D_xy = \dfrac{2x}{(x^2 - 2)^2}$

**7.** $D_xy = 1 - \dfrac{1}{2t}$

**9.** $D_xy = t$

**SEC. 5-10**

**1.** $y' = 5(x - 2x^3)^4(1 - 6x^2)$

**3.** $y' = \dfrac{-6}{x^2}\left(1 + \dfrac{1}{x}\right)^5$

**5.** $y' = 18x(3x^2 + 4)^2$

**SEC. 5-11**

**1.** $y' = \frac{3}{2}$

**3.** $y' = \dfrac{-2y}{x}$

**5.** $y' = \dfrac{1 - 3y}{3x + 2y}$

**7.** $y' = -\dfrac{x + 1}{y - 2}$

**9.** $y' = \dfrac{-y^2}{2xy + 4}$

**11.** $y = -\frac{4}{3}x + \frac{25}{3}$
**13.** $y = 16x - \frac{255}{2}$
**15.** Curves intersect at $x = \pm 2$ and $m_1m_2 = -1$

**SEC. 5-12**

1. $y' = \dfrac{1}{2\sqrt{x}} - \dfrac{1}{2x^{\frac{3}{2}}}$
3. $y' = \dfrac{1}{\sqrt{2x-1}}$
5. $y' = \dfrac{1}{2\sqrt{x}} - \dfrac{1}{x^{\frac{3}{2}}}$
7. $y' = \dfrac{-2x}{3(4-x^2)^{\frac{2}{3}}}$
9. $y' = -\dfrac{1}{x^2}$

## CHAPTER 6

**SEC. 6-1**

1. Rising at $x = 5$
3. Stationary at $x = -2$
5. Rising for $x > 1$; falling for $x < 1$
7. Falling for all $x \neq 0$
9. Rising for $x < -1$ and $x > 1$; falling for $-1 < x < 1$
11. Rising for all $x \neq 0$

**SEC. 6-3**

1. Min at $(2,-1)$
3. Min at $(1,-6)$; max at $(-2,21)$
5. Max at $(0,4)$; min at $(2,0)$
7. Max at $(0,\frac{1}{4})$

**SEC. 6-4**

1. Concave up for all $x$
3. Concave down for $x < 0$; concave up for $x > 0$; no inflection points
5. Concave down for $\dfrac{-1}{\sqrt{3}} < x < \dfrac{1}{\sqrt{3}}$; concave up for $x < \dfrac{-1}{\sqrt{3}}$, $x > \dfrac{1}{\sqrt{3}}$; inflection points at $\left(\pm\dfrac{1}{\sqrt{3}}, \dfrac{-10}{9}\right)$
7. Concave down for $x < -\frac{1}{2}$; concave up for $x > -\frac{1}{2}$; inflection point at $(-\frac{1}{2},17)$

**SEC. 6-5**

1. Min at $(-2,-8)$
3. No max or min
5. Max at $(-1,10)$; min at $(1,-6)$
7. Max at $(-2,56)$; min at $(2,-8)$

**SEC. 6-6**

1. Intercepts: $x = \pm 2$, $y = 8$; symmetric with respect to the $Y$ axis; extent: all $x$, $y \leq 8$; no asymptotes; max at $(0,8)$; curve is always concave down

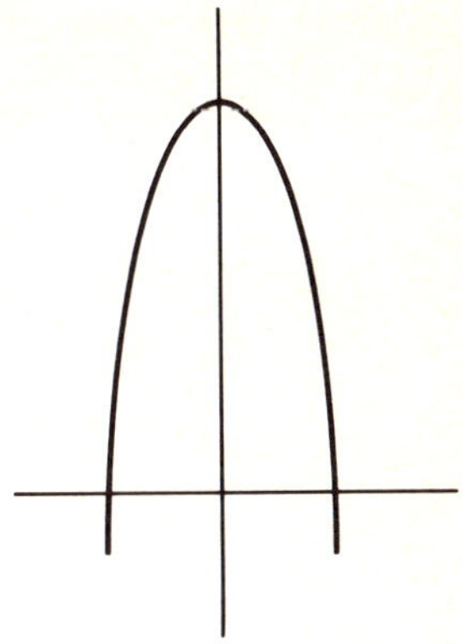

**3.** Intercepts: $x = -\frac{1}{2}$, $x = 1 \pm \sqrt{6}$, $y = -5$; not symmetric with respect to either axis or the origin; extent: all $x$ and all $y$; no asymptotes; max at $(-1,2)$, min at $(2,-25)$; concave down when $x < \frac{1}{2}$, concave up when $x > \frac{1}{2}$; inflection point at $(\frac{1}{2}, -\frac{23}{2})$

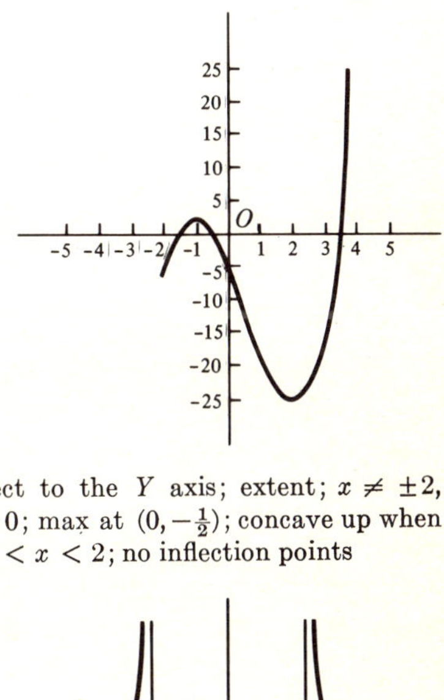


**5.** Intercepts: $y = -\frac{1}{2}$; symmetric with respect to the $Y$ axis; extent; $x \neq \pm 2$, $y \leq -\frac{1}{2}$, $y > 0$; asymptotes: $x = \pm 2$, $y = 0$; max at $(0, -\frac{1}{2})$; concave up when $x > 2$ and $x < -2$, concave down when $-2 < x < 2$; no inflection points

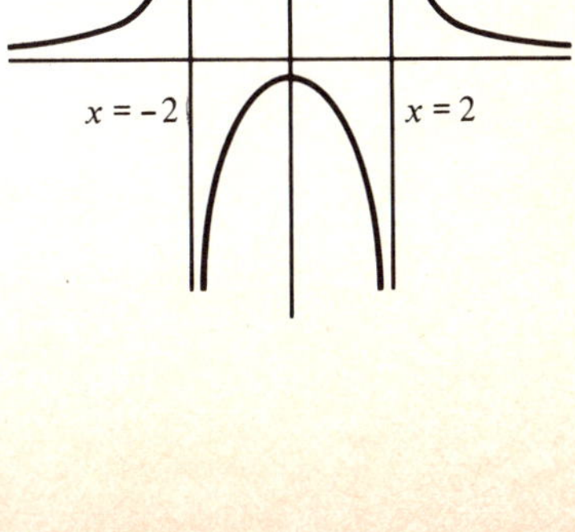

**7.** Intercepts: $y = 0$, $x = 0$, $x = \pm\sqrt{3}$; symmetric with respect to the origin; extent: all $x$ and all $y$; no asymptotes; max at $(-1,2)$, min at $(1,-2)$, inflection point at $(0,0)$; concave down when $x < 0$, concave up when $x > 0$

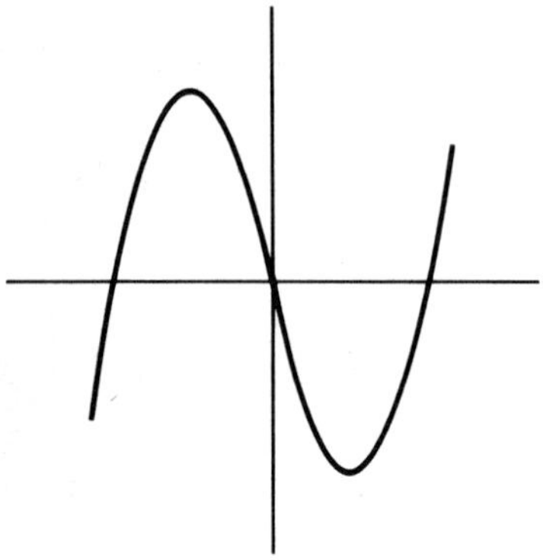

**SEC. 6-7**

**1.** Length $=$ width $= 10$
**3.** Length $=$ width $=$ height $= 2$
**5.** 10 days
**7.** $2 \times 2$
**9.** \$3.25
**11.** Diameter of base $=$ height

**SEC. 6-8**

**1.** $32\pi$ in.$^2$/min
**3.** $\frac{1}{16\pi}$ ft/min
**5.** $\frac{1}{3}$ in./min
**7.** 4.17 ft/min
**9.** $\frac{-18}{5\pi}$ ft/min

**SEC. 6-9**

**1.** $dy = 6x\,dx$
**3.** $dy = -\frac{1}{x^2}\,dx$
**5.** $dy = \frac{dx}{(x+1)^2}$
**7.** $dy = \frac{3x+2}{2\sqrt{1+x}}\,dx$
**9.** $\Delta y = 0.0801$, $dy = 0.08$
**11.** $\sqrt[3]{65} \approx 4.02$

**SEC. 6-10**

**1.** $C = \pm\sqrt{3}$
**3.** $C = \frac{1}{2}$
**5.** $C = \frac{25}{4}$

## CHAPTER 7

**SEC. 7-1**

**1.** $\frac{x^3}{3} + C$
**3.** $\frac{3}{5}x^{\frac{5}{3}} + C$
**5.** $\frac{3}{4}x^{\frac{4}{3}} + C$
**7.** $x^4 - \frac{5}{3}x^3 + \frac{1}{2}x^2 + C$
**9.** $\frac{2}{3}x^3 + \frac{3}{4}x^4 + C$
**11.** $\frac{1}{4}x^4 + x^3 + C$
**13.** $\frac{1}{4}x^4 + x^2 + C$
**15.** $\frac{2}{7}x^{\frac{7}{2}} + C$
**17.** $\frac{1}{3}x^3 - x + C$
**19.** $\frac{1}{3}x^3 + 3x^2 + 9x + C$

**SEC. 7-2**

**1.** $\frac{1}{4}(1 + x^2)^4 + C$
**3.** $-\frac{1}{5}(7 - x^3)^5 + C$
**5.** $\frac{1}{8}(x^4 - x^2)^8 + C$
**7.** $\frac{2}{5}(x^5 - 4x)^{\frac{5}{2}} + C$
**9.** $\frac{1}{3}(4x^3 - 12x)^3 + C$
**11.** $x^4 - \frac{1}{8}x^8 + C$
**13.** $\frac{1}{6}(3x^4 - 4x^3)^{\frac{1}{2}} + C$
**15.** $\frac{1}{3}(x^2 - 4)^{\frac{3}{2}} + C$
**17.** $-\frac{1}{6}(4 - 6x^2)^{\frac{1}{2}} + C$
**19.** $\frac{1}{5}x^5 + \frac{2}{3}x^3 + x + C$

**SEC. 7-3**

1. $-\frac{1}{60}(7 - 4x^3)^5 + C$
3. $-\frac{1}{12}(1 - 4x^2)^{\frac{3}{2}} + C$
5. $-\frac{1}{8}(4 - 2x^3)^{\frac{4}{3}} + C$

**SEC. 7-4**

1. (a) $y = 2x^2 - 2x + C$; (b) $y = 2x^2 - 2x + 5$
3. $y = 2x^2 - x^3 + 7$
5. $y = 2x^3 - 5x + 4$

**SEC. 7-5**

1. $s = t^2 + 4t + 1$
3. (a) $s = -16t^2 + v_0t$; (b) $s = -16t^2 + v_0t + s_0$
5. $-160\sqrt{2}$ ft/sec
7. $t = 2$, $v = -64$ ft/sec

## CHAPTER 8

**SEC. 8-2**

1. 2
3. $\frac{1}{5}$
5. $\frac{1}{2}$
7. 1
9. $\frac{1}{2}$

**SEC. 8-3**

1. $D_xy = 3 \cos 3x$
3. $D_xy = \dfrac{1}{x^2} \sin \dfrac{1}{x}$
5. $D_xy = -6 \cos 3x \sin 3x$
7. $D_xy = -x \sin x + \cos x$
9. $D_xy = \dfrac{2x \cos x + x^2 \sin x}{\cos^2 x}$
11. $D_xy = -\dfrac{4}{x^2} \sin \dfrac{2}{x} \cos \dfrac{2}{x}$
13. Max at $\left(\dfrac{\pi}{4}, \sqrt{2}\right)$; min at $\left(\dfrac{5\pi}{4}, -\sqrt{2}\right)$; inflection points at $\left(\dfrac{3\pi}{4}, 0\right)$ and $\left(\dfrac{7\pi}{4}, 0\right)$

**SEC. 8-4**

1. $D_xy = -2 \csc^2 2x$
3. $D_xy = 3 \sec^2 3x$
5. $D_xy = x \sec^2 x + \tan x$
7. $D_xy = -2 \cot x \csc^2 x$
9. $D_xy = 6 \tan 3x \sec^2 3x$
11. No max or min

**SEC. 8-5**

1. $\frac{1}{5} \sin 5x + C$
3. $\frac{1}{3} \tan 3x + C$
5. $-\frac{3}{2} \cos x^2 + C$
7. $\frac{1}{2} \tan x^2 + C$
9. $\dfrac{\sin^2 x}{2} + C$
11. $\frac{1}{6} \tan^6 t + C$
13. $\dfrac{1}{\cos x} + C$
15. $\dfrac{2}{3} \sin^3 \dfrac{t}{2} + C$
17. $-2(1 + \cos x)^{\frac{1}{2}} + C$
19. $-\frac{1}{2} \csc^2 x + C$

## CHAPTER 9

**SEC. 9-1**

**1.** $3 + 4 + 5 + 6 + 7$
**3.** $2x_3 + 2x_4 + 2x_5 + 2x_6 + 2x_7$
**5.** $\frac{4}{5} + \frac{5}{6} + \frac{6}{7} + \frac{7}{8} + \frac{8}{9} + \frac{9}{10}$
**7.** $\sum_{k=1}^{5} f(x_k)$
**9.** $\sum_{k=1}^{8} k^2$
**11.** 55
**13.** 5,050
**15.** 500,500

**SEC. 9-2**

**1.** $A = 18$
**3.** $A = \frac{4}{3}$

**SEC. 9-4**

**1.** 255
**3.** $\frac{68}{3}$
**5.** $\frac{13}{3}$
**7.** $\frac{38}{3}$
**9.** $\frac{7}{3}$
**11.** 1
**13.** $\frac{\sqrt{3}}{8}$
**15.** $\frac{1}{2}$

**SEC. 9-5**

**1.** $\frac{4}{3}$

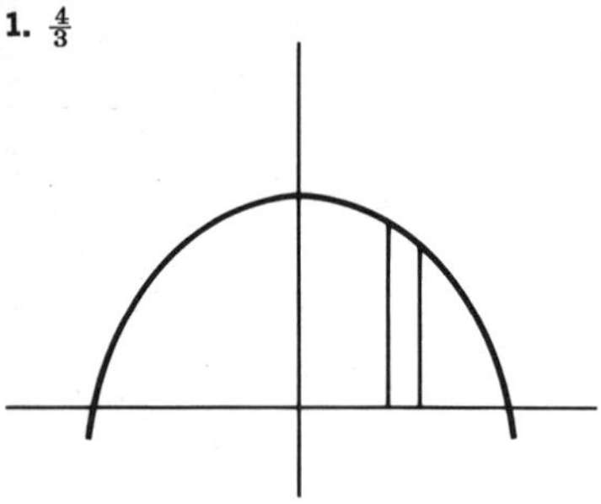

**3.** 2

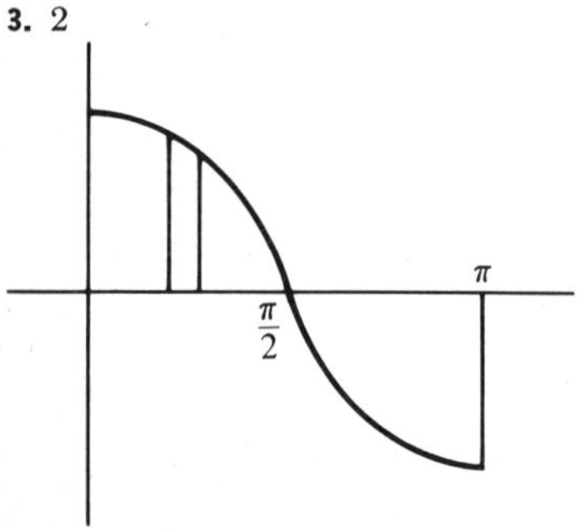


**5.** $\frac{9}{2}$

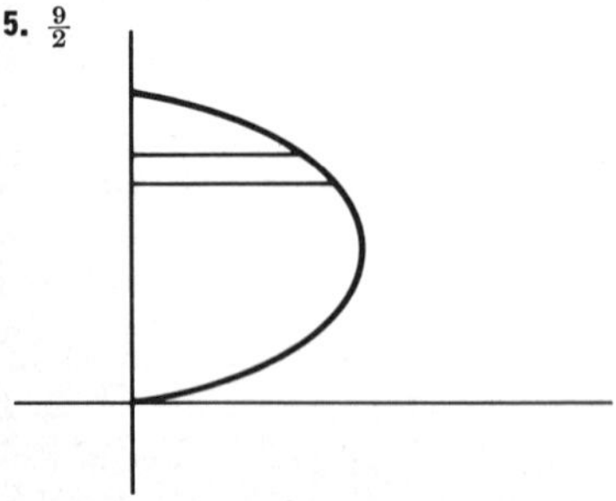

**7.** $\frac{32}{3}$

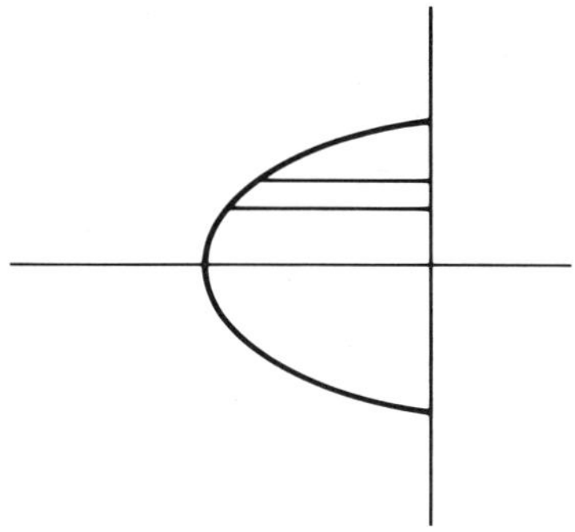

**9.** 4

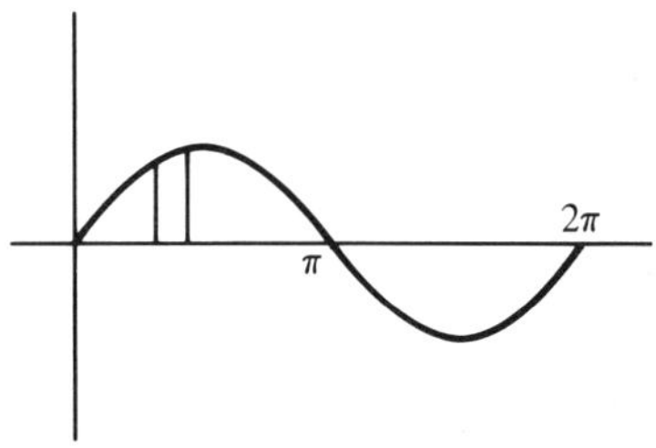


**11.** $\frac{8}{3}$

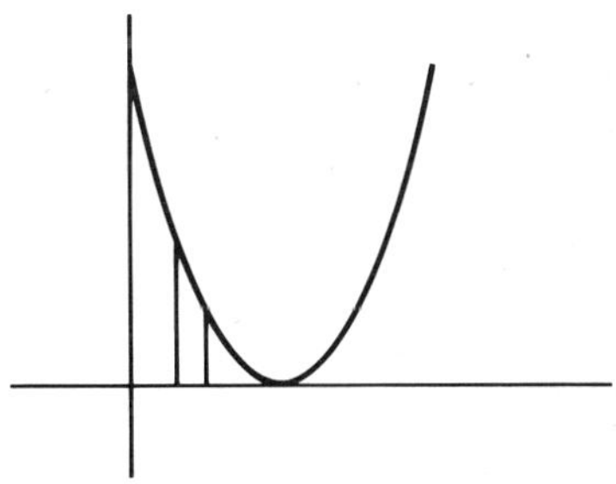

**13.** $\frac{16}{3}$

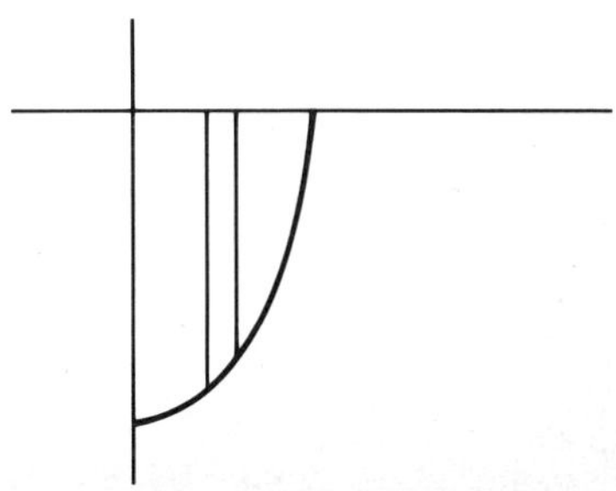

**SEC. 9-6**

1. $\frac{1}{6}$
3. $27$
5. $\frac{32}{3}$
7. $\frac{4}{3}$
9. $\frac{1}{12}$
11. $\frac{21}{8}$

**SEC. 9-7**

1. $\frac{64\pi}{3}$
3. $\frac{16\pi}{3}$
5. $8\pi$
7. $\frac{4}{3}\pi a^3$
9. $\frac{\pi}{2}$
11. $\frac{\pi}{6}$
13. $\frac{512\pi}{15}$

**SEC. 9-8**

1. $512\pi$
3. $8\pi$
5. $\frac{128\pi}{3}$
7. $\frac{8\pi}{15}$
9. $\frac{32\pi}{3}$

**SEC. 9-9**

1. $20.5$
3. $1.11$
5. $\frac{2+\sqrt{3}}{12}\pi$
7. $2.38$

## CHAPTER 10

**SEC. 10-1**

1. $\frac{dy}{dx} = 2\cos 2x$
3. $\frac{dy}{dx} = 3(\cos^2 3x - \sin^2 3x)$
5. $\frac{dy}{dx} = 8\sin 2x \cos 2x$
7. $\frac{dy}{dx} = \sec^3 x + \tan^2 x \sec x$
9. $-\frac{1}{3}\cos 3x + C$
11. $\sec x + C$
13. $\frac{1}{2}\sec 2x + C$
15. $-\tan\frac{1}{x} + C$

**SEC. 10-2**

1. $\frac{-\pi}{6}$
3. $\frac{\pi}{4}$
5. $0$
7. $y = \frac{1}{2}\text{Arcsin } x$
9. $y = \frac{1}{2}\sin x$

**SEC. 10-3**

1. $y' = \frac{1}{\sqrt{-x^2 - x + \frac{3}{4}}}$
3. $y' = \frac{-1}{2\sqrt{x(1-x)}}$
5. $y' = \frac{2x}{\sqrt{1-x^4}}$
7. $y' = \frac{x}{\sqrt{1-x^2}} + \text{Arcsin } x$
9. $y' = \frac{2x^2}{\sqrt{1-x^2}}$

**SEC. 10-5**

1. $x = 2$
3. $x = -2$
5. $x = 5\sqrt{5}$
7. $y = \ln x^2 + 2x$
9. $y = \ln x$

**SEC. 10-6**

1. $y' = \frac{1}{x+1}$
3. $y' = \frac{4x}{x^2-1}$
5. $y' = \cot x$
7. $y' = 1 + \ln x$
9. $y' = \frac{1}{2x\sqrt{\ln x}}$
11. $y' = -\frac{1}{x}$
13. $y' = \sec x \csc x$
15. $y' = -2x - 4x\ln x$
17. $\ln|x+7| + C$
19. $\frac{1}{4}\ln|4x+3| + C$
21. $\frac{4}{3}\ln|3x-4| + C$
23. $-\ln|\cos x| + C$

**SEC. 10-7**

**1.** $y' = 2e^{2x}$

**3.** $y' = \dfrac{2}{x^3} e^{-1/x^2}$

**5.** $y' = e^{1/x}(2x - 1)$

**7.** $y' = \dfrac{e^x}{x^2}(x - 1)$

**9.** $y' = \dfrac{1 - x \ln x}{xe^x}$

**11.** $y' = \cos x e^{\sin x}$

**13.** $\frac{1}{3}e^{3x} + C$

**15.** $e^{-1/x} + C$

**17.** $-e^{\cos x} + C$

**19.** $2(e^3 - e^2)$

**21.** (*a*) No max or min, no inflection points; (*b*) min at $\left(-1, -\dfrac{1}{e}\right)$, inflection point at $\left(-2, \dfrac{-2}{e^2}\right)$; (*c*) max at $\left(1, \dfrac{1}{e}\right)$, inflection point at $\left(2, \dfrac{2}{e^2}\right)$; (*d*) max at $\left(\dfrac{1}{\sqrt{2}}, \dfrac{e^{-\frac{1}{2}}}{\sqrt{2}}\right)$, min at $\left(-\dfrac{1}{\sqrt{2}}, -\dfrac{e^{-\frac{1}{2}}}{\sqrt{2}}\right)$, inflection points at (0,0), $(\sqrt{\frac{3}{2}}, \sqrt{\frac{3}{2}}e^{-\frac{3}{2}})$, $(-\sqrt{\frac{3}{2}}, -\sqrt{\frac{3}{2}}e^{-\frac{3}{2}})$

**SEC. 10-8**

**1.** $y' = 6x^2 + 6x$

**3.** $y' = \dfrac{-3}{2\sqrt{x+1}\,(x-2)^{\frac{5}{2}}}$

**5.** $y' = \dfrac{2x^2 - 4x - 9}{3(x+3)^{\frac{4}{3}}(2x-1)^2}$

**7.** $y' = \left(\dfrac{\sin x}{x} + \cos x \ln x\right) x^{\sin x}$

**SEC. 10-9**

**11.** $y' = 2x \sinh x^2$

**13.** $y' = 2 \cosh (2x + 1)$

**15.** $y' = 2 \sinh 2x - 2 \cosh 2x$

**17.** $y' = -x \operatorname{csch} x \coth x + \operatorname{csch} x$

**19.** $-\sinh (1 - x) + C$

**21.** $\frac{3}{2} \cosh x^2 + C$

**23.** $-\frac{1}{3} \operatorname{sech} 3x + C$

**25.** $\ln (\cosh x) + C$

## CHAPTER 11

### SEC. 11-1

1. $-\frac{1}{9}(1 - x^3)^3 + C$
3. $\frac{1}{2}\ln(1 + x^2) + C$
5. $\frac{1}{6}\sin^3 2x + C$
7. $\frac{1}{3}\tan 3x + C$
9. $\ln|\sin x| + C$
11. $\frac{1}{3}\sec 3x + C$
13. $\frac{1}{4}(1 + e^{x^2})^2 + C$
15. $e^{\sin x} + C$
17. $-\frac{1}{2}e^{(1-x^2)} + C$
19. $\frac{1}{6}(\sin^2 3x) + C$
21. $\ln|x - \cos x| + C$
23. $-\frac{1}{2}e^{2/x} + C$
25. $\frac{1}{3}\ln|1 + x^3| + C$
27. $-\frac{1}{4}(1 - e^x)^4 + C$
29. $-\frac{1}{3}e^{-x^3} + C$

### SEC. 11-2

1. $\frac{1}{2}xe^{2x} - \frac{1}{4}e^{2x} + C$
3. $\frac{1}{2}x^2\ln x - \frac{1}{4}x^2 + C$
5. $-x\cot x + \ln|\sin x| + C$
7. $-e^{-x}(x^2 + 2x + 2) + C$
9. $(x + 2)\ln|x + 2| - x + C$
11. $\frac{1}{3}x^3\ln x - \frac{1}{9}x^3 + C$

### SEC. 11-4

1. (*a*) $\frac{1}{2}e^x(\sin x - \cos x) + C$; (*b*) $\frac{1}{2}e^x(\sin x + \cos x) + C$

### SEC. 11-5

1. $\operatorname{Arctan} x + C$
3. $-\frac{1}{5}(1 - x^2)^{\frac{5}{2}} + C$
5. $\frac{1}{2}x\sqrt{9 + x^2} + \frac{9}{2}\ln|\sqrt{9 + x^2} + x| + C$
7. $\dfrac{1}{16}\dfrac{x}{\sqrt{16 - x^2}} + C$

### SEC. 11-6

1. $\ln\left|\dfrac{(x + 1)^2}{x + 2}\right| + C$
3. $\ln|x - 1| - \dfrac{1}{x - 1} + C$
5. $\ln\left|\dfrac{x^2}{(x + 1)(x + 2)}\right| + C$
7. $\frac{1}{2}\ln(x^2 + 2) + \dfrac{1}{x^2 + 2} + C$
9. $x - \frac{3}{2}\ln|x^2 - 4| + C$

## CHAPTER 12

### SEC. 12-2

1. $C = (-2,3)$, $r = 3$
3. $C = (-10,10)$, $r = 10$
5. $C = (-5,4)$, $r = 7$
7. $(x - 2)^2 + (y + 5)^2 = 9$
9. $x^2 + y^2 = 25$

### SEC. 12-3

1. $F = (2,0)$, $V = (0,0)$, $x = -2$
3. $F = (0,1)$, $V = (0,0)$, $y = -1$
5. $F = (0,6)$, $V = (0,0)$, $y = -6$
7. $F = (6,0)$, $V = (2,0)$, $x = -2$
9. $F = (\frac{3}{2},-3)$, $V = (1,-3)$, $x = \frac{1}{2}$
11. $F = (3,-3)$, $V = (3,-2)$, $y = -1$
13. $y^2 = -12x$
15. $y^2 = 12x$
17. $y^2 = -20x$
19. $(x - 3)^2 = 8(y - 4)$
21. $x^2 = 12(y - 3)$
23. $(y - 2)^2 = 12(x - 1)$

**SEC. 12-4**

**1.** $F = (\pm\sqrt{3},0)$, $V = (\pm 2,0)$
**3.** $F = (0,\pm 3)$, $V = (0,\pm 5)$
**5.** $F = (\pm 12,0)$, $V = (\pm 13,0)$
**7.** $F = (0,\pm\sqrt{5})$, $V = (0,\pm 3)$
**9.** $\dfrac{x^2}{25} + \dfrac{y^2}{16} = 1$
**11.** $\dfrac{y^2}{64} + \dfrac{x^2}{28} = 1$
**13.** $\dfrac{y^2}{64} + \dfrac{x^2}{39} = 1$
**15.** $C = (3,2)$, $F = (3 \pm \sqrt{7}, 2)$, $V = (7,2), (-1,2)$
**17.** $C = (-3,3)$, $F = (-8,3), (2,3)$, $V = (-16,3), (10,3)$
**19.** $C = (2,-3)$, $F = (2, -3 \pm \sqrt{5})$, $V = (2,-6), (2,0)$
**21.** $C = (3,4)$, $F = (-1,4), (7,4)$, $V = (-2,4), (8,4)$
**23.** $\dfrac{(x-3)^2}{100} + \dfrac{(y-4)^2}{36} = 1$
**25.** $\dfrac{(y-5)^2}{9} + \dfrac{(x-5)^2}{5} = 1$
**27.** $\dfrac{(y-4)^2}{9} + \dfrac{(x-4)^2}{5} = 1$

**SEC. 12-5**

**1.** $F = (\pm\sqrt{5},0)$, $V = (\pm 2,0)$
**3.** $F = (0,\pm\sqrt{13})$, $V = (0,\pm 3)$
**5.** $F = (\pm 10,0)$, $V = (\pm 8,0)$
**7.** $F = (0,\pm 5)$, $V = (0,\pm 4)$
**9.** $\dfrac{y^2}{3} - \dfrac{x^2}{4} = 1$
**11.** $\dfrac{x^2}{16} - \dfrac{y^2}{9} = 1$
**13.** $\dfrac{x^2}{36} - \dfrac{y^2}{28} = 1$
**15.** $C = (-2,-3)$, $F = (-2 \pm \sqrt{13}, -3)$, $V = (-4,-3), (0,-3)$
**17.** $C = (-2,4)$, $F = (-2,-1), (-2,9)$, $V = (-2,0), (-2,8)$
**19.** $C = (2,-4)$, $F = \left(2 \pm \dfrac{\sqrt{13}}{6}, -4\right)$, $V = (\frac{3}{2},-4), (\frac{5}{2},-4)$
**21.** $\dfrac{(x-3)^2}{25} - \dfrac{(y-4)^2}{16} = 1$
**23.** $\dfrac{(x-5)^2}{25} - \dfrac{(y-5)^2}{9} = 1$

**25.**

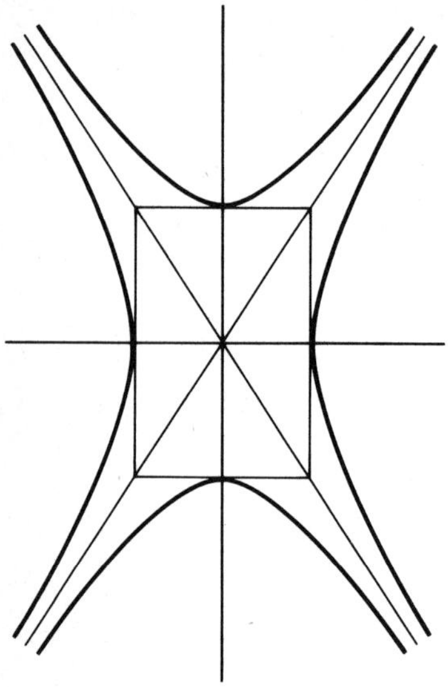

## CHAPTER 13

### SEC. 13-1

**3.** Straight line making an angle of $\frac{\pi}{4}$ with the positive $X$ axis

### SEC. 13-2

**1.** $\left(2\sqrt{2}, \frac{\pi}{4}\right)$

**3.** (5, Arctan $-\frac{3}{4}$)

**5.** $\left(3, \frac{\pi}{2}\right)$

**7.** (13, Arctan $-\frac{5}{12}$)

**9.** $\left(6, \frac{\pi}{6}\right)$

**11.** $\left(6, \frac{3\pi}{2}\right)$

**13.** $\left(\sqrt{10}, \frac{5\pi}{4}\right)$

**15.** $\left(3\sqrt{2}, \frac{7\pi}{4}\right)$

**17.** $\left(-\frac{5}{2}, \frac{5\sqrt{3}}{2}\right)$

**19.** $(1, \sqrt{3})$

21. $(-1,\sqrt{3})$
23. $\left(-\frac{\sqrt{3}}{2}, -\frac{1}{2}\right)$
25. $(-3,3)$
27. $(-4,0)$
29. $x^2 + y^2 - x + 4 = 0$
31. $x^2 + y^2 - 2y - 3x = 0$
33. $xy = 1$
35. $r \sin \theta = 2$
37. $\theta = \frac{\pi}{4}$
39. $r^2(\cos^2 \theta - \sin^2 \theta) = 4$

**SEC. 13-3**

1. Circle of radius 5 with center at the pole

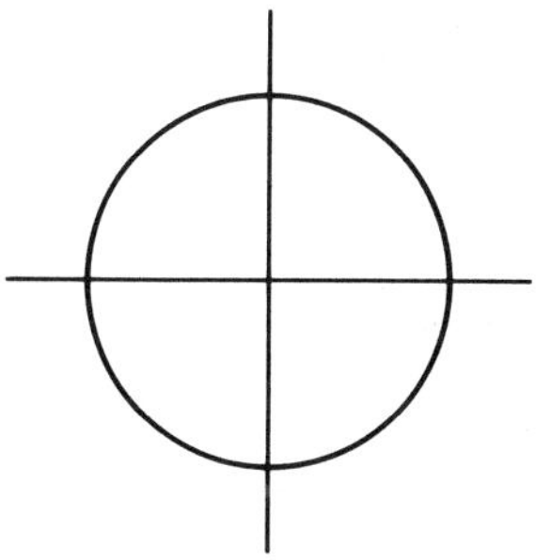

3. Intercepts: $(0,0)$, $\left(0, \frac{\pi}{2}\right)$, $(0,\pi)$, $\left(0, \frac{3\pi}{2}\right)$; symmetric with respect to the polar axis, the 90° axis and the pole; closed curve with $-1 \leq r \leq 1$

**5.** Intercepts: $(1,0)$, $\left(0, \frac{\pi}{2}\right)$, $(-1,\pi)$, $\left(0, \frac{3\pi}{2}\right)$; symmetric with respect to the polar axis; closed curve with $-1 \le r \le 1$

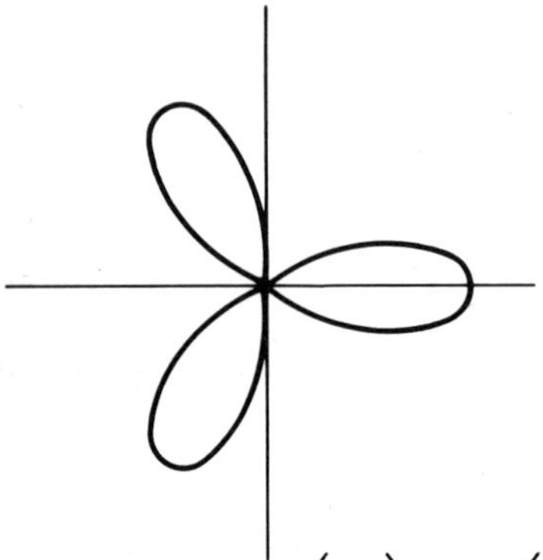

**7.** Intercepts: $(4,0)$, $\left(2, \frac{\pi}{2}\right)$, $(0,\pi)$, $\left(2, \frac{3\pi}{2}\right)$; symmetric with respect to the polar axis; closed curve with $0 \le r \le 4$

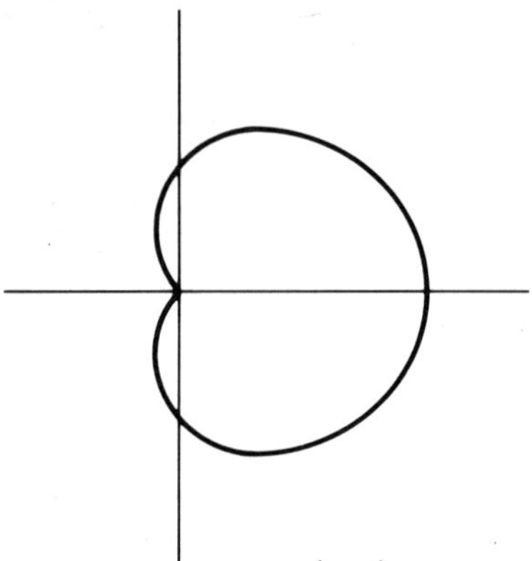

**9.** Intercepts: $(2,0)$, $\left(0, \frac{\pi}{2}\right)$, $(2,\pi)$, $\left(4, \frac{3\pi}{2}\right)$; symmetric with respect to $90°$ axis; closed curve with $0 \le r \le 4$

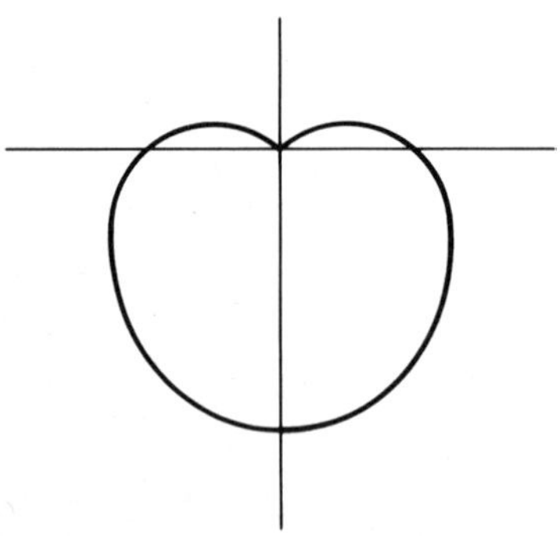

**11.** Intercepts: (3,0), $\left(2, \frac{\pi}{2}\right)$, $(1,\pi)$, $\left(2, \frac{3\pi}{2}\right)$; symmetric with respect to polar axis; closed curve with $1 \leq r \leq 3$

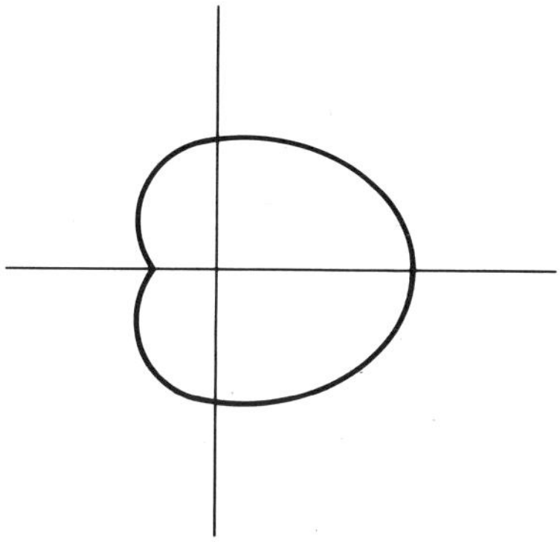

**13.** Intercepts: (0,0), $\left(0, \frac{\pi}{2}\right)$, $(0,\pi)$, $\left(0, \frac{3\pi}{2}\right)$; symmetric with respect to the pole; closed curve with $0 \leq r \leq 2$

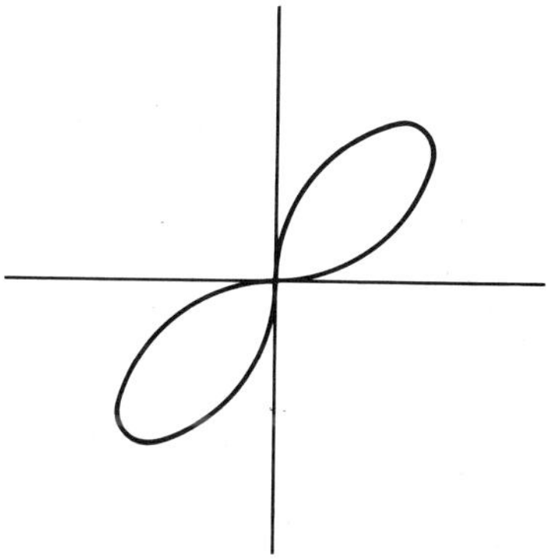

**15.** Intercepts: (3,0), $\left(0, \frac{\pi}{2}\right)$, $(-3,\pi)$, $\left(0, \frac{3\pi}{2}\right)$; symmetric with respect to the polar axis; closed curve with $0 \leq r \leq 3$

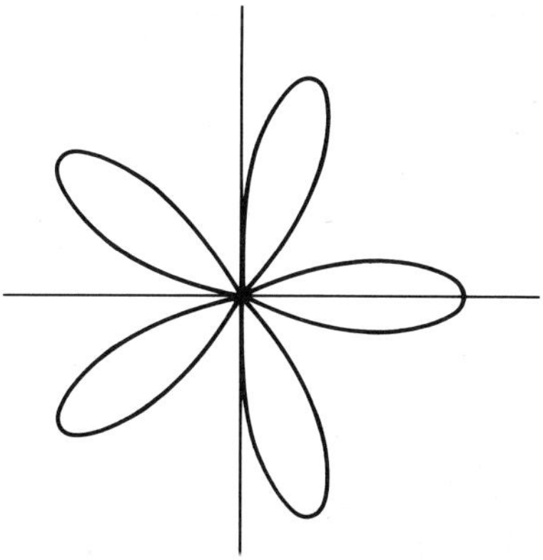

**17.** Intercepts: $(0,0)$, $\left(12, \frac{\pi}{2}\right)$, $(0,\pi)$, $\left(-12, \frac{3\pi}{2}\right)$; symmetric with respect to the 90° axis; closed curve with $0 \le r \le 12$

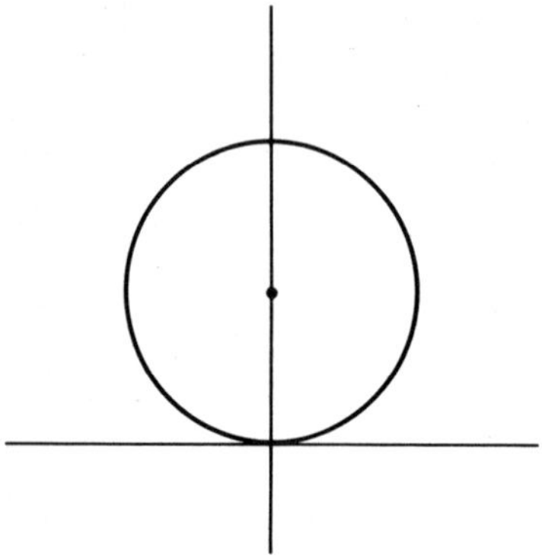

**19.** Intercepts: $(3,0)$, $\left(4, \frac{\pi}{2}\right)$, $(3,\pi)$, $\left(2, \frac{3\pi}{2}\right)$; symmetric with respect to the 90° axis; closed curve with $2 \le r \le 4$

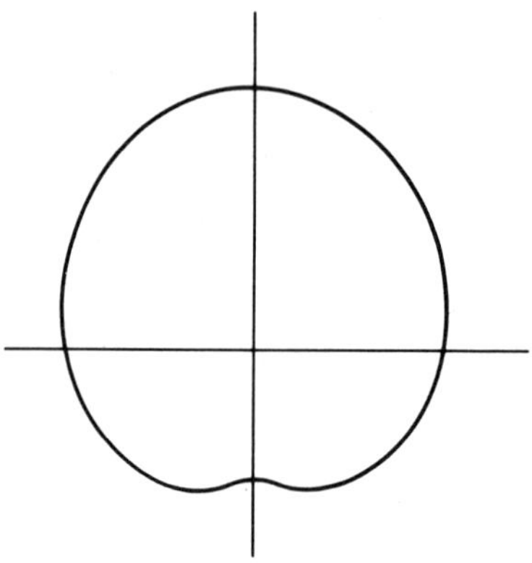

**SEC. 13-4**

**1.** $\frac{\pi}{8}$

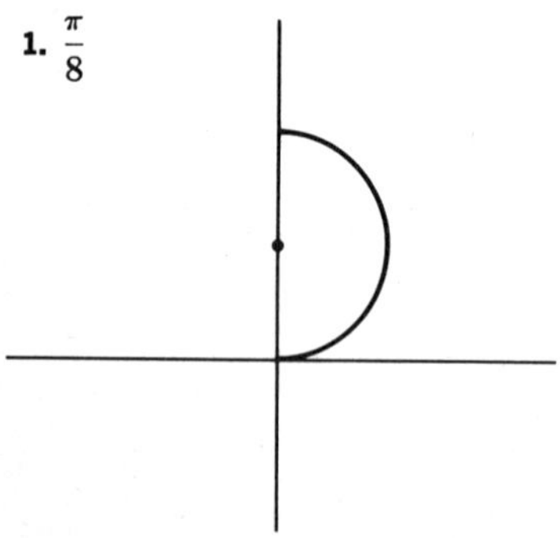

**3.** $16\pi$

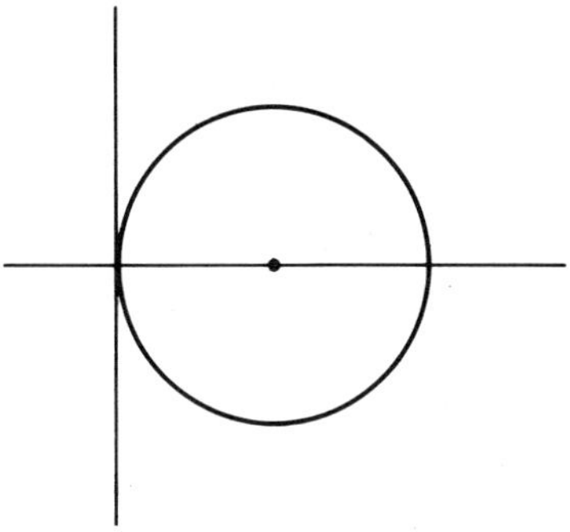

**5.** $\dfrac{\pi a^2}{8}$

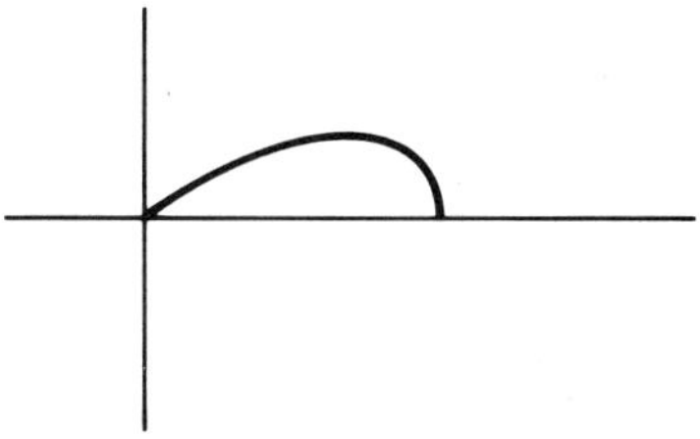

**7.** $\frac{1}{4}(e^{2\pi} - 1)$

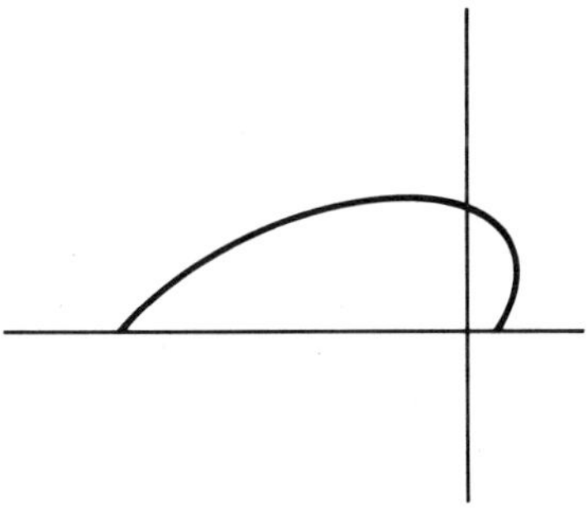

**9.** $2 + \dfrac{\pi}{4}$

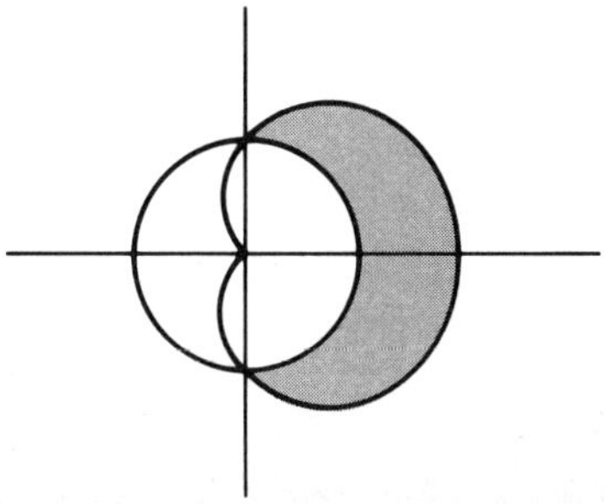

**11.** $\frac{\pi}{2} - 1$

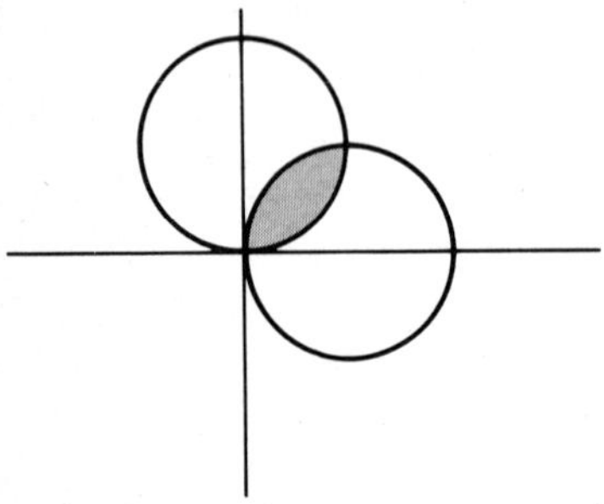

# Index